Undetermined Coefficents

$r(t)$	Attempted solution
$\alpha e^{\beta t}$	$a e^{\beta t}$
$\alpha \cos \omega t + \beta \sin \omega t$	$a \cos \omega t + b \sin \omega t$
$\alpha (\neq 0)$	a
$\alpha + \beta t$	$a + bt$
$\alpha + \beta t + \gamma t^2$	$a + bt + ct^2$
$\alpha + \beta t + \gamma t^2 + \delta t^3$	$a + bt + ct^2 + dt^3$
$\vdots$	$\vdots$

Laplace Transforms

$f(t)$	$\mathcal{L}[f] = F(s)$	Domain of F
1	$\dfrac{1}{s}$	$s > 0$
e^{at}	$\dfrac{1}{s - a}$	$s > a$
$\sin at$	$\dfrac{a}{s^2 + a^2}$	$s > 0$
$\cos at$	$\dfrac{s}{s^2 + a^2}$	$s > 0$
$t^n,\ n = 0, 1, 2, \ldots$	$\dfrac{n!}{s^{n+1}}$	$s > 0$
$t \sin at$	$\dfrac{2as}{(s^2 + a^2)^2}$	$s > 0$
$t \cos at$	$\dfrac{s^2 - a^2}{(s^2 + a^2)^2}$	$s > 0$
$e^{at} \sin bt$	$\dfrac{b}{(s - a)^2 + b^2}$	$s > a$
$e^{at} \cos bt$	$\dfrac{s - a}{(s - a)^2 + b^2}$	$s > a$
$\dfrac{1}{a} \sin at - t \cos at$	$\dfrac{2a^2}{(s^2 + a^2)^2}$	$s > 0$
$\sinh at$	$\dfrac{a}{s^2 - a^2}$	$s > |a|$
$\cosh at$	$\dfrac{s}{s^2 - a^2}$	$s > |a|$
$t^n e^{at}$	$\dfrac{n!}{(s - a)^{n+1}}$	$s > a$
$u_a(t)$	$\dfrac{e^{-as}}{s}$	$s > 0$
$\delta(t - t_0), t \geq t_0$	e^{-st_0}	
$f(at)$	$\dfrac{1}{a} F\left(\dfrac{s}{a}\right),\ a > 0$	Same as F
$e^{at} f(t)$	$F(s - a)$	Depends on F
$f'(t)$	$sF(s) - f(0)$	Same as F
$f''(t)$	$s^2 F(s) - sf(0) - f'(0)$	Same as F
$f(t - a)u_a(t)$	$e^{-as} F(s)$	Same as F

An Introduction to
Differential Equations

Order and Chaos

The author, center, talking with another participant during a coffee break at a conference in Namur, Belgium, in July 1998.

Florin Diacu studied mathematics at the University of Bucharest, Romania, and received his Ph.D. degree from the University of Heidelberg, Germany, with a thesis on the n-body problem of celestial mechanics. He has held positions at the University of Dortmund, Germany, and Centre de Recherches Mathématiques in Montréal, Canada. He is currently a Professor of Mathematics at the University of Victoria, in Victoria, British Columbia, Canada, and the UVic-Site Director of the Pacific Institute for the Mathematical Sciences. His research interests are celestial mechanics, the qualitative theory of dynamical systems, mathematical physics, and the history and philosophy of mathematics. Among researchers he is known primarily for introducing the study of differential equations given by quasihomogeneous potentials, which have applications in astronomy, physics, and chemistry. He has received several awards. He shares two of them with Philip Holmes of Princeton University for their best-selling book *Celestial Encounters—The Origins of Chaos and Stability,* published by Princeton University Press in 1996 and translated into several languages. He has given invitational talks and lectures at prestigious research institutions and universities in Belgium, Brazil, Bulgaria, Canada, France, Germany, Greece, Hungary, Italy, Mexico, New Zealand, Romania, Spain, Switzerland, and the United States. He is fluent in four languages. His hobbies include reading, fitness workouts, and hiking.

An Introduction to Differential Equations

Order and Chaos

Florin Diacu

Professor of Mathematics, University of Victoria,
Victoria, British Columbia

 W. H. Freeman and Company • New York

Publisher: Michelle Julet
Acquisitions Editor: Craig Bleyer
Marketing Manager: Claire Pearson
Development Editor: Frank Purcell
Cover Designer: Michael Jung
Text Designer: Diana Blume
Illustrations: Publication Services
Illustration Coordinator: Bill Page
Photo Researchers: Kathy Bendo, Vikii Wong
Production Coordinator: Susan Wein
Project Management and Composition: Publication Services
Manufacturing: RR Donnelley & Sons Company

Library of Congress Cataloging-in-Publication Data

Diacu, Florin, 1959-
    An introduction to differential equations: order and chaos / Florin Diacu.
        p. cm.
    Includes bibliographical references and index.
    ISBN 0-7167-3296-3
    1. Differential equations. I. Title.

QA372 .D48 2000
515'.35–dc21

00-028837

Printed in the United States of America

First printing 2000

To my former, present, and future students

Among all mathematical disciplines the theory of differential equations is the most important.... It furnishes the explanation of all those elementary manifestations of nature which involve time.

MARIUS SOPHUS LIE
Lobachevskii Prize, 1897

CONTENTS

PREFACE

During the last decade of the 20th century, the teaching of calculus in North America has seen a split between the classical approach and a more intuitive one, commonly known as "reform." Some universities adopted the new point of view; others chose to stick with tradition. This split influenced the teaching of the sophomore differential equations course, which is a natural extension of the freshman calculus curriculum. The standard path in differential equations is based on exact techniques, whereas the modern one emphasizes qualitative methods. This textbook aims to reconcile these two directions and to offer the student a trip along all the main avenues of the field.

Goals

Our first goal was to strike a balance between these two points of view, keeping what is good and efficient from the classical approach, eliminating its nonessential and heavy-going aspects, and developing the qualitative theory as soon as possible. To present the student with a global and realistic image of the methods used today in differential equations, we cover all aspects of the theory: exact, qualitative, numerical, and computer techniques, as well as some elements of modeling.

Our second goal was to emphasize the usefulness of mathematics. Sophomores have reached the stage where their knowledge of algebra, geometry, and calculus bears fruit. They can apply the acquired tools to understand certain phenomena in fields ranging from physics, chemistry, business, and biology, to linguistics, literature, sports, and art. We have therefore gathered here several achievements of the theory, hoping to convince the students that mathematical thinking is a key component in modern society.

Our third goal was to make the textbook attractive and interesting. This was the most demanding aspect of our work. It involved intense library research, an effort of imagination, sparks of inventiveness, plus continuous refining and revision. Any writer knows how difficult it is to get the words right.

Structure

The textbook contains seven chapters. The first has only an introductory section in which we present the object of the theory and an overview of its methods. We also introduce some classification criteria, several equations and systems important in applications, and a few elements of modeling.

Chapter 2 deals with first-order equations. We present the exact methods for separable equations, discuss the variation of parameters for linear equations, and assign the homogeneous, exact, Bernoulli, and Riccati equations as problems. The qualitative methods involve drawing the slope field and the phase line, understanding the behavior of solutions near equilibria, and deciding about the existence and uniqueness of solutions for initial value problems. The method of successive approximations makes the passage from the qualitative to the numerical approach. We present the Euler and second-order Runge-Kutta methods as well as some rudiments of error theory. The computer techniques offer alternative ways of obtaining exact and numerical solutions and of drawing slope fields. At the end we invite the students to use their knowledge and imagination in doing some library research to come up with original ideas in modeling certain problems.

Chapter 3 presents the fundamental theory of linear second-order equations. We introduce the method of reduction, an algorithm for solving linear homogeneous equations with constant coefficients, the variation of parameters, and the method of undetermined coefficients. The qualitative approach reduces to the study of the phase plane for autonomous equations. We use the characteristic equation to establish the nature of the flow near an equilibrium and provide some elements of structural stability and bifurcation theory. The numerical methods generalize the ones presented in the previous chapter. The computer techniques give some new options for obtaining exact and numerical solutions and for drawing vector and direction fields. We end up with some modeling experiments.

Chapter 4 extends the results of Chapter 3 to linear systems. We introduce some elements of linear algebra, using a formalism that avoids operations with matrices. We present most results in terms of two- and three-dimensional systems, an approach that allows obvious generalization to n dimensions. The computer methods extend the ones of the previous chapter and also include programs for drawing the flow. These computer techniques can also be applied to nonlinear systems. We close the chapter with a few modeling experiments.

Chapter 5 deals with qualitative methods for nonlinear systems. We start with the linearization method and the study of the flow near equilibria based on the Hartman-Grobman theorem. We then investigate simple periodic orbits and cycles for two-dimensional systems in terms of polar coordinates, and we describe the connection between gradient and Hamiltonian systems and their reciprocal flows. We also introduce the notion of Liapunov stability, showing how the Liapunov function method can succeed when linearization fails. The chapter ends with a description of chaos in the language of symbolic dynamics and with some modeling experiments.

Chapter 6 covers differential equations and systems with the help of the Laplace transform. We first introduce the Laplace transform and its inverse and give an existence criterion. We then compute the Laplace transform of elementary and step functions and present a three-step method for solving differential equations and systems. The computer techniques can either apply this method directly or help with carrying it out step by step. We end the chapter with a few modeling experiments.

Chapter 7 considers exact and approximate power series solutions for differential equations. We first introduce a method that provides solutions near regular points and then use the Frobenius theorem to obtain solutions near regular singular points. The computer techniques implement these methods step by step. We end the chapter with some modeling experiments.

Applications

An important selection was the applications. To stress the importance of the theory of differential equations and to make the presentation attractive and interesting, we chose examples from various fields of human activity: anthropology, astronomy, population biology, brewing, business, chemistry, cooking, cosmology, rock climbing, ecology, economics, electronics, engineering, epidemiology, finance, mechanics, medicine, meteorology, oceanography, pharmaceuticals, physics, politics, space science, and sports. Some models are well established. Others are mere didactic toys.

We use differential equations to understand the motion of celestial bodies (pp. 6–9), model prices in a free-market economy (pp. 9–11, 196–197), compute the interest of investments (pp. 27–28), date the Shroud of Turin (pp. 28–29), cook a salmon (pp. 29–30), make a pharmaceutical drug (pp. 36–38), estimate the growth of the cougar population on Vancouver Island (pp. 46–47), follow the landing of Apollo 11 (pp. 47–48), study the swings of Galileo's pendulum (pp. 97–98, 256–258), and participate in some maglev transportation experiments (pp. 107–109). We also analyze the oscillations of water in a pipe (pp. 99–100), understand simple electric circuits (pp. 131–132), shed some light on the Tacoma Narrows Bridge disaster (pp. 116–119), determine the motion of a bungee jumper (pp. 137–139), and study the vibrations of a cantilever beam (pp. 139–141). We follow some chemical reactions in the search for an AIDS vaccine (pp. 183–184), determine the mixtures in a brewing technique (pp. 184–186), find the optimal shape of a rock-climbing tool (pp. 188–189), describe the evolution of two fish populations in the Tasmanian Sea (pp. 195–196), study epidemics with quarantine (pp. 203–204), and test the strength of buffer springs between the cars of trains (pp. 204–206). We further draw conclusions about the change of wolf and fox populations in northern Canada (pp. 232–233), explain the temperature variation of an engine and its coolant (pp. 239–240), see how lobsters scavenge (pp. 248–249), investigate why long-term weather forecasts are unreliable (pp. 260–261), determine the elasticity of a pole-vaulting pole (pp. 317–319), and study the escalation of expenditures in an arms race (pp. 319–320). Though far from displaying the entire spectrum of this theory, we can at least glimpse the variety of phenomena it describes.

Computers

A difficult choice was that of the computer environments. In a field like computer science, in which textbooks become obsolete soon after publication, opinions are changing fast, so it is impossible to satisfy everybody. In the end we decided to go for the three *M*'s: Maple, *Mathematica*, and MATLAB, which are popular in colleges and universities, have better chances of survival, and whose designers promote constructive upgrading. In Sections 2.7, 3.7, 4.6, 6.4, and 7.4 we present Maple, *Mathematica*, and MATLAB separately. This allows instructors a lot of freedom. They can teach one or all of them, treat them as independent entities or use them interactively while covering the other sections, or assign them as homework in connection with a computer project or modeling experiment.

Modeling

This is a difficult and time-consuming issue, which if stressed is done so at the expense of the core material on differential equations. Though we briefly discuss the modeling problem in most of our applications, we decided to emphasize this aspect at the end of every chapter and give instructors the option of assigning lab experiments to the students. Each modeling exercise has as a final goal the writing of an essay, which may contribute to the final grade. In Section 2.8 we deal with money investments, a model of the memory, the landing of Apollo 11, and a population dynamics experiment. Section 3.8 considers Galileo's pendulum, a model for bungee jumping, suspension bridges, and a simple electric circuit. In Section 4.7 we propose free-market models, a system describing malignant tumors and metastasis, another epidemic with quarantine, and a model for a decelerating train. Section 5.6 refers to chaotic aspects of the van der Pol, Düffing, and Lorenz equations, and to the three-body problem of celestial mechanics. In Section 6.5 we model car suspensions, electric circuits with ramped forcing, and instant shocks on harmonic oscillators. Finally, Section 7.5 deals with pole vault and arms race models and electric circuits with variable resistance and capacitance.

Historical Remarks

We mention names, dates, and nationalities for the mathematicians whose results we present, sometimes adding brief historical remarks. Some books use separate notes for this purpose. We chose to include the historical facts in the text in order to convey the feeling that mathematics is a cultural edifice built through collective human efforts.

Style

We tried to be direct and concise. Bombarded with information, students have no time and no desire to read more than they need. So we attempted to follow a geodesic toward our goal, keeping theory to a minimum. But we were generous with metaphors, figures, and examples, which give the text a friendly look, allowing a better and faster understanding. In most cases we took the route metaphor-theory-example, but sometimes we favored a more heuristic approach. We also aimed to strike a balance between rigor and intuition, relying on the latter whenever the former endangered clarity.

Teaching

As mentioned earlier, the text presents several methods—analytic, numerical, qualitative, and computer-based—emphasizing that each has its merits, depending on the circumstances. All these aspects can be covered in one term. But unless instructors have at their disposal 16 or 17 weeks, they will be unable to teach all the material. In the fall of 1998 and the spring of 1999, we covered Chapters 1 through 6 during the usual 13-week term at the University of Victoria, leaving aside the computer techniques, for which we used only 1 hour of demonstrations. Alternatively, in the summer of 1999, Cristina Stoica replaced Chapter 6 with Chapter 7. But there are many other choices an instructor can make with regard to chapters and even sections. For a more classical approach, Chapters 1, 2, 3, 4, 6, and 7 would be adequate. The computer sections can be intensively used or totally ignored. Somebody uninterested in numerical methods can simply avoid them. The only rule to follow is represented in the diagram below, which explains the logical construction of the textbook. Chapter 5, for example, can be taught only after going through Chapters 1, 2, 3, and 4. However, instructors have a lot of flexibility in what they can choose to teach and skip.

$$
\begin{array}{c}
6 \\
\uparrow \\
1 \rightarrow 2 \rightarrow 3 \rightarrow 4 \rightarrow 5 \\
\downarrow \\
7
\end{array}
$$

Acknowledgments

Many people helped me along the way. First I would like to thank my graduate student and friend Bogdan Verjinschi, the main author of the solutions manual, who followed the development of this project from the very beginning. He taught the tutorials for this course in the fall of 1998; spring, summer, and fall of 1999; and spring of 2000. He corrected many mistakes, proposed some problems, and suggested several of the physics applications. I would also like to thank my undergraduate students and friends Jacquie Gregory and Heather Seymour, who collaborated with Bogdan and myself in the summer of 1999, solved many of the problems, checked the remaining ones, and together with Cara Stewart and Matthew Theobald helped produce the camera-ready test for the solutions manual. Also I want to address my gratitude to Frank Purcell, a highly experienced free-lance mathematics editor, who read the manuscript in detail and suggested many changes and didactical improvements. Also, many thanks to the undergraduate students Daniel Flath, Alex Hertel, Philipp Hertel, Ryan Johnson, Karl Kappler, Rodney Marseille, Shane Ryan, and Francis Vitek, who corrected many typos in an earlier version of the manuscript. And a warm word of thanks to all the students that took this course with me in the fall of 1998, spring of 1999, and spring of 2000.

Several colleagues and friends at universities in Canada and the United States read parts of the manuscript or led discussions with me, made corrections, or suggested improvements. I would like to thank here Edoh Amiran, Jacques Bèlair, Pauline van den Driessche, Jinqiao Duan, Ramesh V. Garimella, Tomas Gedeon, Mathias Kawski, Melvin D. Lax, Douglas B. Meade, Jim Meiss, William G. Newman, Robert

McOwen, Rick Moeckel, Dan Offin, Donald Saari, David A. Sanchez, Carl Simon, Cristina Stoica, John Travis, Richard D. West, Jane Ye, and all the anonymous reviewers of the initial proposal. Of course, any remaining mistakes are entirely due to the author.

The editors and staff at W.H. Freeman and Company played an important role in this project. I would like to thank Richard Bonacci, who contacted me in the spring of 1997 and asked me to write an introductory textbook in differential equations. Without his enthusiasm and encouragement I might never have started. In addition, I would like to thank Michelle Russel Julet, the publisher of this text and an excellent short-time replacement after Richard's departure, and Craig Bleyer, the new acquisitions editor, who worked closely with me during the intense part of finalizing the project and seeing the book into print. I would also like to thank the editorial assistants Tim Solomon, Tim Johnson, Melanie Mays, and Christina Powell, who collaborated with Richard, Michelle, and Craig.

Several people helped with producing the book form of this text and with its marketing. I would like to acknowledge here Tami Craig, the manuscript preparation coordinator; Mary Louise Byrd, the project editor; Vikii Wong and Kathy Bendo, the photo researchers; Diana Blume, the book designer; Maria Eps, the art director; Claire Pearson, the marketing manager; Susan Wein, the production coordinator; Bill Page, the illustration coordinator; Kris Engberg, Jason Brown, David Mason, and everybody else at Publication Services in Champaign, Illinois; and all those responsible with manufacturing at RR Donnelley & Sons. Also thank you to George Lobell, the mathematics editor at Prentice Hall, for his interest in the initial project. I would like to apologize to anyone who has been involved and whom I forgot to mention.

Last but not least, I would like to thank Marina and Razvan for all their support, patience, and understanding. And I shouldn't forget Nina, our Siberian Husky pup, who, tired after our evening run, would doze at my feet while I was writing.

1 Introduction

Mathematics has, in modern times, brought [the idea of] order into greater and greater prominence.

BERTRAND RUSSELL
Nobel Prize for Literature, 1950

On November 7, 1940, some astounding pictures made the headlines of the North American television channels. They showed a man struggling to reach a car abandoned on a bridge that was wildly waving in the storm. After several unsuccessful attempts he gave up and, with visible efforts, returned to the shore. This proved a wise decision. The Tacoma Narrows Bridge near Seattle, a suspension structure more than a mile long, collapsed minutes later, witnessed by the helpless eyes of those who had dared to approach it. How could a construction of iron and concrete wave for days like a flag in the wind and then break all of a sudden? The answer is difficult (see Section 3.4), but some insight can be given through the theory of differential equations.

The history of science enumerates many achievements of this theory. The first is due to the English mathematician and physicist Isaac Newton (1642–1727), who in addition to being its cocreator, used it to show that the force that keeps the moon on its orbit is the same as the one that makes objects fall to the ground.

Shortly after this resounding success, the astronomer Edmond Halley (1656–1742), a friend of Newton, noticed the similarity of four cometary orbits observed in 1456, 1531, 1607, and 1682 and wondered whether they represent the same periodic trajectory. Using Newton's theory, Halley computed that the comet would return in 1758. He did not live to see the event, but his prediction proved accurate, and Halley's Comet has appeared in earth's skies on schedule three times more since then.

The discovery of the planet Neptune through numerical computations, performed independently by the French astronomer Jean Joseph Le Verrier (1811–1877) and the English astronomer John Couch Adams (1819–1892), was another significant success for the theory of differential equations. The observed orbit of Uranus had disagreed with the one predicted by theory. The two scientists argued that the discrepancy was due to the existence of some unknown planet. Using numerical methods, they computed the orbit of this hypothetical object, which was then observed on September 18, 1846.

First applied to the physical sciences, the theory of differential equations has later extended to other human activities ranging from engineering and biology to medicine, business, history, sports, and arts. The goal of this textbook is to introduce you to the main methods, ideas, techniques, and applications of this branch of mathematics, whose strength lies in its large applicability.

The Object of Study

A *differential equation* relates an unknown function and one or more of its derivatives. For example, the equation

$$x' = x \tag{1}$$

relates the function $x = x(t)$ and its derivative $x' = dx(t)/dt$. Unlike the unknowns of algebraic equations, which are numbers, the unknowns of differential equations are functions. Solving a differential equation means finding all its *solutions*, i.e., all functions that satisfy the equation. For example, $x(t) = e^t$ is a solution of equation (1) because $(e^t)' = e^t$. Can you find another solution?

The theory of differential equations has three main branches, which involve exact, numerical, and qualitative methods. Let us briefly describe them.

The *exact methods* are those meant to obtain all the solutions of a given equation. They first appeared more than three centuries ago at the same time as calculus. Though fundamental for understanding and developing further concepts, the exact methods have a narrow range of applications, because only a few classes of equations can be completely solved.

The *numerical methods* are designed to obtain, with some reasonable accuracy, particular solutions of a given equation. They have thrived during recent decades due to the invention of modern computers. Today these methods are widely used in practical problems ranging from physics and engineering to psychology and art, but, for reasons we will understand later, they offer good approximations only locally, i.e., on small intervals of the solution's domain. Therefore, in practical time-dependent problems, long-term predictions are difficult to achieve with this approach.

The *qualitative methods* are used to investigate properties of solutions without necessarily finding those solutions. For example, questions regarding existence and uniqueness, stability, or chaotic or asymptotic behavior can be answered with the help of these methods. Except for existence and uniqueness theorems, which appeared early in the development of the theory, the mainstream qualitative methods began to be developed toward the end of the 19th century, mainly through the work of the French mathematician Henri Poincaré. These methods are successful in understanding fundamental issues of the theory of differential equations.

We could add to this classification the process of *modeling*, which deals with obtaining the equations that describe certain phenomena and with interpreting the results of their analysis within the framework of the model. However, this is a much larger subject that goes beyond the theory of differential equations. We will present two examples of modeling later in this section and deal with this aspect in many of our applications.

The study of differential equations is a difficult task. Only a combination of quantitative, numerical, and qualitative methods brings insight toward understanding most problems. Mastering this theory at the research level requires knowledge in several branches of mathematics as well as a taste for applications. At the introductory level it asks for a solid background in mathematics, which includes the basic notions and techniques of calculus, algebra, and geometry.

Classification There are several ways of classifying differential equations, of which we will consider here four criteria.

Domain of the unknown The main distinction is between *ordinary differential equations* (ODEs) and *partial differential equations* (PDEs). ODEs involve functions of one variable and their derivatives, whereas PDEs concern functions of several variables and their partial derivatives.

> **EXAMPLE 1**   The equations
>
> $$x' = 2x^2, \tag{2}$$
>
> $$u' = 2u - t^3, \tag{3}$$
>
> $$v'' - 2tv' + v - 6 = 0 \tag{4}$$
>
> are ODEs. For simplicity, the argument t of the functions $x, u,$ and v is omitted. The equation
>
> $$\frac{\partial u}{\partial t} = 2\left(\frac{\partial u}{\partial x}\right)^2 - 3xy\left(\frac{\partial u}{\partial y}\right)^3, \tag{5}$$
>
> where the unknown function is $u(t, x, y)$, is a PDE. In this textbook we will deal only with ODEs.

Number of the unknowns We distinguish between *single differential equations* and *systems of differential equations*.

> **EXAMPLE 2**   All previous ODE examples have been single equations. The following ones are systems:
>
> $$\begin{cases} x' = -y \\ y' = x, \end{cases} \tag{6}$$
>
> $$\begin{cases} x_1' = x_1^2 + 5x_2^3 - x_3 \\ x_2' = \frac{1}{2}tx_1 - 3 \\ x_3' = x_1x_2x_3 \sin t. \end{cases} \tag{7}$$
>
> The first system is two-dimensional and the second is three-dimensional.

Structure of the equation   We distinguish between *linear differential equations* and *nonlinear differential equations*. Linear equations are those whose left- and right-hand sides are linear functions (i.e., polynomials of degree 1) with respect to the unknown and its derivatives, whereas nonlinear ones do not satisfy this property. Linear systems are those formed by linear equations only, whereas nonlinear ones are those involving at least one nonlinear equation.

EXAMPLE 3   The equations

$$P' = t^2 P, \tag{8}$$

$$u'' = (\sin \theta)u' + (\cos \theta)u \tag{9}$$

are linear in spite of having nonlinear coefficients (in t and θ, respectively). The equations

$$x' = 3x^2 + t, \tag{10}$$

$$y'' = -y' + yy' \tag{11}$$

are nonlinear. Also, system (6) in Example 2 is linear, whereas (7) is nonlinear.

Order of the equation   We say that an equation has *order k* if the highest derivative involved in the equation has order k.

EXAMPLE 4   The equations

$$w' = -4w + 3, \tag{12}$$

$$2X' - 5X'' = X + 7t^4, \tag{13}$$

$$\xi^2 x''' = 6x' + x'' - 2\xi x \tag{14}$$

have order 1, 2, and 3, respectively. Systems (6) and (7) in Example 2 both have order 1.

Applications   The examples above have been chosen artificially, in the sense that they do not necessarily describe natural phenomena. But since the theory of differential equations is mainly concerned with those equations that have applications in other fields of human activity, we will present some examples from physics, astronomy, meteorology, chemistry, biology, anthropology, medicine, economics, and engineering. The area of applications is much larger than what we show here.

Some of the equations below are easy to solve, and we will solve them later; others continue to defy our attempts at obtaining explicit solutions. Progress in understanding them has been slow so far. Such differential equations can take the life-work of several generations of mathematicians and still remain poorly understood.

(i) The equation

$$x' = k \cdot x \qquad (15)$$

models growth or decay problems. It says that, at a certain time, the rate at which a given quantity is changing is proportional to the amount existing at that time. This equation is used, for example, to determine the half-life of a radioactive substance or the doubling time of a money investment with continuously compounding interest. The well-known carbon dating method used in anthropology is based on this simple equation. In this case x represents the quantity of radioactive substance and k is a constant characteristic of the substance. The value of the constant can be determined through practical experiments and measurements. We will study this equation in detail in Section 2.2.

(ii) Newton's law of cooling or heating differs slightly from (15),

$$T' = k(T - \tau), \qquad (16)$$

where k and τ are constants. It models the cooling or heating of a body immersed in a medium of constant temperature τ, where T is the unknown temperature function of the body, and k is a constant depending on the body. This equation describes a known physical phenomenon: that the rate at which the temperature of a body is changing is proportional to the difference between the temperatures of the medium and the body. We will study this equation in detail in Section 2.2.

(iii) The equation

$$u' = k(a - u)(b - u), \qquad (17)$$

where k, a, and b are constants, models mixing problems in chemistry.

(iv) The logistic equation

$$p' = \lambda p(\alpha - p) \qquad (18)$$

is a special case of equation (17), where α and λ are constants; it is used in biology and medicine for simple population and epidemiological models. We will study this equation in Section 2.4 in connection with a model that describes the evolution of the cougar population on Vancouver Island.

(v) The equation

$$x'' + bx' + kx = \gamma \sin \omega t, \qquad (19)$$

where b, k, γ, and ω are constants, models a simple electric circuit or the motion of a damped spring with a periodic forcing term. It is connected to many practical problems, which range from bungee jumping to the collapse of the Tacoma Narrows Bridge. We will consider this equation in Section 3.4.

(vi)   A special form of van der Pol's equation is

$$x'' + \alpha(x^2 - 1)x' + x = \beta \cos \omega t, \tag{20}$$

where α, β, and ω are constants. This equation models an electric circuit with a triode, the resistive properties of which change with the current.

(vii)   A version of Düffing's equation considered in 1979 by the British-American engineer and mathematician Philip Holmes,

$$x'' + \delta x' - x + x^3 = \gamma \cos \omega t, \tag{21}$$

describes the dynamics of a buckled beam or plate that vibrates in one direction. Here x represents the displacement of the beam from the equilibrium position, and δ, γ, and ω are constants. This equation is used in mechanical engineering, for example, in modeling the action of leaf springs on trucks. We will consider this system in Sections 3.6 and 5.5.

(viii)   The Lorenz equations

$$\begin{cases} x' = \sigma(y - x) \\ y' = \rho x - y - xz \\ z' = -\beta z + xy, \end{cases} \tag{22}$$

where $\sigma, \rho, \beta > 0$ are constants, describe the fluid convection in a two-dimensional layer heated from below. This set of equations has important applications in meteorology. Proposed in 1963 by the American meteorologist Edward Lorenz, it is little understood today. It has become famous because of the so-called chaotic character of its solutions—a property that, though discovered by Poincaré in 1889 in tackling the three-body problem, has become mainstream research mainly due to Lorenz's, Düffing's, and van der Pol's equations. We will discuss the notion of chaos in Chapter 5 and will deal with the Lorenz system in Section 5.5.

(ix)   The celebrated *three-body problem* of celestial mechanics models the motion of three bodies of masses m_1, m_2, m_3 under the influence of gravitation, such as the system of the sun, earth, and moon. The equations are given by a nine-dimensional second-order system,

$$\begin{cases} q_{1i}'' = Gm_2 \dfrac{q_{2i} - q_{1i}}{r_{21}^3} + Gm_3 \dfrac{q_{3i} - q_{1i}}{r_{31}^3} \\[2mm] q_{2i}'' = Gm_1 \dfrac{q_{1i} - q_{2i}}{r_{12}^3} + Gm_3 \dfrac{q_{3i} - q_{2i}}{r_{32}^3} \qquad (i = 1, 2, 3), \\[2mm] q_{3i}'' = Gm_1 \dfrac{q_{1i} - q_{3i}}{r_{13}^3} + Gm_2 \dfrac{q_{2i} - q_{3i}}{r_{23}^3} \end{cases} \tag{23}$$

in which G is the gravitational constant, q_{ji} $(j, i = 1, 2, 3)$ represents the ith coordinate of the jth body in a three-dimensional frame, and r_{ji} $(j, i = 1, 2, 3)$ is the distance between the jth and ith bodies. In spite of three centuries of efforts, in which 10 generations of mathematicians have written thousands of research papers, this system has revealed very few of its mysteries. In the following subsection we will show how to derive the equations describing the gravitational motion of two bodies, a particular case of the above problem, which is better understood today.

(x) The system

$$\begin{cases} x' = au(t)y \\ y' = [kv(t) - u(t)]y, \end{cases} \tag{24}$$

where a, k are constants and u, v are known functions, has applications in economics for diverse planning problems.

Modeling　　Many natural phenomena can be modeled mathematically. In general, modeling is a difficult task that requires interdisciplinary knowledge and skills. In this textbook we will learn a few modeling techniques by working through some accessible examples. We start with two problems that lead to systems of differential equations.

The Kepler problem　　Although the three-body problem is very complicated, the gravitational motion of two bodies is well understood. The Kepler problem of celestial mechanics, named after the German astronomer Johannes Kepler

Astronomer Johannes Kepler.

Figure 1.1.1. The coordinates of the point of mass M in the Kepler problem.

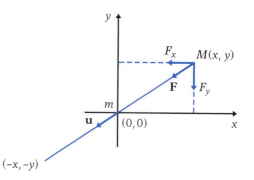

(1571–1630), was proposed by Newton as a model for the motion of the earth around the sun, ignoring the moon's influence.

Assume that two bodies (point masses) moving in a plane are mutually attracted by a force that is directly proportional to the product of their masses and inversely proportional to the square of their distance, where the proportionality factor is the gravitational constant G. Let the bodies of mass m and M be situated at $(0, 0)$ and (x, y), respectively (see Figure 1.1.1). Then the distance between them is $(x^2 + y^2)^{1/2}$, so the magnitude of the attraction force exerted by m on M is $GmM/(x^2 + y^2)$. But this force acts in the direction $(-x, -y)$. The unit vector of this direction is

$$\mathbf{u} = \frac{(-x, -y)}{(x^2 + y^2)^{1/2}}.$$

(Can you explain why it has length 1?). Thus, to show the direction of the force $\mathbf{F}$ acting on M, we must multiply the magnitude of this force by the unit vector $\mathbf{u}$, which means that

$$\mathbf{F} = \frac{(-x, -y)GmM}{(x^2 + y^2)^{3/2}}.$$

If we write this force on the two components F_x and F_y along the coordinates, we obtain

$$\begin{cases} F_x = -\dfrac{GmMx}{(x^2 + y^2)^{3/2}} \\ F_y = -\dfrac{GmMy}{(x^2 + y^2)^{3/2}}. \end{cases} \tag{25}$$

On the other hand, Newton's second law states that *force = mass × acceleration*. In physics the acceleration is taken to be the second derivative of the position with respect to time. So the components F_x and F_y of the force $\mathbf{F}$ exerted by m on M are

$$\begin{cases} F_x = Mx'' \\ F_y = My''. \end{cases} \tag{26}$$

Figure 1.1.2. Possible orbits in the Kepler problem.

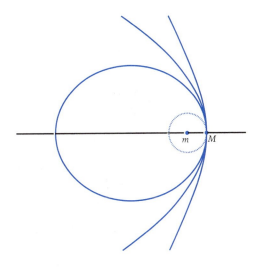

Comparing (25) and (26), we obtain

$$\begin{cases} x'' = -\dfrac{Gmx}{(x^2 + y^2)^{3/2}} \\ y'' = -\dfrac{Gmy}{(x^2 + y^2)^{3/2}}, \end{cases} \qquad (27)$$

a system of two second-order equations that describe the relative motion of M with respect to m.

In his masterpiece *Principia* (1687), Newton used a geometrical approach to derive these equations, so their formal appearance was different from (27). The explicit solution was given in 1710 by the Swiss mathematician Johann Bernoulli (1667–1748), who showed that the motion of M with respect to m can be an ellipse (a circle in particular), a parabola, a hyperbola, or a straight line (see Figure 1.1.2).

Prices in a free-market economy

In the second part of the 18th century, the Scottish economist Adam Smith (1723–1790) asserted that in a free-market economy governed by the law of demand and supply, prices will eventually tend to an equilibrium. Let us derive a system that models the price change for one commodity (i.e., a product or service).

If d and s are the time-dependent functions that denote the *demand* and the *supply* of a given commodity, then $r = d - s$ is called *excess demand*. Let p denote the price of this commodity (which is also a function of t) and assume that p and r are differentiable. The law of demand and supply works as follows:

(1) If at a given time, demand is less than supply, prices decrease, so the demand increases; therefore the excess demand increases.

(2) If at a given time, demand is greater than supply, prices increase, so the supply increases; therefore the excess demand decreases.

(3) If at a given time, demand equals supply, prices are constant, so the excess demand remains constant, namely 0.

Economist Adam Smith.

Retaining only the desired information and assuming that r and p are differentiable functions of t, we can rewrite the above laws as follows:

(L1) $r(t) < 0 \Rightarrow p'(t) < 0 \Rightarrow r'(t) > 0$;
(L2) $r(t) > 0 \Rightarrow p'(t) > 0 \Rightarrow r'(t) < 0$;
(L3) $r(t) = 0 \Rightarrow p'(t) = 0 \Rightarrow r'(t) = 0$.

We can thus ignore the change in demand and supply and deal only with the change in excess demand. According to Adam Smith, an economy is a *free market* if the laws L1, L2, and L3 are the only ones that rule the variations in price and excess demand.

We can now write the equations that govern a free market. From the first implication in L1, L2, and L3, we see that a possible way to express these laws is that the rate of change in price is proportional to the excess demand, so in a first approximation $p' = \alpha r$, where $\alpha > 0$ is a constant. Looking at the relation between r and r' in L1, L2, and L3, we similarly write that $r' = -\beta r$, where $\beta > 0$ is a constant. Combining the two equations, we obtain the system

$$\begin{cases} p' = \alpha r \\ r' = -\beta r \end{cases} \qquad \alpha, \beta > 0. \tag{28}$$

This is of course only a first approximation, since we do not know whether a proportionality law truly describes the free market.[1] Though the equations in (28) are in agreement with L1, L2, and L3, the market might be more complicated. A more accurate description of a free-market economy should involve a system of the type

$$\begin{cases} p' = \alpha f(r) \\ r' = -\beta h(r), \end{cases} \tag{29}$$

where f and h are functions with certain properties. What would be the natural conditions to impose on f and h such that the law of demand and supply remains true? The choice is, of course, not unique and the experience of economists can now be of help. A possibility is

(C1) $f, h < 0$ for $r < 0$,
(C2) $f, h > 0$ for $r > 0$,
(C3) $f(0) = h(0) = 0$,
(C4) $\lim_{r \to 0} f'(r) = 0$.

Conditions C1, C2, and C3 are such that system (29) agrees with L1, L2, and L3, whereas condition C4 tells us that when the difference between demand and supply is small, the variation in price is also small. Without C4, a function like $f(r) = 1/r$ would satisfy the conditions C1–C3, but such a function has nothing to do with prices. Though we could have replaced C4 with, say, the condition that f is increasing, such a requirement is more restrictive, and we prefer to keep the options as general as possible.

---

[1]Recall that if a quantity x increases proportionally with a quantity y, then there is a constant $k > 0$ such that $x = ky$.

In Section 4.4 we will study system (28) and show that the solutions tend to equilibria. The same idea used for the linear model can then be applied to system (29) when f and h satisfy the conditions C1–C4. Moreover, it will be interesting to notice that if other forces occur, i.e., if the system changes to

$$\begin{cases} p' = \alpha f(r) + u_0 \\ r' = -\beta h(r) + v_0, \end{cases}$$

then, in general, prices do not reach an equilibrium.

The field of mathematical economics has developed many models regarding prices, and some of them exhibit chaotic behavior, as defined in Section 5.5. These models, however, are beyond the scope of this textbook.

The following chapters will include exact, numerical, and qualitative methods as well as results from the theory of ordinary differential equations. After studying this textbook in detail, you will have not only a good image of the importance, complexity, and beauty of this mathematical theory, but also a background for further theoretical or applicative studies and research.

PROBLEMS

In each of Problems 1 to 8, tell whether the given equations are linear or nonlinear and determine their order.

1. $3x'x = 2t^2$

2. $(\sqrt{3} - 1)t^3 x'' + t^2 x' - 3x + \cos t = 0$

3. $\frac{2}{3}x''' - 7x'' + x' + 9x = 0$

4. $y'' + \frac{17}{3}yy' + \ln t = \sqrt{3}$

5. $x' + 2e^x = 0$

6. $x'' = \dfrac{\sqrt{2}}{2}\cos t$

7. $uu't + u'u''t^2 + u''ut^3 = 0$

8. $v^{(4)} - xv''' + x^2 v'' - x^4 v' + x^8 v + x^{16} = 0$

9. Characterize each of the equations (15) through (24) from the point of view of the criteria given in the subsection "Classification."

10. Show how Newton's gravitational law leads to the equations (23), which model the three-body problem.

11. Molecular physics deals with models in which molecules repel each other with a force proportional to the seventh power of their distance (or ninth power for some types of molecules). They are called van der Waals (pronounced "fahn der vahlss") forces, after the Dutch physicist Johannes Diderik van der Waals (1837–1923), who won the Nobel Prize for Physics in 1910. Write the equations that describe the motion of two molecules for each of the above forces.

12. Using $\rho = s - d$ instead of the excess demand function, where s is the supply and d is the demand, write the system modeling the price of a commodity in a free-market economy.

13. Consider two commodities that depend on each other in the sense that one is always a component of the other but can also be commercialized independently. The price of one commodity is approximately double that of the other. Find a model that describes the change in prices in a free-market economy.

14. In 1798 the British scientist Thomas Robert Malthus (1766–1834), a pioneer of population science and economics, stated that populations grow faster than the means that can sustain them. One of his arguments was that the rate of change in a given population is directly proportional to the size of the population. Assuming that a population can be approximated by a differentiable function,

find a model that agrees with Malthus's theory. Can you obtain a solution of the corresponding equation?

15. Unable to explain the moon's motion accurately, Newton thought of replacing the inverse-square force with one in which the attraction is of the type $\alpha/r^2 + \beta/r^3$, where α and β are positive constants and r is the distance between bodies. He published some results on this gravitational law in the first edition of *Principia* (1687) but kept most of them in manuscript form; they can be found today in the Portsmouth Collection at Trinity College, Cambridge, England. Unaware of Newton's results, the Bulgarian physicist Georgi Manev found in the 1920s a physical motivation for this law if $\alpha = Gm_1m_2$ and

$$\beta = \frac{3G^2}{c^2}m_1m_2(m_1 + m_2),$$

where m_1, m_2 are the masses of the bodies, G is the gravitational constant, and c is the speed of light. Manev's law explains several astronomical phenomena (such as the perihelion advance of Mercury) with the same accuracy as the theory of general relativity. Write the equations of the Manev two-body problem.

16. The Lennard-Jones potential is used in chemistry in the theory of crystals. One of its goals is to explain why certain configurations lead to stable crystal formations while others do not. The Lennard-Jones force is of the form $-a/r^7 + b/r^{13}$, where a and b are positive constants and r is the distance between molecules. Write the Lennard-Jones equations for two molecules, each assumed to have unit mass.

17. *Quasihomogeneous forces* appear in the theory of particle systems and have the form $\alpha/r^a + \beta/r^b$, where α, β, a, and b are constants (with $a, b > 0$) and r is the distance between particles. Write the equations of motion for two and then for three particles, each assumed to have unit mass.

2 First-Order Equations

> The possibility of using differential equations is, to my mind, the final justification for the use of quantitative models in science.
>
> RENÉ THOM
> Fields Medal, 1958

2.1 Equations and Solutions

In this section we will define the concepts of first-order ordinary differential equation, vector field, solution, initial value problem, and solution of an initial value problem. We will also introduce the notions of general solution and singular solution and present several examples. Many of the basic concepts and techniques acquired in your calculus courses will appear from now on; therefore, we suggest that you refer to a calculus textbook every time you feel unsure about those aspects we assume known.

First-Order Equations

The *general form* of a first-order ordinary differential equation is given by

$$F(t, x, x') = 0,$$

where F is a function that relates the variables $t, x,$ and x'. This allows us to represent the whole class of first-order equations without writing any particular equation down.

EXAMPLE 1   For the equation

$$x' = x,$$

the function F can be defined as $F(t, x, x') = x' - x$. Notice that F is independent of t.

EXAMPLE 2   For the equation

$$x' = -5t^2,$$

the function F can be defined as $F(t, x, x') = x' + 5t^2$. In this case F is independent of x.

EXAMPLE 3   The function F defining the equation

$$3tx' = 6x + \sin t$$

is $F(t, x, x') = 3tx' - 6x - \sin t$.

It is often inconvenient to work with the general form. Many equations encountered in applications can be expressed in the *normal form,*

$$x' = f(t, x).$$

F is now $F(t, x, x') = x' - f(t, x)$. We call f the *vector field,* t the *independent variable,* and x the *dependent variable,* since it depends on t. So the unknown x is a function.

Comment on the notation. As in equations encountered in algebra, x denotes the unknown, except that instead of being a number, it is now a function. The use of t as the variable of the function x suggests that many differential equations model phenomena that change in time. (However, to stimulate flexibility in handling differential equations, we will use alternative notations in many of the exercises.)

EXAMPLE 4   For the equation

$$x' = x,$$

the vector field is $f(t, x) = x$. Notice that f is independent of t.

EXAMPLE 5   For the equation

$$x' = -5t^2,$$

the vector field is $f(t, x) = -5t^2$. Here f is independent of x.

EXAMPLE 6   For the equation

$$x' = \frac{5t^2 - 1}{2t} x,$$

the vector field is

$$f(t, x) = \frac{5t^2 - 1}{2t} x.$$

Notice that f is undefined at $t = 0$.

As the terminology suggests, the *general form* of a first-order equation is more general than the *normal form.* The *general form* can represent *any* first-order equation, even monstrosities like $(x')^5 - \sin(x' + x^2) + t^3 x = 0$. (Such an equation, however, cannot be written in normal form. Can you explain why?)

Vector Fields   As we will see in Chapters 3 and 4, the terminology *vector field* is inspired from second-order equations and two-dimensional systems, for which the function f generates vectors in a plane. For first-order differential equations the notion of *vector line* would be better suited. For example, the vector field $f(x) = \frac{1}{2}x$, which defines the equation

$$x' = \frac{1}{2}x,$$

Figure 2.1.1. The line of vectors generated by the vector field $f(x) = \frac{1}{2}x$.

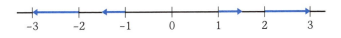

can be represented on the x-axis as in Figure 2.1.1. At $x = 1$, $f(1) = \frac{1}{2}$, so we draw a vector of length $\frac{1}{2}$ based at 1, oriented to the right. For $x = -1$, $f(-1) = -\frac{1}{2}$, so we obtain a vector of length $\frac{1}{2}$ based at -1 and oriented to the left. Analogously, at $x = 0$ we have the zero vector, i.e., a point. In this way we generate a line of vectors.

The General Solution The line of vectors suggests more than terminology. Suppose a car is speeding along a road (as in Figure 2.1.1) and each vector represents its velocity at a given point. Let us find a formula that describes the position of the car at every instant.

This is equivalent to finding a *solution* of the equation $x' = \frac{1}{2}x$, i.e., a function $x(t)$ (the position of the car at time t) whose derivative $x'(t)$ (the velocity of the car at time t) takes the values specified by the vector, i.e., $\frac{1}{2}x(t)$. With this metaphor in mind, let us define what a *solution* means.

DEFINITION 2.1.1

A differentiable function φ that satisfies the relation

$$\varphi'(t) = f(t, \varphi(t)) \tag{1}$$

for every t in some open interval is called a *solution* of the equation $x' = f(t, x)$. Similarly, we define a solution of the equation $F(t, x, x') = 0$ by substituting relation (1) with

$$F(t, \varphi(t), \varphi'(t)) = 0. \tag{2}$$

EXAMPLE 7 For the equation $x' = \frac{1}{2}x$, the function $\varphi(t) = e^{t/2}$ is a solution. Indeed, φ is differentiable and $\varphi'(t) = \frac{1}{2}\varphi(t)$. This shows that if the distance increases proportionally with the velocity, the car moves exponentially fast. Another solution is the zero function $\psi(t) = 0$. In this case the car remains forever at rest.

Let us play a little bit with this model. Since we can choose any units we like, suppose that x is measured in miles east or west of home and x' is measured in miles per hour. So if we are 10 miles west of home, we are driving west at 5 mph. The solution of the equation will tell us where the car is at any time if we know its position at some particular time. If at time 0 we are at home (i.e., $x(0) = 0$), our speed is $x'(0) = x(0) = 0$, so we are stuck at home forever. This is the zero-function solution $\psi(t) = 0$. For the solution $\varphi(t) = e^{t/2}$, $\varphi(0) = 1$, so at $t = 0$ we are one mile west of home. Now that we have the solution φ, we can determine our distance from home at any time. Ten hours after time 0 we are $\varphi(10) = e^5 \approx 148$ miles west of home, driving at a speed of 74 mph. Notice that since we are now west of home, we could not have started from home. If we had, we would have been stuck there forever, as the zero solution $\psi(t) = 0$ proves. So we must always have been somewhere west of home. For example, 24 hours before time 0 we were $\varphi(-24) = e^{-12}$ miles west of home, which is a bit less than half an inch. What was our speed then? Where were we one week before time 0?

EXAMPLE 8   For the equation $x' = -5t^2$, a solution is $v(t) = -\frac{5}{3}t^3$. Indeed, x is differentiable and $v'(t) = -5t^2$. Straightforward integration leads to infinitely many solutions: functions of the form $-\frac{5}{3}t^3 + c$, where c is a real constant. (Can you check this?)

EXAMPLE 9   For $x' = 3t^2x$, $u(t) = e^{t^3}$ is a solution. Indeed, u is differentiable, $u'(t) = 3t^2e^{t^3}$, and $3t^2u(t) = 3t^2e^{t^3}$, so u satisfies the equation.

For some of the above equations we guessed more than one solution, even infinitely many in one case. The set of *all* solutions of a differential equation is called the *general solution*, to distinguish it from single (particular) solutions like the ones we have checked in most examples. In Sections 2.2 and 2.3 we will present some methods that provide the general solutions for certain classes of equations.

Comment on the notation. We will usually denote by x the general solution[1] of an equation $x' = f(t, x)$. In this case x represents a set of (infinitely many) functions. When picking up a particular solution, i.e., a single function of this set, we denote it by a different letter, say, φ, ϕ, u, v, etc.

Singular Solutions

The class of *singular solutions* is formed by the constant functions that cancel the vector field. They are particular solutions contained in the general solution. We can formally define them as follows.

DEFINITION 2.1.2

A function $\varphi(t) = \gamma$, where γ is a constant such that $f(t, \gamma) = 0$ for all t, is called a *singular solution* of equation (1).

Notice that any function with the properties in Definition 2.1.2 is a solution as required by Definition 2.1.1. (Can you explain why?) Let us provide some examples of singular solutions.

EXAMPLE 10   The equation

$$x' = 2x^2$$

has the zero function, $\varphi(t) = 0$, as a singular solution. Indeed, φ is constant and the vector field $f(x) = 2x^2$ cancels for $x = 0$, i.e., $f(0) = 0$.

[1]If the general solution cannot be written as a single formula, as in Example 15, x may denote less than the general solution, but still a set of infinitely many solutions.

EXAMPLE 11   The equation

$$x' = -3t(x - 5)$$

has $\psi(t) = 5$ as a singular solution. Indeed, ψ is constant and the vector field $f(t, x) = -3t(x - 5)$ cancels at $x = 5$ for any t, i.e., $f(t, 5) = 0$.

EXAMPLE 12   The equation

$$x' = 4x^2 - 1$$

has the constant functions $u(t) = \frac{1}{2}$ and $v(t) = -\frac{1}{2}$ as singular solutions. Indeed, the vector field $f(x) = 4x^2 - 1$ cancels at $\frac{1}{2}$ and $-\frac{1}{2}$.

EXAMPLE 13   The equation

$$x' = \frac{1}{x}$$

has no singular solutions, because there is no constant that cancels the vector field $f(x) = 1/x$.

As we will see in Sections 2.2 and 2.3, it often happens that the singular solutions are particular cases of a formula obtained using a certain method, which provides infinitely many solutions. But sometimes this formula fails to include one or more singular solutions.

EXAMPLE 14   The method of Section 2.2 gives for the solutions of the equation

$$x' = x^{2/3}$$

the formula $x(t) = \left(t/3 + c\right)^3$. But the singular solution $\psi(t) = 0$ cannot be recovered from this formula, since there is no real value of c for which $\left(t/3 + c\right)^3 = 0$ for all t.

EXAMPLE 15   As we have seen, $u(t) = -\frac{1}{2}$ and $v(t) = \frac{1}{2}$ are singular solutions of the equation

$$x' = 4x^2 - 1.$$

In Section 2.2, Example 6, we will show that infinitely many solutions are given by the formula

$$x(t) = \frac{1 + ce^{4t}}{2(1 - ce^{4t})},$$

where c is a real constant. Only v can be recovered from this formula (for $c = 0$). There is no constant c such that

$$\frac{1 + ce^{4t}}{2(1 - ce^{4t})} = -\frac{1}{2}.$$

(Can you show this?)

The equations in Examples 14 and 15 are *autonomous*, which means that the vector field does not "explicitly" depend on t (i.e., the normal form of the equation is $x' = f(x)$). Can it happen that the general formula fails to include singular solutions in the *nonautonomous* case (i.e., for equations in which the vector field depends on t explicitly)? The example below shows that it can.

EXAMPLE 16   Infinitely many solutions of the equation

$$x' = -tx^2$$

are obtained from the general formula $x(t) = 1/t^2 + c$ (as it can be shown with the method of Section 2.2). The singular solution $u(t) = 0$, however, is impossible to recover from this formula. (Can you prove this?)

The last three examples teach us that even if we are finding a formula by some method, we also have to determine all singular solutions, and, if they exist, we must check whether they are included in the formula. Therefore, the general solution often consists of a formula and one or more singular solutions.

Initial Value Problems

Sometimes we are interested only in a particular solution that satisfies a certain condition, say, $x(t_0) = x_0$. In geometric terms this is like seeking the curve that represents a solution in the tx-plane and passes through the point of coordinates (t_0, x_0) (see Figure 2.1.2).

For example, the equation $x' = -5t^2$ has the general solution $x(t) = -\frac{5}{3}t^3 + c$, where c is a real constant. The particular solution that satisfies the initial condition $x(1) = 2$ is the one whose graph passes through the point of coordinates $(t, x) = (1, 2)$. To find the expression of this solution, we need to determine a suitable constant c in the general solution. The formula gives $x(1) = -\frac{5}{3} + c$, whereas the condition is $x(1) = 2$, so $c = \frac{11}{3}$. Therefore, the particular solution we were seeking is $\varphi(t) = -\frac{5}{3}t^3 + \frac{11}{3}$.

DEFINITION 2.1.3

An initial value problem for a first-order differential equation in normal form is given by

$$x' = f(t, x), \qquad x(t_0) = x_0.$$

A *solution* φ of this initial value problem is a solution of the equation $x' = f(t, x)$ (i.e., $\varphi'(t) = f(t, \varphi(t))$) that also satisfies the initial condition (i.e., $\varphi(t_0) = x_0$).

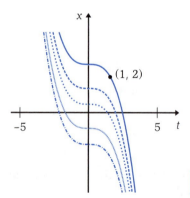

Figure 2.1.2. The graphs of some solutions for the general solution of the equation $x' = -5t^2$.

EXAMPLE 17   Consider the initial value problem

$$x' = x, \qquad x(0) = 1.$$

We have seen that $\varphi(t) = e^t$ is a solution of the equation $x' = x$, but it also satisfies the initial condition, because $\varphi(0) = e^0 = 1$. So $\varphi(t) = e^t$ is a solution of the initial value problem.

Figure 2.1.3. The points of coordinates $(0, 1)$ and $(1, e)$ belong to the graph of the function $\varphi(t) = e^t$.

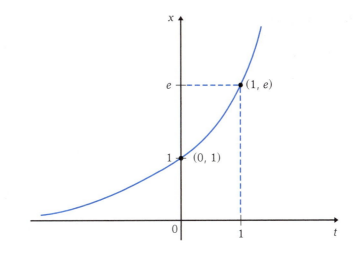

EXAMPLE 18 Let us consider the same equation with a different initial condition, say,

$$x' = x, \qquad x(1) = e.$$

Again, the function $\varphi(t) = e^t$ satisfies both the equation and the initial condition: $\varphi(1) = e^1 = e$. This shows that we can have the same solution for different initial value problems. This happens when the points we choose correspond to the graph of the same solution (see Figure 2.1.3).

EXAMPLE 19 Let us consider once more the same equation with a different initial condition,

$$x' = x, \qquad x(0) = 0.$$

Obviously, $\varphi(t) = e^t$ fails to satisfy this initial condition. Recall, however, that the zero function, $\psi(t) = 0$, is also a solution of the equation. Moreover, it satisfies the above initial condition, since $\psi(0) = 0$, so $\psi(t) = 0$ is a solution of the initial value problem. Can you find a different initial condition attached to the equation $x' = x$ such that the zero function is a solution of this new initial value problem?

It is natural to ask whether all initial value problems have solutions, and if so, whether they are unique. We will answer these questions in Section 2.5. But until then, let us notice that there is a close connection between these questions and that of asking whether the graphs of the solutions fill the tx-plane. By this we mean that for every point (t_0, x_0) there is a solution φ for which $\varphi(t_0) = x_0$.

EXAMPLE 20 The family of functions $x(t) = ce^t$, where c takes all real values, form the general solution of the equation $x' = x$. (Can you check that every member of the family is a solution?) To show that the graphs of the family fill the tx-plane (see Figure 2.1.4), we must prove that for every point (t_0, x_0) of the plane there exists a function φ

Figure 2.1.4. The graphs of the family of functions $x(t) = ce^t$, where c takes all real values, fill the tx-plane.

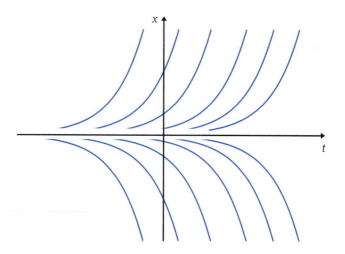

in the family such that $\varphi(t_0) = x_0$. Since φ must be of the form ce^t for some c, to find φ we must solve the equation $ce^{t_0} = x_0$ for c. This has the solution $c = x_0 e^{-t_0}$, so the corresponding function of the family is $\varphi(t) = x_0 e^{-t_0} e^t$. Since for every value of t_0 and x_0 such a function exists, the graphs fill the plane.

PROBLEMS

1. Give an example of an equation in general form that cannot be written in normal form.

2. From the point of view of the set in which the variable t is defined, are normal forms more restrictive than general forms? If so, give an example that shows this. If not, explain why not.

For each of the equations in Problems 3 through 10, find a function F that defines the general form. Is your choice of F unique? Then, if possible, write the equation in normal form and determine the vector field.

3. $2x' - x + t = \sqrt{t}$

4. $\dfrac{t}{3}x' + x^2 = \dfrac{tx}{2}$

5. $\dfrac{\sqrt{3}-1}{2}\theta'\theta = t^{3/2}$

6. $X'\sin(t+1) - X'X = 2e^t + 1$

7. $6w' + \dfrac{3}{2}w'xw^2(2x-3) = \dfrac{\sqrt{7}-4}{\sqrt{7}+4}$

8. $(y')^2 = y^2 + t^2$

9. $(v')^3 - 1 = 0$

10. $u' + \sin(u') - \ln(uu') = 0$

In Problems 11 through 16, determine whether the given equations have the corresponding functions as solutions.

11. $x' = \dfrac{2x}{t}$, $x(t) = 5t^2$

12. $u' = \dfrac{u}{t} + t$, $u(t) = t^2 + 3t$

13. $v' + v = 0$, $v_1(x) = 3\sin x - 4\cos x$,
     $v_2(x) = 3\sin x + 4\cos x$

14. $x' - x = 0$, $x_1(t) = ce^t$, $x_2(t) = e^{kt}$, where c and k are real constants

15. $y' + y = 0$, $y_1(t) = te^t$, $y_2(t) = t^2 e^t$

16. $\theta' = \dfrac{1}{\cos\theta} - \theta$,

     $\theta_1(t) = t\sin t + (\cos t)\ln(\cos t)$, $\theta_2(t) = t\sin t$

For each of the equations in Problems 17 through 22, show that the corresponding function gives a family of solutions. Do the graphs of these functions fill the plane? In each case, find all singular solutions of the given equation and see whether they are contained in the corresponding formula.

17. $tx' - 2x = 0$, $x(t) = ct^2$

18. $y' = x\sqrt{y}$, $y(x) = \left(\dfrac{x^2}{4} + c\right)^2$

19. $uu' + u^2 = 0, \qquad u(x) = -ce^{-x}$

20. $(v')^2 = v^4, \qquad v(t) = -\dfrac{1}{t+c}$

21. $\tan\left(\dfrac{\pi\, yww'}{4}\right) = 1, \qquad w(y) = \sqrt{2\ln y + c}$

22. $(\theta')^3 = t^3\theta^3, \qquad \theta(t) = ce^{t^2/2}$

For each of the functions in Problems 23 through 30, find a differential equation that has the given function (or family of functions) as a solution.

23. $x(t) = t$

24. $x(z) = 6z^2$

25. $u(t) = \sqrt{2t}$, with $t \geq 0$

26. $\varphi(t) = -\sqrt{-5t}$, with $t \leq 0$

27. $v(x) = Ke^x$, with K a real constant

28. $y(x) = ce^{2x}$, with c a real constant

29. $z(t) = C\cos t$, where C is a real constant

30. $w(\theta) = k\sin^2\theta$, where k is a real constant

31. Show that the formula $x(t) = -\tanh(t-c)$ gives the solutions of the equation $x' = x^2 - 1$, where $\tanh z = (e^z - e^{-z})/(e^z + e^{-z})$. Then find the singular solutions and check whether they are contained in the above

formula. Graph some of the functions expressed by the formula and then graph the singular solution.

32. Check that the formula $y(x) = (k - 2x)^{-1/2}$ gives the solutions of the equation $y' = y^3$, then show that $\varphi(x) = 0$ is a singular solution. Is the zero function included in the above formula? Graph some of the functions given by the formula and then graph the singular solution. Do the graphs of all solutions fill the xy-plane?

For each of the initial value problems in Problems 33 through 38, determine whether the given function is a solution.

33. $x' = x^3,\ x(1) = 1; \qquad x(t) = \dfrac{1}{\sqrt{3-2t}}$

34. $v' = \frac{1}{2}v + 4,\ v(0) = 2; \qquad v(x) = 10e^{x/2} - 8$

35. $w' = 9w^2,\ w(0) = 0; \qquad w(z) = 0$

36. $x' = 6yx,\ x(1) = 3; \qquad x(y) = 3e^{3(y^2-1)}$

37. $x' = -(\sin t)x,\ x\left(\dfrac{\pi}{2}\right) = 2; \qquad x(t) = 2\cos t$

38. $y' = (\cos z)y,\ y\left(\dfrac{3\pi}{2}\right) = \dfrac{1}{e}; \qquad y(z) = e^{\sin z}$

2.2 Separable Equations

The simplest differential equations are of the type $x' = f(t)$, where f is a given function of t alone and x is the unknown function. Such equations can be solved by direct *integration*, as we know from calculus. For example, the general solution of the equation

$$x' = (1+t)^2$$

is $x(t) = \int (1+t)^2\, dt = (1+t)^3/3 + c$, where c can take any real value. Similarly, the general solution of the equation

$$x' = \frac{1}{t}, \qquad \text{for } t > 0,$$

is $x(t) = \int (1/t)\, dt = \ln|t| + c$, where c can take any real value. The method of integration, however, can be applied to the larger class of *separable equations*, which have the form

$$x' = g(t)h(x), \tag{1}$$

where g and h are functions and h is defined in a set in which $h(x) \neq 0$. Such differential equations were first solved over three centuries ago by the founders of the theory: the English mathematician and physicist Isaac Newton (1642–1727) and the German mathematician and philosopher Gottfried Wilhelm Leibniz (1646–1716).

G. W. Leibniz.

Isaac Newton.

EXAMPLE 1 The equation

$$x' = \frac{1}{2}t^3 x^2$$

is separable, since $g(t) = \frac{1}{2}t^3$ and $h(x) = x^2$. Is this the only choice for g and h?

EXAMPLE 2 The equation

$$x' = \frac{e^t}{x}$$

is separable, since we can define $g(t) = e^t$ and $h(x) = 1/x$.

EXAMPLE 3 The equation

$$x' = \frac{\sin^2 x}{t} + 1$$

is not separable, since there are no functions g and h such that $g(t)h(x) = (\sin^2 x)/t + 1$.

Method for Solving Separable Equations

A separable equation is like a mixture of oil and water: The liquids part naturally. The idea of the method is to break up the variables x and t and then integrate the new equation. This technique is fully justified by a calculus theorem: the change of variable under the integral. The method works as follows.

Step 1. Separate the variables and obtain the equivalent equation

$$\frac{x'}{h(x)} = g(t).$$

Step 2. Since x and x' are functions of t, both sides of the equation depend on t, so apply the integral and write

$$\int \frac{x'}{h(x)} dt = \int g(t) dt.$$

Step 3. Use the change-of-variable theorem for the left integral and formally write $x'dt = dx$, so the above equation becomes

$$\int \frac{dx}{h(x)} = \int g(t) dt.$$

Step 4. After integration obtain the general formula in *implicit form* (i.e., an equation involving x but unsolved for x),

$$H(x) = G(t) + c,$$

where H and G are differentiable functions such that $dH/dx = 1/h$ and $dG/dt = g$, and c is the combination of the left and right integration constants.

Step 5. Solve the above equation for x and obtain the general formula in *explicit form* (i.e., solved for x as a function of t).

Step 6. Find all singular solutions, check whether they are represented in the general formula, and then write the general solution.

EXAMPLE 4   Let us solve the equation $x' = x$ with the above method.

Step 1. Separate the variables and obtain

$$\frac{x'}{x} = 1.$$

Step 2. Apply the integral and write

$$\int \frac{x'}{x} dt = \int dt.$$

Step 3. Use the change-of-variable theorem and formally write $x'dt = dx$. The above equation becomes

$$\int \frac{dx}{x} = \int dt.$$

Step 4. Compute each integral and write the general solution in implicit form,

$$\ln |x(t)| = t + c,$$

where c is a real constant.

Step 5. To solve for x, first obtain $|x(t)| = e^c e^t$. This is equivalent to $x(t) = e^c e^t$ if $x > 0$ and to $x(t) = -e^c e^t$ if $x < 0$, or $x(t) = ke^t$, where $k = e^c$ if $x > 0$ and $k = -e^c$ if $x < 0$. Since c takes all real values, k also takes all real values, except for the value 0. So this leads to the general formula in explicit form

$$x(t) = ke^t,$$

where k is a real constant, $k \neq 0$.

Step 6. The only singular solution is the zero function and appears in the above formula for $k = 0$. So the general solution is given by the formula obtained at Step 5 for all real values of k.

The graphs of the functions that form the general solution fill the tx-plane with exponential curves, as shown in Figure 2.1.4 of Section 2.1.

"How does a mathematician catch a rabbit? He catches two and then sets one free." This joke is not far from the truth. As we will see in our next example, the usual way of solving an initial value problem is to find the general solution, which is far more than we need, and then keep only the particular solution we are interested in.

EXAMPLE 5   To solve the initial value problem

$$x' = \frac{2x}{t}, \quad x(1) = 3,$$

first ignore the initial condition and proceed as before. Separate the variables and obtain

$$\frac{x'}{x} = \frac{2}{t},$$

which, if integrated, yields the implicit general solution

$$\ln|x| = 2\ln|t| + c,$$

where c is a constant. This leads to the general solution

$$x(t) = kt^2,$$

where k takes all the real values. Notice that the zero solution, which is the only singular one, appears in the above formula for $k = 0$. Also notice that since solutions must be defined on intervals and since $t \neq 0$, each particular solution is defined either on $(-\infty, 0)$ or on $(0, \infty)$.

Figure 2.2.1. The graphs of the family of functions $x(t) = kt^2$, where k takes all real values, fill the tx-plane except for the line $t = 0$. The particular solution $\varphi(t) = 3t^2$ passes through the point of coordinates $(1, 3)$.

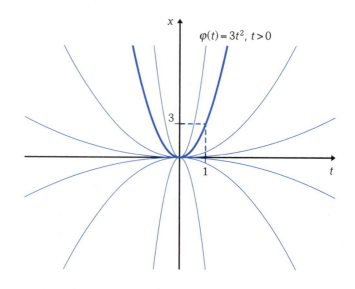

To determine the particular solution φ that satisfies the initial condition $\varphi(1) = 3$, notice from the general solution that $\varphi(1) = k \cdot 1^2 = k$, so $k = 3$, and therefore the solution of the initial value problem is $\varphi(t) = 3t^2$, defined on $(0, \infty)$ (see Figure 2.2.1).

EXAMPLE 6 To solve the equation $x' = 4x^2 - 1$, take $g(t) = 1$ and $h(x) = 4x^2 - 1$. Separation of variables leads to

$$\frac{x'}{4x^2 - 1} = 1,$$

which further yields

$$\int \frac{dx}{4x^2 - 1} = \int dt.$$

Recall from calculus that such an integral can be solved by first decomposing the expression $1/(4x^2 - 1)$ in partial fractions. This can be done by finding numbers A and B so that

$$\frac{A}{2x - 1} + \frac{B}{2x + 1} = \frac{1}{4x^2 - 1}.$$

This is equivalent to

$$\frac{2(A + B)x + A - B}{4x^2 - 1} = \frac{1}{4x^2 - 1},$$

which then implies that $A = \frac{1}{2}$ and $B = -\frac{1}{2}$. So the above integral equality reduces to

$$\int \frac{dx}{2(2x - 1)} - \int \frac{dx}{2(2x + 1)} = t + c,$$

where c is a constant. This leads to

$$\frac{|2x - 1|}{|2x + 1|} = ke^{4t}, \tag{2}$$

where k is a constant. To solve for x, we first remark that

$$|2x - 1| = \begin{cases} 1 - 2x, & \text{for } x \text{ in } \left(-\infty, \dfrac{1}{2}\right) \\[2mm] 2x - 1, & \text{for } x \text{ in } \left(\dfrac{1}{2}, \infty\right), \end{cases}$$

and

$$|2x + 1| = \begin{cases} -2x - 1, & \text{for } x \text{ in } \left(-\infty, -\dfrac{1}{2}\right) \\[2mm] 2x + 1, & \text{for } x \text{ in } \left(-\dfrac{1}{2}, \infty\right), \end{cases}$$

which implies

$$\frac{|2x - 1|}{|2x + 1|} = \begin{cases} \dfrac{1 - 2x}{2x + 1} & \text{for } x \text{ in } \left(-\dfrac{1}{2}, \dfrac{1}{2}\right] \\[2mm] \dfrac{2x - 1}{2x + 1} & \text{for } x \text{ in } \left(-\infty, -\dfrac{1}{2}\right] \text{ or } \left(\dfrac{1}{2}, \infty\right). \end{cases}$$

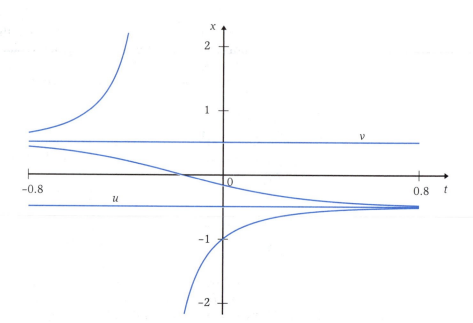

Figure 2.2.2. The graphs of the family of functions

$$x(t) = \frac{1 + ke^{4t}}{2(1 - ke^{4t})},$$

where k takes all real values, together with the singular solutions $u(t) = -\frac{1}{2}$ and $v(t) = \frac{1}{2}$, fill the tx-plane. For example, if $k = 3$, the graph has two branches: one in the upper half-plane, defined for $t < \frac{1}{4} \ln \frac{1}{3}$, and the other in the lower half-plane, defined for $t > \frac{1}{4} \ln \frac{1}{3}$. Since in Definition 2.1.1 we are asking that solutions be defined on intervals, each branch of the curve represents a different solution.

Solving now equation (2) first in $(-\frac{1}{2}, \frac{1}{2}]$ and then in $(-\infty, -\frac{1}{2}] \cup (\frac{1}{2}, \infty)$, we obtain

$$x(t) = \frac{1 - ke^{4t}}{2(1 + ke^{4t})} \qquad \text{and} \qquad x(t) = \frac{1 + ke^{4t}}{2(1 - ke^{4t})},$$

respectively. However, since k takes all real values, the above are two different representations of the same formula, so we can choose to write the general solution as

$$x(t) = \frac{1 + ke^{4t}}{2(1 - ke^{4t})}.$$

As we have already seen in Example 12 of Section 2.1, there are two singular solutions: $u(t) = -\frac{1}{2}$ and $v(t) = \frac{1}{2}$, of which only v is reflected in the above formula (for $k = 0$). The general solution is thus described by the above formula and by the singular solution $u(t) = -\frac{1}{2}$ (see Figure 2.2.2).

The method of solving separable equations is simple. The only difficulties appear with integration (calculus) and with solving the implicit equation (algebra). Often these steps cannot be completed. When they can, we may save time by using modern technology. Computer environments such as Maple, *Mathematica*, or MATLAB are endowed with built-in programs able to solve such equations (see Section 2.7). When the steps cannot be completed by straightforward integration, we can still find approximate solutions by *numerical methods*. We will discuss them in Section 2.6.

Applications

Growth and decay problems

Let us now consider the equation

$$x' = kx. \tag{3}$$

If x is a quantity that changes in time, this equation indicates that the rate of change is proportional to x; i.e., the quantity increases (decreases) like kx if $k > 0$ ($k < 0$, respectively). The variable x can represent the size of a population (in which case $k > 0$ is the birth rate; the more people, the more births), the quantity of a radioactive substance (in which case $k < 0$ is the rate of decay; the more atoms in a given space unit, the more of them break in a unit of time), the amount of invested or borrowed money, etc. Before presenting specific examples, let us solve equation (3) using the above method. Separate the variables and obtain

$$\frac{x'}{x} = k,$$

which after integration yields the general solution that depends on the constant C,

$$x(t) = Ce^{kt}. \tag{4}$$

MONEY INVESTMENTS   Investments can be with simple or compound interest, usually paid at equal intervals of time. At the end of each interval, simple interest is paid as a fixed percentage of the original amount, whereas compound interest is paid on the existing amount. Compound interest is usually applied annually or semiannually. However, since we regard time as a continuous variable, it seems more natural to compound the interest continuously. To have some idea of the difference between these types of investments, let us compute the yield after 5 years for a $10,000 investment made at 5% annual interest, compounded continuously.

Solution.   Since the rate of change of the amount x of money is proportional to the amount itself, with proportionality constant 0.05, the growth equation is

$$x' = 0.05x,$$

i.e., equation (3) with $k = 0.05$. Its general solution is given by (4), i.e.,

$$x(t) = Ce^{0.05t}.$$

The initial condition $x(0) = 10,000$ leads to the particular solution $x(t) = 10,000e^{0.05t}$. The amount of money after 5 years is therefore $x(5) = 10,000e^{0.25} = \$12,840.25$. For comparison, a simple-interest investment produces $12,500, whereas an annually compounded one yields $12,762.81. If you take examples in which the interest rate is higher and the time to maturity is longer, you will see that the difference between these types of investments changes significantly (see Experiment 2.8.1 in Section 2.8).•

THE SHROUD OF TURIN   Believed by some to have been the burial garment of Jesus Christ, the Shroud of Turin has been preserved since 1578 in the royal chapel of the Cathedral San Giovanni Battista in Turin, Italy. In 1988 three independent dating tests revealed that the quantity of carbon-14 in the shroud's linen cloth was between 99.119% and 99.275% of that found in new cloth. When was the shroud made?

Solution. Radioactive substances have a decay rate proportional to their quantity at every instant. For the carbon-14 isotope, the proportionality constant is approximately -0.00001216. So equation (3) with $k = -0.00001216$ describes the decay process of a quantity x of carbon 14. The solution is given by (4), i.e., $x(t) = Ce^{-0.00001216t}$. If the initial quantity of radioactive substance is $x(0) = x_0$, then $x_0 = x(0) = Ce^0 = C$,

The Shroud of Turin.

so the variation of the initial quantity of substance in time is $x(t) = x_0 e^{-0.00001216t}$.

We would like to determine first how long it takes until the quantity x_0 changes to $0.99119 x_0$. For this we have to solve the equation

$$0.99119 x_0 = x_0 e^{-0.00001216t},$$

which has the solution $t = -(\ln 0.99119)/0.00001216 = 727.7168996$, i.e., approximately 728 years. Similarly, 99.275% corresponds to 598.3888957, i.e., approximately 598 years. This means that the shroud was made sometime between 1260 and 1390, about 13 centuries after the death of Jesus.

Since this discovery, several magazine articles have claimed that new research has shown how the tests were affected by carbon-14-accumulating microorganisms living in the cloth. But no such research has yet been published in serious, peer-reviewed scientific journals. To this day, the scientific community remains firm about the above carbon-dating conclusions.●

Newton's law of cooling or heating

Newton's law of cooling or heating tells that the rate of change of the temperature T of a body in a medium of constant temperature τ is proportional to $T(t) - \tau$. This process is described by the equation

$$T' = k(T - \tau), \qquad k < 0. \tag{5}$$

To solve this equation, separate the variables and obtain $T'/(T - \tau) = k$. Integration then yields the general solution

$$T(t) = \tau + Ce^{kt}. \tag{6}$$

Notice that heating occurs for $C < 0$, whereas cooling occurs for $C > 0$.

HOW TO COOK A SALMON? In the late 1950s, chefs like Jean and Pierre Troisgros, Paul Bocuse, and Michel Guérard came up with a new cooking philosophy, *la nouvelle cuisine*, whose main achievement was the recognition of chefs as creative artists. Chefs at famous restaurants all over the world today may earn higher salaries than the senior executives who dine there. The secret of their success rests with detailed and intense research on the amount and mixture of ingredients, the cooking time, etc. We can imagine a simple example of how the law of heating can help.

A salmon fillet, initially at 50°F, is cooked in an oven at the constant temperature of 400°F. After 10 minutes the temperature of the fillet is measured to be 150°F. Since the fish is thin and tender, we assume in a first approximation that its temperature is uniform. How long does it take until the salmon is considered rare, say, at 200°F?

Solution. The cooking process is described by equation (5) with $\tau = 400$ and k unknown. However, k as well as C can be found using the information: $T(0) = 50$ and $T(10) = 150$. Indeed, the first condition and (6) yield

$$50 = T(0) = 400 + Ce^0 = 400 + C,$$

which means that $C = -350$, and the second condition and (6) give

$$150 = T(10) = 400 - 350e^{10k},$$

from which $k = \frac{1}{10} \ln \frac{25}{35} \approx -0.034$. Thus, the particular solution of the differential equation describing our problem is $T(t) = 400 - 350e^{-0.034t}$. To determine when the fish reaches 200°F, we have to solve for t the equation

$$200 = 400 - 350e^{-0.034t},$$

whose solution is $t = (-\frac{1}{0.034}) \ln \frac{4}{7}$, i.e., approximately 16.5. So, under the assumption of uniform temperature, the salmon would reach 200°F after about 16 and a half minutes.●

PROBLEMS

Find the general solution of each of the separable equations in Problems 1 through 6. If you cannot complete Step 5 of the method, leave the solution in implicit form. If you wish, check your answers using Maple, Mathematica, or MATLAB, as indicated in Section 2.7.

1. $x' = 2x^2$

2. $u' = \frac{1}{2}u \sin t$

3. $\sqrt{2}v' = \sin v$

4. $y' = e^{2x+5y}$

5. $\dfrac{\theta'}{\ln r} = \dfrac{7\theta}{3}$

6. $x' = \dfrac{t^2 - 1}{x^2 + 1}$

Solve the initial value problems in Problems 7 through 12. If you wish, check your answers using Maple, Mathematica, or MATLAB, as indicated in Section 2.7.

7. $x' = tx + t$, $x(0) = 0$

8. $y' = [(t-1)y]^{-1/3}$, $y(0) = 1$

9. $u' = \dfrac{u}{\tan x}$, $u\left(\dfrac{\pi}{4}\right) = \dfrac{\pi}{4}$

10. $z - rz' = 1 + r^2z'$, $z(1) = 1$

11. $w' = (2 + w)\cot x$, $w\left(\dfrac{\pi}{3}\right) = 1$

12. $r' = -\dfrac{\sin^2 r}{\cos^2 t}$, $r\left(\dfrac{\pi}{4}\right) = 1$

Find the general solutions of the equations in Problems 13 through 18 in implicit or explicit form.

If you wish, check your answers using Maple, Mathematica, or MATLAB, as indicated in Section 2.7. Find the singular solutions and see whether they are reflected in the formula you have found.

13. $x' = x^2 - 4$

14. $y' = 6 - 5y + y^2$

15. $w' = (w - 2)(w - 3)$

16. $u' + \frac{1}{9}u^2 = 1$

17. $2v' = (2v - 1)^2$

18. $\frac{1}{3}r' = r^3 - 8r^2 + 8r - 1$

In Problems 19 through 24, find (if possible) a separable equation that has the given function as a particular solution. Is it always possible to find such an equation?

19. $\varphi(t) = 2t^2 + 7.3$

20. $\varphi(x) = 8x^2 - \frac{2}{3}x + 7$

21. $\varphi(\theta) = \ln(2\theta - 4) + \sqrt{2\theta^3}$

22. $\varphi(t) = 15e^{2t}$

23. $\varphi(r) = \sin r + \cos r$

24. $\varphi(\omega) = \tan^2 \omega$

25. Assume that the amount x was invested in a savings account in which the interest is continuously compounded at a constant rate of 5.5% per year. Write and solve the equation that describes the growth rate. If $5000 was initially invested, what is the amount after 3 years? How long does it take for the initial amount to double? Does the doubling time depend on the value of the initial amount?

26. In 1997 the inflation rate in Canada was 1.2%. The economists predicted that this rate would remain about the same for the next few years. If they are correct, what is the price of a basket of goods in the year 2000 if the 1997 price is $199.85?

27. In 1960, the year in which the American scientist Willard Frank Libby (1908–1980) received the Nobel Prize for Chemistry for his work in archaeological dating techniques, a group of specialists from the Royal British Museum in London checked whether an art object found in Tutankhamon's tomb had been made during the pharaoh's time or belonged, as some historians claimed, to an earlier period. Knowing that Tutankhamon died about 1352 B.C., what should have been the object's carbon-14 percentage if it had been made during Tutankhamon's time?

28. A curator with the Smithsonian Institution in Washington, D.C., the oldest scientific foundation in the United States, was asked to check the age of an Egyptian papyrus manuscript that the museum intended to purchase from an antiquities dealer. The curator ordered a carbon-dating test and found out that the manuscript contained about 60% of the normal quantity of carbon-14. He concluded that this was indeed a valuable document from the time papyrus writing had been invented. When was the manuscript produced?

29. What is the half-life of krypton-85 (i.e., the time until the element loses half of the initial quantity), knowing that its decay rate is 6.3% per year?

30. The French microbiologist Louis Pasteur (1822–1895), who, among many remarkable achievements, proved that microorganisms cause fermentation and disease, noticed that a certain culture of bacteria grows in milk by 2% per hour. He wanted to know the size of the culture after 48 hours. If the initial size was unity, can you answer Pasteur's question?

31. The temperature of a liquid is 90°F and the room temperature is 65°F. If the liquid cools to 84°F in 5 minutes, what is its temperature after 15 minutes?

32. A forensic specialist took the temperature of the victim's body at a murder scene at 3:20 A.M. and found it to be 85.7°F. By 3:50 A.M. the temperature was 84.8°F. When was the murder committed, if the air temperature during the night was 55°F?

33. In 1929, the American astronomer Edwin Powell Hubble (1889–1953) stated that galaxies observed at a particular time, say, t_0, recede from each other such that the ratio between speed and distance is a constant H that depends on t_0. Since then, cosmology has shown that the law of motion of galaxies is more complicated than this and that Hubble's statement offers only a first approximation. Can you find the formula that describes the change of distance between galaxies according to Hubble's initial result?

Homogeneous Equations *The equations in Problems 34 through 40, called homogeneous equations, have the form*

$$x' = h\left(\frac{x}{t}\right),$$

where h is a continuous function. For each equation, consider the change of variable $u = x/t$; this means substitute ut instead of x and $u + tu'$ instead of x', which reduces each homogeneous equation to the separable equation

$$u' = \frac{h(u) - u}{t}.$$

Solve each separable equation, obtain a solution $u = u(t)$, and then return to the change of variable and express the solution as a function $x = x(t)$. If you wish, check your answers using Maple, Mathematica, or MATLAB, as indicated in Section 2.7.

34. $x' = \dfrac{x + t}{t}$

35. $x' = \dfrac{x}{t} + \dfrac{t}{x}$

36. $x' = \dfrac{x^2 - t^2}{xt}$

37. $x' = \tan\dfrac{x}{t} + \dfrac{x}{t}$

38. $x' = \dfrac{t^2 + 2tx + x^2}{t^2}$

39. $x' = \dfrac{x}{t} + e^{x/t}$

2.3 Linear Equations

In this section we will study equations of the form

$$x' = f(t)x + g(t), \tag{1}$$

called *linear*, where f and g are known functions. If g is not the zero function, equation (1) is called *linear nonhomogeneous*, whereas if g is the zero function (i.e., $g(t) = 0$ for all t), it is called *linear homogeneous*. Notice that every linear homogeneous equation is separable and has the zero function as the only singular solution. The general solution of linear nonhomogeneous equations was obtained in 1743 by the Swiss mathematician Leonhard Euler (1707–1783).

EXAMPLE 1   The equation

$$x' = tx + t^2$$

is linear nonhomogeneous, because $f(t) = t$ and $g(t) = t^2$.

EXAMPLE 2   The equation

$$x' = x/t + \sin t$$

is linear nonhomogeneous, because $f(t) = 1/t$ and $g(t) = \sin t$.

EXAMPLE 3   The equation

$$x' = (\cos t)x$$

is linear homogeneous, because $f(t) = \cos t$ and g is the zero function.

Variation of Parameters

Linear nonhomogenous equations can be solved by the method of *variation of parameters*. Since the nonhomogeneous equation $x' = f(t)x + g(t)$ and its homogeneous counterpart[1] $X' = f(t)X$ differ only by an additive function, we expect that their general solutions are somewhat alike. Therefore the idea of the method is to seek for the nonhomogeneous equation a solution similar to that of the homogeneous one, but whose constant c is replaced by a to-be-determined function $c(t)$. In other words, we change the constant into a varying parameter. This method was proposed in 1774 by the French mathematician Joseph Louis Lagrange (1736–1813). It works as follows.

Step 1. Solve the first homogeneous equation

$$\frac{dX}{dt} = f(t)X,$$

and obtain the general solution

$$X(t) = ce^{\phi(t)},$$

where ϕ is any function with $\phi' = f$.

Joseph Louis Lagrange.

---

[1] By "the homogeneous counterpart" we mean that we leave $g(t)$ out of the nonhomogeneous equation. We use the notations x and X for the unknowns, to make clear that each equation has a different solution.

Step 2. For the nonhomogeneous equation, seek a solution of the form

$$x(t) = c(t)e^{\phi(t)}.$$

Step 3. Substitute x and x' into the nonhomogeneous equation and eventually see that

$$c'(t) = g(t)e^{-\phi(t)}.$$

Step 4. Integrate and obtain the family of functions

$$c(t) = \int \left(g(t)e^{-\phi(t)} \right) dt.$$

Step 5. Substitute this c into u and obtain the general solution of the nonhomogeneous equation,

$$x(t) = [G(t) + k]e^{\phi(t)},$$

where ϕ is any function such that $\phi' = f$, G is any function such that $G' = ge^{-\phi}$, and k is constant.

This summary, which will guide our steps through exercises, leaves out the detail between Steps 2 and 3. To fill it in, substitute $x(t) = c(t)e^{\phi(t)}$ into the nonhomogeneous equation and using the product and the chain rules, obtain

$$x'(t) = \frac{d}{dt}(c(t)e^{\phi(t)}) = c(t)\phi'(t)e^{\phi(t)} + c'(t)e^{\phi(t)}.$$

But since $x(t) = c(t)e^{\phi(t)}$ with $\phi'(t) = f(t)$, then $x'(t) = f(t)x(t) + c'(t)e^{\phi(t)}$. Comparing this with the initial homogeneous equation $x'(t) = f(t)x(t) + g(t)$, it follows that $c'(t)e^{\phi(t)} = g(t)$, so $c'(t) = g(t)e^{-\phi(t)}$, as claimed at Step 3.

EXAMPLE 4 Let us use the method of variation of parameters to solve the equation

$$x' = 2tx + e^{t^2}.$$

Step 1. Solve first the homogeneous equation

$$\frac{dX}{dt} = 2tX,$$

which is separable and has the general solution $X(t) = ce^{t^2}$.

Step 2. Seek those functions c for which

$$x(t) = c(t)e^{t^2}$$

is a solution of the linear nonhomogeneous equation.

Step 3. Substitute x and x' into the nonhomogeneous equation and find that

$$c'(t)e^{t^2} + 2c(t)te^{t^2} = 2tc(t)e^{t^2} + e^{t^2},$$

which if solved for $c'(t)$ yields

$$c'(t) = 1.$$

Step 4. Integrate this equation and obtain the family of functions

$$c(t) = t + k,$$

where k is a constant.

Step 5. The function u with c as above provides the general solution

$$x(t) = (t + k)e^{t^2}$$

of the nonhomogeneous equation.

EXAMPLE 5   To solve the initial value problem

$$x' = \frac{\sin t}{t} - \frac{2x}{t}, \qquad x\left(\frac{\pi}{2}\right) = 1,$$

solve first the homogeneous equation $X' = -2X/t$, which has the solution $X(t) = c/t^2$. So for the nonhomogeneous equation seek a solution of the form $x(t) = c(t)/t^2$. Substituting this into the nonhomogeneous equation gives

$$\frac{c'(t)t - 2c(t)}{t^3} = \frac{\sin t}{t} - \frac{2c(t)}{t^3},$$

which yields $c'(t) = t \sin t$. Integration by parts leads to $c(t) = \sin t - t \cos t + k$, so the general solution is

$$x(t) = \frac{\sin t - t \cos t + k}{t^2}.$$

To determine the constant k that suits the initial condition, write

$$1 = x\left(\frac{\pi}{2}\right) = \frac{\sin(\pi/2) - (\pi/2)\cos(\pi/2) + k}{(\pi/2)^2} = \frac{1 + k}{\pi^2/4},$$

which leads to $k = \pi^2/4 - 1$. So the solution of the initial value problem is

$$\varphi(t) = \frac{4 \sin t - 4t \cos t + \pi^2 - 4}{4t^2}.$$

Notice that the graph of this function passes through the point $(\pi/2, 1)$ (see Figure 2.3.1).

Figure 2.3.1. The graph of the solution to the initial value problem in Example 5 passes through the point of coordinates $(\pi/2, 1)$. Notice that though the function is defined for all $t \neq 0$, our solution must be defined in an interval; the largest we can find is $(0, \infty)$.

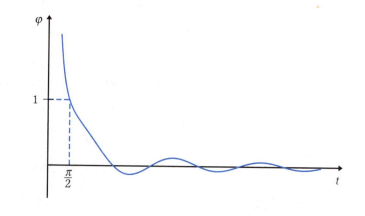

Remark 1. To make sure that the answer is correct, we can check the solution into the initial equation. (Can you do this for Examples 4 and 5?)

Remark 2. Notice that when finding $c'(t)$, the terms containing $c(t)$ always cancel. So in case our result involves terms with $c(t)$, we have made a mistake somewhere. This is a good checkpoint of the correctness of our computations on the way to $c'(t)$.

The only difficulty when solving linear nonhomogeneous equations is to integrate f and $ge^{-\phi}$. If this works, the method is successful; otherwise it fails. Again, Maple, *Mathematica,* or MATLAB can help you with the computations (see Section 2.7).

Applications

A falling object   A 5-kg box containing scientific instruments is dropped from a helicopter above the Antarctic. Due to excessive dryness and cold in the atmosphere, the box encounters a specific air resistance during the fall. Taking into account the atmospheric conditions, a meteorologist estimates that within the time and duration of the mission this force would be well approximated by the function

$$F_{\text{resistance}} = -\frac{2}{5(t+1)}v,$$

where v is the velocity of the object in meters/second and t is the time in seconds. (The negative sign shows that the resistance force acts upward, opposed to gravitation.) For the safety of the scientific instruments, the impact with the ground must be mild. To know the height from which to drop the object, the team designing the mission needs to find the formula of the object's velocity during the fall. What is this formula?

Solution. Newton's second law states that force = mass × acceleration, or

$$F = ma.$$

Since we are interested in the velocity of the object, we can write the acceleration as v'. The force in this case is $F = F_{\text{weight}} + F_{\text{resistance}}$. Since $F_{\text{weight}} = mg$, where $g = 9.8$ m/s^2 is the approximate gravitational acceleration, $m = 5$ kg, and

$$F_{\text{resistance}} = -\frac{2}{5(t+1)}v,$$

the equation describing the velocity is $5v' = 5 \cdot 9.8 - 0.4v/(t+1)$. After simplification this equation takes the linear nonhomogeneous form

$$v' = 9.8 - 0.08\frac{v}{t+1}.$$

The corresponding linear homogeneous equation is

$$V' = -0.08\frac{V}{t+1},$$

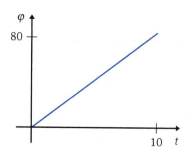

Figure 2.3.2. The graph of the solution $\varphi(t) = 9.074[t + 1 - (t + 1)^{-0.08}]$ is practically indistinguishable from a straight line. This means that under the given circumstances the velocity of the falling object is changing almost linearly in time. Under different atmospheric conditions and for higher velocities this is no longer true.

which has the general solution $V(t) = c(t + 1)^{-0.08}$. According to the method of variation of parameters, we seek solutions of the form $v(t) = c(t)(t + 1)^{-0.08}$. After substituting v into the nonhomogeneous equation and reducing the terms, we obtain $c'(t) = 9.8(t + 1)^{0.08}$, which implies that $c(t) = \frac{9.8}{1.08}(t + 1)^{1.08} + k$, where k is a constant. Thus, with good enough approximation, the general solution of the nonhomogeneous equation is

$$v(t) = 9.074(t + 1) + k(t + 1)^{-0.08}.$$

Because at the time it was dropped from the helicopter, the box had zero velocity with respect to the ground, the initial condition is $v(0) = 0$. This leads to $k = -9.074$ and consequently to the particular solution

$$\varphi(t) = 9.074[t + 1 - (t + 1)^{-0.08}],$$

which is the velocity formula computed for this mission. The graph of this function is very close to a line, as Figure 2.3.2 shows. This means that at least for the first few seconds, for which the model is valid, the velocity increases almost linearly in time. This conclusion will help the research team compute the highest altitude at which the box can be safely dropped.●

The making of a drug

During production of a certain drug at the Bayer pharmaceutical company in Germany, an insoluble, sandlike substance S must be washed by a mixture of two liquids A and B, whose volume ratio changes continuously. A tank is initially filled with 1000 m^3 of the liquid A. Through a pipe, a solution of 9 volume parts liquid A and 1 volume part liquid B flows into the tank at a rate of 30 m^3 per hour and mixes instantly with the liquid in the tank. Substance S, which immediately settles to the bottom, comes through the same pipe at a rate of 2 m^3 per hour. Through an outflow pipe, the new liquid is pumped out of the tank at a rate of 32 m^3 per hour. The process ends when substance S fills the tank, but several times the inflow and outflow must be simultaneously stopped. These breaks are determined by the quantity of liquid B in the mixture. What is the formula that describes this quantity at every instant?

Solution. Let x denote the volume of liquid B. The tank initially contains only liquid A, so $x(0) = 0$. The rate of change of x in time is

$$x'(t) = r_1(t) - r_2(t),$$

Figure 2.3.3. A sketch of the tank and pipes in the mixture problem.

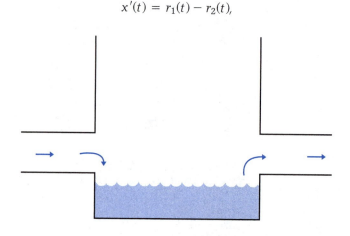

where r_1 is the rate at which liquid B enters and r_2 the rate at which it leaves the tank. The concentration of liquid B in the volume of inflow liquid is $\frac{1}{10}$, so

$$r_1(t) = 30 \cdot \frac{1}{10} = 3 \text{ m}^3 \text{ liquid } B \text{ per hour,}$$

which is constant. The outflow rate is $r_2 = 32C(t)$, where C is the concentration of liquid B in the tank. The concentration C is given by $C(t) = x(t)/V(t)$, where $V(t) = 1000 - 2t$ is the volume of the mixture of liquids in the tank at time t (without the substance S, which settles at the bottom). So

$$r_2(t) = \frac{32x(t)}{1000 - 2t}.$$

Thus the rate of change of liquid B in the tank is $x'(t) = 3 - 32x(t)/(1000 - 2t)$, so the initial value problem to solve is

$$x' = 3 - \frac{32x}{1000 - 2t}, \qquad x(0) = 0,$$

given by a linear nonhomogeneous equation. To use the method of the variation of parameters, first consider the corresponding separable equation

$$X'(t) = -\frac{32X(t)}{1000 - 2t},$$

which has the general solution $X(t) = c(1000 - 2t)^{16}$, c constant. For the linear nonhomogeneous equation, seek solutions of the form

$$x(t) = c(t)(1000 - 2t)^{16}.$$

Substituting this into the nonhomogeneous equation leads to

$$c'(t) = 3(1000 - 2t)^{-16},$$

which, if integrated, yields

$$c(t) = \frac{1}{10}(1000 - 2t)^{-15} + k,$$

where k is a constant. Thus, the solution of the nonhomogeneous equation is

$$x(t) = \frac{1}{10}(1000 - 2t) + k(1000 - 2t)^{16}.$$

The initial condition $x(0) = 0$ implies that $k = -10^{-46}$, so the desired particular solution is

$$\varphi(t) = 10^{-1}(1000 - 2t) - 10^{-46}(1000 - 2t)^{16}.$$

The graph of this solution (see Figure 2.3.4) shows that the quantity of liquid B reaches a maximum after about 100 hours and then decreases to 0 at the time the tank is filled with substance S, after 500 hours. We can

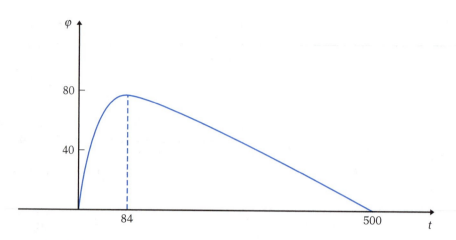

Figure 2.3.4. The graph of the function φ that gives the quantity of liquid B in the tank at any time.

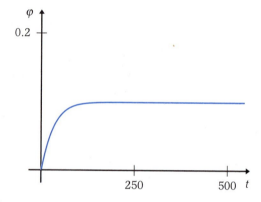

Figure 2.3.5. The graph of the function

$$\phi(t) = \frac{\varphi(t)}{1000 - 2t}$$

gives the concentration of the liquid B in the tank at any time.

also draw the graph of the function $\phi(t) = \varphi(t)/(1000 - 2t)$ (see Figure 2.3.5), which gives the concentration of the liquid B in the tank at any time. This shows that after $t = 100$, the concentration of liquid B in the tank increases very slowly.•

PROBLEMS

Find the general solution of the linear nonhomogeneous equations in Problems 1 through 8. If you wish, check your answers using Maple, Mathematica, or MATLAB, as indicated in Section 2.7.

1. $x' = 2x - 3t + e^{-t}$

2. $u' = \dfrac{u}{t} + 2t$

3. $p' = \dfrac{r-1}{r}p + r^2$

4. $x' = tx + 6te^{-t^2}$

5. $v' = (\cot x)v + \sin x$

6. $x' = (\tan t)x - \cos t$

7. $tw' = w + \dfrac{1}{t}$

8. $tx' = -\dfrac{x}{t} + \dfrac{e^{1/t}}{t^3}$

Solve the initial value problems in Problems 9 through 16. If you wish, check your answers using Maple, Mathematica, or MATLAB, as indicated in Section 2.7.

9. $x' = tx + 2t, \ x(1) = 2$

10. $v' = t^2 v - 3t^{-2}, \ v(0) = 3$

11. $w' = (\tan\theta)w + 1, \ w(\pi) = 1$

12. $r' = (\cot 2\phi)r - \cos 2\phi, \ r(\pi/4) = 0$

13. $x' = (\cot t)x + \csc t, \ x(\pi/4) = 10$

14. $y' = -(\cot x)y + 7\csc x, \ y(\pi/4) = 0$

15. $x' = x + e^{2t}, \ x(0) = 2$

16. $t\theta' = (t+2)\theta + 3t, \ \theta(\ln 2) = 1$

17. Can linear nonhomogeneous equations have singular solutions? Under what circumstances?

18. At Herculane (a spa in southwestern Romania, named after the legendary Hercules) a pool for rheumatic treatment, when half filled, contains 50,000 gallons of natural hot spring-water that has 5000 pounds of Herculane salt. Fresh water is poured into the pool at the rate of 2000 gallons per hour. Assuming that the fresh water mixes instantly with the hot springwater and that the solution leaves the tank at a rate of 1000 gallons per hour, what is the salt concentration when the tank is completely filled?

19. An electric circuit contains a capacitor, a resistor, and a voltage source. The voltage of the source is given by the periodic function $V(t) = \sin t$, whereas the resistance R and the capacitance C are constant. The voltage v across the capacitor satisfies the equation

$$v' = -\frac{1}{RC}v + \frac{V(t)}{RC}.$$

If the resistance is 5 ohms (Ω) and the capacitance is $2 \cdot 10^{-6}$ farads (F), determine the voltage across the capacitor at every moment if it is 0 at $t = 0$.

20. Because of evaporation (on one hand proportional to the surface area, on the other hand dependent on the increasing temperature), the radius of a raindrop changes during the fall as $r_0 - 0.002t$, where $r_0 = 0.01$ m is the initial radius. Assuming that friction forces are negligible, Newton's second law yields the equation

$$[m(t)v(t)]' = F_W,$$

where m is the function that gives the variable mass of the raindrop and F_W is the weight force. Knowing that the initial velocity is zero, find the formula that gives the velocity v of the raindrop at every instant. Graph this function and decide whether the model is realistic.

21. An elastic cable, 30 m long if unstretched, is bound to a heavy box on the ground, and the other end is attached to a helicopter, initially at 30 m above the ground. When the helicopter starts flying up slowly with a constant velocity of 3 m/s, a cufflike device around the cable, initially at 10 m above the ground, shifts up along the cable with a velocity $v(t) = \frac{1}{2}t$ m/s with respect to the cable. This device is part of the system that tightens the box securely when transported in the air. When the device reaches the helicopter, the string system ceases to be elastic and the box is lifted from the ground. At what height is the helicopter when the box is lifted? (*Hint:* First find the expression of the velocity of the device with respect to the ground.)

Bernoulli Equations *The equations in Problems 22 through 27, called Bernoulli equations after the Swiss mathematician Jakob Bernoulli (1654–1705), have the form*

$$x' = f(t)x + g(t)x^\alpha,$$

where f and g are known functions and $\alpha \neq 0, 1$ is a constant. For each equation below consider the change of variables $u = x^{1-\alpha}$; this means substituting $u^{1/(1-\alpha)}$ for x and $[1/(1-\alpha)]u^{\alpha/(1-\alpha)}u'$ for x', which reduce the Bernoulli equation to the linear nonhomogeneous equation

$$u' = (1-\alpha)f(t)u + (1-\alpha)g(t).$$

Solve each linear nonhomogeneous equation, obtain a solution $u = u(t)$, then return to the change of variable and express the solution as a function $x = x(t)$. If you wish, check your answers using Maple, Mathematica, or MATLAB, as indicated in Section 2.7.

22. $x' = \dfrac{4x}{t} + t\sqrt{x}$

23. $x' = -2\dfrac{x}{t} + \dfrac{x^3}{t^2}$

24. $x' = -tx^2 + \dfrac{x}{t}$

25. $x' = 2x - 3x^2$

26. $x' = x - x^3$

27. $x' = ax - bx^3, \qquad a > 0, \ b > 0$

Riccati Equations *The equations in Problems 28 through 33, called Riccati equations after the Venetian mathematician Jacopo Francesco Riccati (1676–1754), have the form*

$$x' = f(t)x^2 + g(t)x + h(t),$$

where f, g, and h are known functions. For each equation below, find attached a particular solution φ. Use it to make the change of variable $u = x - \varphi$; this means substituting $u + \varphi$ for x and $u' + \varphi'$ for x', which reduces each Riccati equation to the Bernoulli equation:

$$u' = [g(t) + 2f(t)\varphi(t)]u + f(t)u^2.$$

Solve each Bernoulli equation (see Problems 21 through 26), obtain a solution $u = u(t)$, then return to the change of variable and express the solution as a function $x = x(t)$. (This method was proposed in 1760 by Leonhard Euler.) If you wish, check your answers using Maple, Mathematica, or MATLAB, as indicated in Section 2.7.

28. $x' = -x^2 + 3x - 2; \ \varphi(t) = 1$

29. $x' = -x^2 + t^2 + 1; \ \varphi(t) = t$

30. $x' = x^2 - \dfrac{x}{t} - \dfrac{1}{t^2}; \ \varphi(t) = \dfrac{1}{t}$

31. $x' = \dfrac{x^2}{\cos t} - x\tan t + \cos t; \ \varphi(t) = \sin t$

32. $x' = (x - t)^2 + 1; \ \varphi(t) = t$

33. $x' = -x^2 + 2tx - t^2 + 5; \ \varphi(t) = t - 2$

Exact Equations *The equations in Problems 34 through 39, called exact, have the form*

$$x' = -\frac{P(t, x)}{Q(t, x)},$$

where $\partial P/\partial x = \partial Q/\partial t$. They can be reduced to the implicit equation

$$W(t, x) = c,$$

where c is a constant, by finding W from the equations

$$\frac{\partial W}{\partial t}(t, x) = P(t, x) \qquad and \qquad \frac{\partial W}{\partial x}(t, x) = Q(t, x).$$

To do this, integrate the first equation above with respect to t. This leads to $W(t, x) = \int P(t, x)dt = R(t, x) + k(x)$. (Indeed, $k(x)$ is a constant with respect to t.) To find k, substitute W thus obtained into the second equation.

Example. To solve the equation

$$x' = -\frac{5t + 8x}{8t - 3x},$$

first notice that $P(t, x) = 5t + 8x$, $Q(t, x) = 8t - 3x$ and $\partial P(t, x)/\partial x = \partial Q(t, x)/\partial t = 8$, so the equation is exact. To find W from the equation $\partial W(t, x)/\partial t = 5t + 8x$, we are led to $W(t, x) = \int (5t + 8x)dt + k(x) = \frac{5}{2}t^2 + 8tx + k(x)$. We can determine the function k from the equation

$$\frac{\partial W}{\partial x}(t, x) = Q(t, x),$$

which in our case is $8t + dk(x)/dx = 8t - 3x$, so $k(x) = -\frac{3}{2}x^2$. Thus the implicit-form solution of the given equation is $-\frac{3}{2}x^2 + 8tx + 5t^2 = c$, where c is a constant—a formula from which $x(t)$ can be computed explicitly.

For the equations in Problems 34 through 39, check first that they are exact, then determine the function W and see whether you can obtain the general solution in explicit form. If you wish, check your answers using Maple, Mathematica, or MATLAB, as indicated in Section 2.7.

34. $x' = -\dfrac{t + x}{t + 2x}$

35. $x' = \dfrac{2t - 3x}{3t + x}$

36. $x' = \dfrac{t^3 - 3tx^2 + 2}{3t^2x - x^2}$

37. $x' = -\dfrac{t^2 + \ln t + t + x}{t}$

38. $x' = \dfrac{2e^{tx}\sin 2t - 2t - xe^{tx}\cos 2t}{te^{tx}\cos 2t - 3}$

39. $x' = \dfrac{6t^2 + x}{t(2 - \ln t)}$

2.4 Qualitative Methods

Henri Poincaré.

In this section we will start the qualitative study of differential equations. We introduce the notions of slope field, phase line, and equilibrium solution of an autonomous equation, classify the equilibria, and explain the phase-line portrait. In the end we apply these qualitative methods to understand the evolution of the cougar population of Vancouver Island and the landing dynamics of the Apollo 11 command module.

Since exact methods apply only to a few classes of equations, we will now present some ways of finding properties of solutions even when these solutions are unknown. Questions such as *stability* (will the moon always revolve around the earth on its present orbit?) or *asymptotic behavior* (does an ever-expanding universe tend to a certain geometrical configuration?), can be investigated with the help of qualitative methods.

This research direction was opened by the Swiss mathematician Charles Sturm (1803–1855), who in 1836 published a paper dealing with a certain class of second-order equations. But the qualitative approach established itself as fundamental only toward the end of the 19th century through the work of the French mathematician and philosopher Henri Poincaré (1854–1912). Today, more than a century later, this is a very active research area.

The Slope Field

We can guess the shape of the solutions' graphs by drawing the *slope field*. If φ is a solution of the equation

$$x' = f(t, x)$$

and $\varphi(t_0) = x_0$, the *slope* of the tangent to the graph of φ at (t_0, x_0) is $f(t_0, x_0)$ (see Figure 2.4.1). Since we can compute $f(t_0, x_0)$ for every (t_0, x_0), we can find the slope of the tangent to the graph of a solution at every point.

Figure 2.4.1. The slope $f(t_0, x_0)$ of the tangent to the graph of φ at (t_0, x_0).

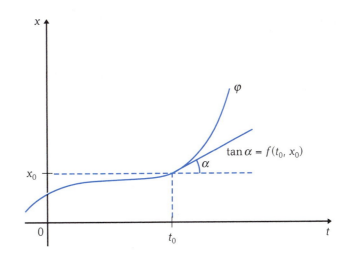

Figure 2.4.2. The slope field and some solution graphs of the equation $x' = -x + 2t$.

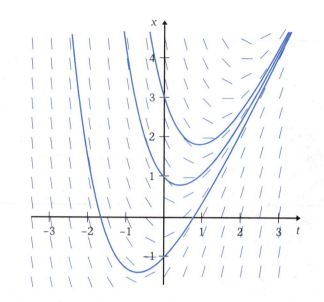

EXAMPLE 1   The vector field of the equation

$$x' = -x + 2t$$

is $f(t, x) = -x + 2t$. The slope of the tangent to the graph of a solution at some point, say $(2, 1)$, is given by $f(2, 1) = -1 + 2 \cdot 2 = 3$. Taking a mesh of points in some region of the plane and computing the slope at every point, we obtain a slope field as in Figure 2.4.2, in which we have also drawn the graphs of some solutions.

Let us take a look at the solution graph passing through the point $(1.5, 2)$ (see Figure 2.4.3(a)). If we zoom in, we can better see the tangency point (see Figure 2.4.3(b)). It becomes clear that it is difficult to draw the graph by eye if looking only at the slope field. We may need a denser grid, because the slopes are changing faster around the origin.

EXAMPLE 2   The vector field of the autonomous equation

$$x' = x^2 - 2x$$

is $f(x) = x^2 - 2x$. At the point $(-2, 2)$, for example, the slope of the tangent to the graph is $f(-2, 2) = 0$. In Figure 2.4.4(a) we have drawn the slope field and a few solution graphs.

Computing slope fields by hand is tedious. Fortunately, computers are now of great help. We can use one of the programs in Section 2.7 for this purpose and instantly obtain slope fields as dense as we like.

Observe in Figure 2.4.4(a) that the slope vectors having the same x-coordinate are parallel. Moreover, the slope is 0 at each point (t, x) whose x-coordinate cancels the vector field, i.e., makes $f(x) = 0$ (see the lines $x = 0$ and $x = 2$ in Figure 2.4.4(a)). These two properties are, in fact, true for the slope field of every autonomous equation $x' = f(x)$. (Can you explain why?)

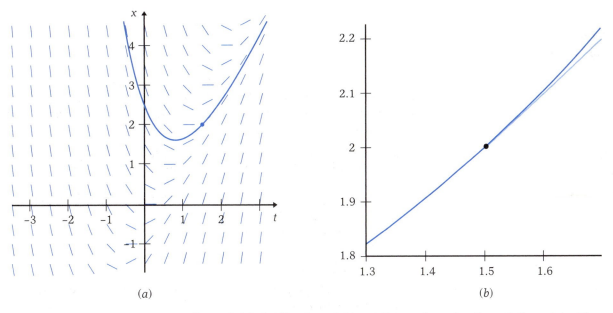

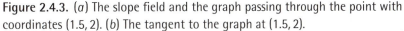

(a)

(b)

Figure 2.4.3. (a) The slope field and the graph passing through the point with coordinates (1.5, 2). (b) The tangent to the graph at (1.5, 2).

Figure 2.4.4. (a) The slope field and some solution graphs of the equation in Example 2. (b) The phase line of the same equation, drawn vertically.

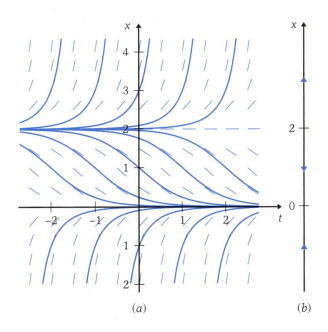

(a)

(b)

We will further consider in this section only *autonomous* equations. Recall from Section 2.1 that the constant solutions canceling the vector field are called *singular*. For autonomous equations, however, we do not have to impose the constancy condition, because it follows as a consequence. Indeed, since $x' = f(x)$ and $f(x) = 0$, we necessarily have $x(t) = c$ (constant). The value of c depends on the initial condition. Singular solutions of autonomous equations are also called *equilibria*. They play an important role in understanding the qualitative behavior of solutions.

Figure 2.4.5. The phase line
for the equation in Example 2,
drawn horizontally.

The Phase Line

The slope field in Figure 2.4.4(a) also suggests that, relative to their asymptotic behavior, the solutions in Example 2 can be grouped into four classes:

(a) equilibrium solutions: $x(t) = 0$ and $x(t) = 2$,

(b) solutions x with $x > 2$ for which $\lim_{t \to -\infty} x(t) = 2$ and $\lim_{t \to +\infty} x(t) = +\infty$,

(c) solutions x with $0 < x < 2$ for which $\lim_{t \to -\infty} x(t) = 2$ and $\lim_{t \to +\infty} x(t) = 0$,

(d) solutions x with $x < 0$ for which $\lim_{t \to -\infty} x(t) = -\infty$ and $\lim_{t \to +\infty} x(t) = 0$.

We can plot these classes of solutions on a *phase line*, which indicates the direction of the graphs of all solutions with respect to the x-axis (see Figure 2.4.4(b), in which the phase line is drawn vertically, and Figure 2.4.5, in which—for spacing reasons—the phase line is drawn horizontally). This portrait represents the solutions that correspond to the vectors on the vector line defined in Section 2.1.

We can think of the phase line in Figure 2.4.5 as a channel with a *source* and a *sink*. Water flows out of the source at 2 in both directions and into the sink at 0 from both directions.

Drawing the Phase Line

Step 1. For the autonomous equation $x' = g(x)$, first find the domain of the vector field g and mark it on the oriented line.

Step 2. Determine the equilibrium solutions, i.e., find the solutions of the equation

$$g(x) = 0,$$

then mark those points on the line.

Step 3. Study the sign of the function g in the intervals between the equilibria and (if applicable) between equilibria and points at which g is not defined, by computing the value of g at some point in each interval.

Step 4. In those intervals in which $g > 0$, draw an arrow to the right.[1]

Step 5. In those intervals in which $g < 0$, draw an arrow to the left.

EXAMPLE 3   For the equation

$$x' = x^2(x^2 - 1),$$

the vector field $g(x) = x^2(x^2 - 1)$ is defined everywhere and the equilibria occur at $-1, 0,$ and 1. Computing g at some point in each of the

---

[1]This is because $g(x) > 0$ implies that $x' > 0$, so x is increasing.

Figure 2.4.6. The phase-line portraits of the equations:
(a) $x' = x^2(x - 1)(x + 1)$
and (b) $x' = [(x - 1)^3 \times (3 - x)]/[x^2(x - 2)^4]$.

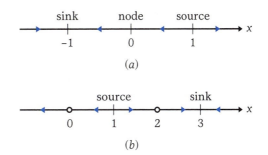

intervals $(-\infty, -1), (-1, 0), (0, 1),$ and $(1, +\infty)$, say at $-2, -0.5, 0.5,$ and 2, we obtain $g(-2) > 0, g(-0.5) < 0, g(0.5) < 0,$ and $g(2) > 0$, so the phase-line portrait is as in Figure 2.4.6(a).

EXAMPLE 4   For the equation

$$x' = \frac{(x - 1)^3(3 - x)}{x^2(x - 2)^4},$$

the vector field $g(x) = [(x - 1)^3(3 - x)]/[x^2(x - 2)^4]$ is undefined at 0 and 2. Equilibria occur at 1 and 3. Computing g at some point in each of the intervals $(-\infty, 0), (0, 1), (1, 2), (2, 3),$ and $(3, +\infty)$, we obtain the phase-line portrait in Figure 2.4.6(b). Notice that the points 0 and 2 *do not* belong to the phase line.

Equilibrium Solutions   Since equilibria play a crucial role in drawing the phase line, let us take a closer look at the behavior of solutions that are close to them. We first classify equilibria as follows.

DEFINITION 2.4.1

An equilibrium solution x_0 of an autonomous equation $x' = g(x)$, where g is differentiable, is called a

    (a) *source* if $g'(x_0) > 0$,
    (b) *sink* if $g'(x_0) < 0$,
    (c) *node* if $g'(x_0) = 0$ or if $g'(x_0)$ does not exist.

The phase-line behavior of solutions near sinks and sources is independent of the equation. In case of a source x_0, the arrows always point away from x_0; in other words, for $x < x_0$ and close to $x_0, g(x) < 0$, and for $x > x_0$ and close to $x_0, g(x) > 0$. The point 1 in Example 3 and the point 1 in Example 4 represent sources (see Figures 2.4.6(a) and (b)). In case of a sink x_0, the arrows always point toward x_0; this means that for $x < x_0$ and close to $x_0, g(x) > 0$, and for $x > x_0$ and close to $x_0, g(x) < 0$. The point -1 in Example 3 and the point 3 in Example 4 represent sinks (see Figures 2.4.6(a) and (b)). Nodes, however, are equation-dependent. The behavior of solutions near a node can be the same as for sources, or sinks, or just different (see Figure 2.4.7).

Figure 2.4.7. Possible behaviors of solutions near a node. The last phase line is filled with nodes.

EXAMPLE 5   The equation

$$x' = x^3$$

has 0 as a node. Indeed, the vector field is $g(x) = x^3$, so $g'(0) = 0$. Since $g(x) < 0$ for $x < 0$ and $g(x) > 0$ for $x > 0$, the phase-line portrait is sourcelike, as in Figure 2.4.7(a).

EXAMPLE 6   The equation

$$x' = -2x^5$$

has 0 as a node. Indeed, the vector field is $g(x) = -2x^5$, so $g'(0) = 0$. Since $g(x) > 0$ for $x < 0$ and $g(x) < 0$ for $x > 0$, the phase-line portrait is sinklike, as in Figure 2.4.7(b).

EXAMPLE 7   The equation

$$x' = x^2$$

has 0 as a node. Indeed, the vector field is $g(x) = x^2$, so $g'(0) = 0$. Since $g(x) > 0$ for $x \neq 0$, the phase line is as in Figure 2.4.7(c).

EXAMPLE 8   The equation

$$x' = 0$$

is an example of degeneracy, in which every point on the line is a node (see Figure 2.4.7(d)).

Applications

Population dynamics In 1998, the British Columbia Ministry of Environment had to determine whether under normal conditions the population of cougars in the provincial Strathcona Park on Vancouver Island would increase, decrease, or remain about the same during the first decade of the 21st century. A group of undergraduate students from the University of Victoria, working for the Ministry during their co-op term, suggested the use of differential equations. The rate of change p' of a population p is proportional to p if p is small (i.e., the population increases), but proportional to $1 - p/n$ if p is large (i.e., the population decreases), where n is the maximum size a population can maintain for a longer time in a given

Figure 2.4.8. The phase-line portrait for the logistic equation $x' = kp(1 - p/n)$.

environment. An equation that agrees with these assumptions is the *logistic equation*

$$p' = kp\left(1 - \frac{p}{n}\right),$$

where $k > 0$ is a proportionality constant. Using statistical data from the last sixty years, the students estimated that in their case $k = 2$ and $n = 127$. Knowing that the population of cougars in Strathcona Park in 1998 was 83, what was the students' prediction?

Solution. The phase-line portrait answers the question posed by the Ministry. The vector field $g(p) = 2p(1 - p/127)$ is defined for all p, and the equilibrium solutions are $p_1(t) = 0$ and $p_2(t) = 127$. Since $g'(0) = 2 > 0$ and $g'(127) = -2 < 0$, it follows from Definition 2.4.1 that 0 is a source and 127 is a sink (see Figure 2.4.8). Therefore, the population of cougars is likely to increase in the near future.

We can also draw other conclusions from the phase line. For example we can tell that it would be unwise to import 200 cougars. Due to the saturation number of 127, most of the cougars would die until the population reached the optimal size.

Remark. It is important to note that time does not pass uniformly while we move on the phase line. The closer to the equilibrium, the slower we move. This becomes pretty clear if we compare Figures 2.4.4(*a*) and (*b*). With respect to the evolution of the cougar population, this means that the closer to the saturation number of 127, the slower the change in the number of individuals.●

The landing of Apollo 11   On July 24, 1969, the command module of the Apollo 11 mission splashed into the Pacific Ocean about 1500 km southwest of Honolulu, Hawaii. When close to the earth, the capsule, which carried home the American astronauts Neil Armstrong and Edwin Aldrin (the first men to walk on the moon), opened its parachutes to make the landing smoother. Knowing that the velocity of the module when entering the atmosphere had been approximately 11 km/s and that the resistance force of the atmosphere for large objects and high speeds is proportional to the square of the velocity, determine whether the velocity of the module increased, decreased, remained constant, or was a combination of these before the opening of the parachutes.

Solution. According to Newton's second law,

$$mv' = F_{\text{weight}} + F_{\text{resistance}},$$

where m is the mass of the module, $F_{\text{weight}} = mg$, g is the gravitational acceleration, and $F_{\text{resistance}} = -kv^2$, $k > 0$, is a force opposed to the di-

Apollo 11 astronauts in raft.

rection of fall. So the velocity is described by the autonomous equation

$$v' = g - \frac{k}{m}v^2.$$

The phase-line portrait will answer the above question. The equilibria are at $-\sqrt{mg/k}$ and $\sqrt{mg/k}$. If f is the vector field, then $f'(v) = -(2k/m)v$, $f'(-\sqrt{mg/k}) = 2\sqrt{kg}/\sqrt{m} > 0$, and $f'(\sqrt{mg/k}) = -2\sqrt{kg}/\sqrt{m} < 0$. Thus $-\sqrt{mg/k}$ is a source and $\sqrt{mg/k}$ is a sink (see Figure 2.4.9).

Ignoring negative velocities, we see that the answer to the question depends on the velocity of the module when entering the atmosphere, which is about 11,000 m/s. The gravitational acceleration varies little in the neighborhood of the earth, so we assume it constant, $g = 9.8$ m/s^2. The constant k depends on the size and shape of the module and on the density of the air and is of the order of 10^1. The mass of the module is of the order of 10^3 to 10^4 kg. With these rough estimates, the value of the equilibrium $v = \sqrt{mg/k}$ is of the order 10^2 m/s at most, i.e., significantly smaller than the initial velocity. Obviously then, the module strongly decelerates when entering the atmosphere.

Again, this example shows us how important it is to determine equilibrium solutions. The entire dynamics of the landing depends on the equilibrium $v = \sqrt{mg/k}$, also called the *terminal velocity*. The module strongly decelerates during the first few seconds after entering the atmosphere and tends to reach a velocity close to the terminal one later on.

Figure 2.4.9. The phase line of the equation $v' = g - (k/m)v^2$.

PROBLEMS

For the nonautonomous equations in Problems 1 through 8, draw the slope fields and the approximate graphs of some solutions. Then group the solutions according to their asymptotic behavior. (Use one of the programs in Section 2.7.)

1. $x' = x + 2t$

2. $x' = x - 2t$

3. $x' = x^2 + t$

4. $x' = x^2 - t$

5. $x' = x + \sin x + t^2$

6. $x' = x + \cos x + t^2$

7. $x' = x^3 + \ln t$

8. $x' = -x^3 + \ln t$

For the autonomous equations in Problems 9 through 16, draw their slope fields and then project them onto the x-axis to obtain the phase-line portraits. Compare the pictures of the equations that have similar forms. (Use one of the programs in Section 2.7.)

9. $x' = x^2 - 3x + 2$

10. $x' = x^2 - 5x + 6$

11. $x' = (x - 1)(x - 2)(x - 3)$

12. $x' = (1 - x^2)(x + 2)$

13. $x' = x^2(\tan x - 1)$

14. $x' = 2x^3 \cot x$

15. $x' = \dfrac{x^2 - 1}{\sqrt{x}|x + 2|}$

16. $x' = \dfrac{x^3 - 1}{\sqrt{x}|x - 2|}$

For the autonomous equations in Problems 17 through 24, find the equilibrium solutions and classify them as sources, sinks, or nodes. Then draw the phase-line portraits and an approximate graph of the corresponding vector field. Compare the pictures of the equations that have similar forms. (Use, if you wish, one of the programs in Section 2.7.)

17. $x' = x^2 - 6x - 16$

18. $x' = 7x - 7x^2$

19. $x' = x \tan x$

20. $x' = x \cot x$

21. $x' = x^2 \sin x$

22. $x' = -x^2 \cos x$

23. $x' = \dfrac{1}{x^2 - 1}$

24. $x' = \dfrac{1}{(x - 1)(x + 2)}$

25. Find an equation $x' = g(x)$ that has -1 and 1 as nodes and in which $g(x) < 0$ for all x different from -1 and 1. Is the choice of the equation unique?

26. Find an equation $x' = g(x)$ that has $-2, 0,$ and 1 as equilibria and in which -2 and 0 are sources and 1 is a sink. Can you find an equation as above that has no more than three equilibria and for which the vector field is defined everywhere?

27. Find an autonomous equation $x' = g(x)$ (where g is differentiable) that has an equilibrium solution x_0 such that $g'(x_0)$ does not exist.

For Problems 28 through 31, use the given graphs of the vector fields to represent the phase-line portraits.

28.

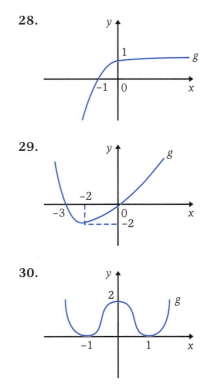

29.

30.

31.

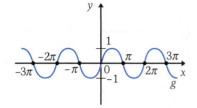

32. Let P denote a population of wolves that inhabit a large area of the Yukon Territory. If P is below a threshold n_1, the wolves cannot find each other during the mating period, so the population decreases as $P/n_1 - 1$. If P is larger than n_1 but smaller than n_2 $(n_1 < n_2)$, the population grows proportionally to its size until it reaches the threshold n_2. If P is larger than n_2, due to limited resources the population decreases as $1 - P/n_2$. Under what circumstances can the population maintain a relatively constant number of individuals?

33. The population of tuna fish of the American East Coast increases or decreases according to a law similar to the one in Problem 32:

$$P' = k\left(\frac{P}{n_1} - 1\right)\left(1 - \frac{P}{n_2}\right) - N,$$

where N represents the number of fish removed by fishing. Describe the dynamics of the fish population.

34. A patient receives glucose into the bloodstream at a constant rate of r grams/minute. The patient's body converts the glucose and removes it from the bloodstream at a rate proportional to the amount of injected glucose, with a constant c of proportionality. The evolution in time of the amount of glucose in the bloodstream can be described by the logistic equation. Write this equation and show that this amount tends to an equilibrium.

35. After its launch from Cape Canaveral, a rocket encounters an air-resistance force that varies proportionally to the square of its velocity with a factor $T - t$, where t is the time variable and T is the instant at which the air-resistance force becomes zero. Considering the mass and propulsion force constant, write the equation that describes the variation of the velocity, draw the slope field, and describe the possible situations of motion that can occur during the flight.

36. Show that if a body moving with velocity v encounters a resistance force of the form $v' = -v^{3/2}$, then, whatever its initial velocity, the body will travel only a finite distance before stopping.

37. In the class of motions given by resistance forces of the type $v' = -kv^\alpha$, $k, \alpha > 0$, where v is the velocity, find those values of α for which, independently of the initial velocity, the body can travel only a finite distance when the time t tends to infinity.

2.5 Existence and Uniqueness

Most equations are difficult to solve by exact methods. Because some algebraic equations do not have solutions at all, it is natural to ask under what circumstances differential equations or initial value problems possess solutions. In this section we will present two existence and uniqueness theorems for initial value problems, point out the importance of such results for the theory of differential equations, and in the end describe the method of successive approximations.

EXAMPLE 1   The equation in general form

$$(x')^2 = -x^2 - 1$$

has no real solutions. Indeed, the left-hand side is positive or zero, whereas the right-hand side is always negative.

EXAMPLE 2 The initial value problem

$$x' = \frac{2x^3 - \sin t}{t}, \qquad x(0) = 1$$

has no solution, because the vector field is undefined at $t = 0$.

The question of uniqueness for the solution of an initial value problem

$$x' = f(t, x), \qquad x(t_0) = x_0$$

is also important. Imagine that a problem of this type describes a nuclear power experiment and that the initial condition $x(t_0) = x_0$ might trigger two possible outcomes: one that leads to the desired experimental result and another that causes a nuclear explosion. Which one will happen after the button is pushed? Fortunately this scenario is fictitious, because most phenomena around us, including nuclear reactions, are modeled by equations having the uniqueness property. There exist, however, initial value problems with two or more solutions.

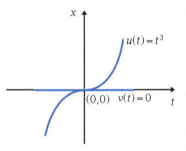

Figure 2.5.1. The graphs of $u(t) = 0$ and $v(t) = t^3$ have the point $(0, 0)$ in common.

EXAMPLE 3 The initial value problem

$$x' = 3x^{2/3}, \qquad x(0) = 0$$

has at least two solutions: $u(t) = 0$ and $v(t) = t^3$ (see Figure 2.5.1).

An Existence Theorem

The above examples are pathological, uncommon to natural phenomena. We are mostly interested in those equations whose initial value problems admit existence and uniqueness of solutions. To distinguish those that do from those that do not, we need some general criteria. The first result we present in this sense is an existence theorem published in 1890 by the Italian mathematician Giuseppe Peano (1858–1932).

The Existence Theorem of Peano

If the vector field $f(t, x)$ is continuous with respect to t and x in some rectangle $\{(t, x)|t_1 < t < t_2, \ x_1 < x < x_2\}$ and if (t_0, x_0) is a point inside this rectangle, then there exists an $\epsilon > 0$ and a function φ defined on $(t_0 - \epsilon, t_0 + \epsilon)$ that is a solution of the initial value problem

$$x' = f(t, x), \qquad x(t_0) = x_0.$$

This result tells us that if in some rectangle containing (t_0, x_0) the slope field changes continuously, then there exists at least one curve passing through (t_0, x_0) that represents the graph of a solution (i.e., is tangent to the corresponding direction vectors). If the direction of the vector field has abrupt changes, as for the equation

$$x' = \begin{cases} 0 & \text{if } t < 0 \\ 1 & \text{if } t \geq 0, \end{cases}$$

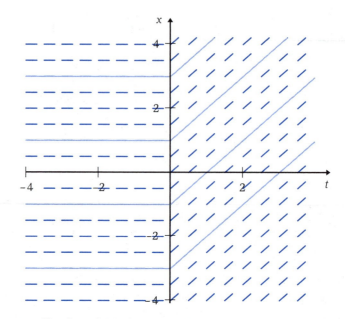

Figure 2.5.2. The slope field of the equation

$$x' = \begin{cases} 0 & \text{if } t < 0 \\ 1 & \text{if } t \geq 0 \end{cases}$$

changes abruptly at $t = 0$, so the expected solutions are not differentiable functions and therefore are not solutions in the sense of Definition 2.1.1.

then we may lack solutions in the sense of Definition 2.1.1. The graphs of the would-be solutions with initial conditions of the form $x(0) = x_0$ consist of two joined half-lines of slopes 0 and 1, respectively (see Figure 2.5.2). Functions of this type are not differentiable at $t = 0$, so they cannot be solutions as required by the definition. (In Chapter 6 we will extend Definition 2.1.1 to include solutions that are not differentiable, but doing so leads to complications we would like to avoid at this stage.)

Also notice that the theorem ensures the existence of the solution only locally (see Figure 2.5.3). This means that the solution may not exist outside a small interval containing t_0.

EXAMPLE 4   The initial value problem

$$x' = 50\pi(x^2 + 1), \qquad x(0) = 0$$

has the solution defined only on a small interval. If we treat the equation as separable, we obtain the solution $\varphi(t) = \tan 50\pi t$. The argument $50\pi t$ of the tangent function must be contained in $(-\pi/2, \pi/2)$, so t is in the interval $(-0.01, 0.01)$. Because it becomes unbounded when t approaches -0.01 or 0.01, φ cannot be extended to a larger interval (see Figure 2.5.4).

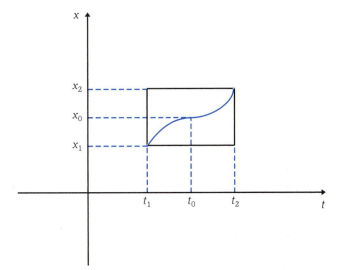

Figure 2.5.3. Peano's theorem ensures only the local existence of a solution.

This example shows that there are situations when the solution is defined only locally (i.e., near t_0), and it explains why Peano's theorem cannot guarantee global existence. In many cases, however, the solution is defined on a large interval.

IDEA OF PROOF

This theorem can be proved by choosing a particular sequence of functions φ_k, $k = 1, 2, \ldots$, called a sequence of *successive approximations*, and by showing that when $k \to \infty$, the sequence tends to a function φ that

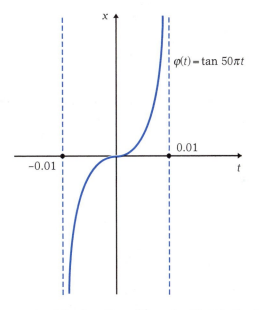

Figure 2.5.4. The graph of the function $\varphi(t) = \tan 50\pi t$ in the interval $(-0.01, 0.01)$.

is a solution of the initial value problem. The *method of successive approximations* first appeared in 1890 in a paper published by the French mathematician Émile Picard (1856–1914), though it was earlier considered in particular cases by two other French mathematicians, Augustin Louis Cauchy (1789–1857) and Joseph Liouville (1809–1882).

An Existence and Uniqueness Theorem

The following existence and uniqueness theorem has its roots in some results obtained by Cauchy in the 1820s.

The Existence and Uniqueness Theorem of Cauchy[1]

If the vector field $f(t, x)$ is continuous with respect to t and x in some rectangle $\{(t, x) \mid t_1 < t < t_2,\ x_1 < x < x_2\}$, if the partial derivative of f with respect to x exists and is continuous, and if (t_0, x_0) is a point inside this rectangle, then there exists an $\epsilon > 0$ and a unique function φ defined on $(t_0 - \epsilon, t_0 + \epsilon)$ that is a solution of the initial value problem

$$x' = f(t, x), \qquad x(t_0) = x_0.$$

Augustin Louis Cauchy.

In contrast to Peano's criterion, this theorem ensures the uniqueness property of the solution by asking that the directions of the slope field do more than avoid abrupt changes: They must also change "smoothly," a property expressed by the differentiability of the vector field with respect to the dependent variable.

EXAMPLE 5   Cauchy's theorem does not apply to initial value problems of the form

$$x' = 3x^{2/3}, \qquad x(t_0) = 0$$

for any real t_0. The reason is that the vector field $f(x) = 3x^{2/3}$ is not differentiable at $x = 0$, i.e., the change of slopes fails to be "smooth" when the slope field passes through the t-axis. Unlike discontinuity, which implies sudden changes of the slope as in Figure 2.5.2, the lack of "smoothness" is hard to see from the picture. For the equation $x' = 3x^{2/3}$, the change of slopes when passing through the t-axis looks quite "smooth" to the eye (see Figure 2.5.5).

Cauchy's theorem can be proved using the method of successive approximations, but it also needs some more sophisticated mathematical tools, which are beyond the scope of this textbook.

---

[1]Some sources call this result the Theorem of Picard; though Picard contributed to it later than Cauchy, he was the first to connect it to the method of successive approximations.

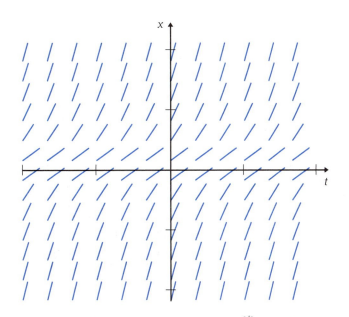

Figure 2.5.5. The slope field of the equation $x' = 3x^{2/3}$ looks "smooth" to the eye, but the vector field is not differentiable at $x = 0$.

Remark 1. Like Peano's theorem, Cauchy's criterion has a local character; i.e., the solution might be defined only in a very small interval around t_0.

Remark 2. Both results above offer only *sufficient conditions* for the uniqueness or existence of solutions to initial value problems. This means that if the hypotheses of the theorems are satisfied, we certainly have uniqueness or existence. If they are violated, however, uniqueness or existence may or may not take place. In Examples 3 and 5, the vector field $f(x) = 3x^{2/3}$ is not differentiable at $x = 0$, and the corresponding initial value problem fails to have a unique solution. However, for the initial value problem

$$x' = |x|, \qquad x(0) = 0,$$

the vector field $g(x) = |x|$ is not differentiable at $x = 0$ either, but the initial value problem has the zero function as a unique solution.[2]

Remark 3. If a differential equation $x' = f(t, x)$ satisfies the hypotheses of Cauchy's theorem for any initial value problem and the graphs of two distinct solutions intersect at one point, then they agree on an interval. Indeed, if the graphs of the solutions φ_1 and φ_2 have a common point (t_0, x_0), then $\varphi_1(t_0) = \varphi_2(t_0) = x_0$, so by Cauchy's theorem $\varphi_1 = \varphi_2$ at least on some interval $(t_0 - \epsilon, t_0 + \epsilon)$.

---

[2]The proof of uniqueness can be obtained using the Cauchy–Lipschitz theorem, published in 1876 by the German mathematician Rudolph Lipschitz (1832–1904).

The Method of Successive Approximations

To gain some insight into the method of successive approximations, which is essential not only for the above theorems but also for the direction of thinking we will deal with in our next section, let us see how this method works. The idea is to use an equivalent form of the initial value problem

$$x' = f(t, x), \qquad x(t_0) = x_0,$$

namely the integral form[3]

$$x(t) = x_0 + \int_{t_0}^{t} f(s, x(s))\, ds, \tag{1}$$

as a *self-feeding machine*. That is, introduce some "raw material" into it and get a product; then feed the machine with this product and obtain a better one; repeat the procedure and achieve a still finer output; and so on. Let us illustrate this idea with the following examples.

EXAMPLE 6   To solve the initial value problem

$$x' = x + t, \qquad x(0) = 1,$$

proceed as follows.

Step 1. As a first (very rough) approximation of the solution, take the function

$$\varphi_0(t) = 1,$$

which obviously satisfies the initial condition.

Step 2. Using equation (1), define the second term in the sequence of approximations,

$$\varphi_1(t) = x_0 + \int_{t_0}^{t} f(s, \varphi_0(s))\, ds = 1 + \int_{0}^{t} (1 + s)\, ds = 1 + t + \frac{t^2}{2},$$

which also satisfies the initial condition.

Step 3. Using equation (1) again, define the third term in the sequence of approximations,

$$\varphi_2(t) = x_0 + \int_{t_0}^{t} f(s, \varphi_1(s))\, ds$$

$$= 1 + \int_{0}^{t} \left(1 + s + \frac{s^2}{2} + s\right) ds = 1 + \frac{t^2}{2} + t + \frac{t^3}{6} + \frac{t^2}{2}.$$

---

[3]The integral form can be obtained by using a version of the fundamental theorem of calculus, which says that if f is a continuous function in the interval $[a, b]$, then the function $F(t) = \int_{a}^{t} f(u)\, du$ is differentiable and $F' = f$; moreover, $F(a) = 0$. Indeed, using this theorem, we can differentiate equation (1) and obtain $x'(t) = f(t, x(t))$; then, taking $t = t_0$ in (1), we obtain $x(t_0) = x_0$.

Step 4. Repeating the procedure, obtain the next approximation,

$$\varphi_3(t) = x_0 + \int_{t_0}^{t} f(s, \varphi_2(s)) \, ds$$

$$= 1 + \int_{0}^{t} \left(1 + s + \frac{s^2}{2} + s + \frac{s^3}{3!} + \frac{s^2}{2}\right) ds$$

$$= 1 + \frac{t^2}{2} + t + \frac{t^3}{3!} + \frac{t^2}{2} + \frac{t^4}{4!} + \frac{t^3}{3!}$$

$$= \left(1 + t + \frac{t^2}{2!} + \frac{t^3}{3!}\right) + \left(1 + t + \frac{t^2}{2!} + \frac{t^3}{3!} + \frac{t^4}{4!}\right) - t - 1.$$

Step 5. The above form suggests that the $(n+1)$st approximation of the solution is

$$\varphi_n(t) = -1 - t + \sum_{k=0}^{n} \frac{t^k}{k!} + \sum_{k=0}^{n+1} \frac{t^k}{k!}.$$

Step 6. Since $\sum_{k=0}^{\infty} t^k/k! = e^t$, it follows that when $n \to \infty$, φ_n tends to φ, where

$$\varphi(t) = -1 - t + 2e^t.$$

Step 7. A straightforward computation shows that φ satisfies the initial value problem.

The graphs of the first four approximations of the solution, as well as the exact solution, are drawn in Figure 2.5.6.

EXAMPLE 7   To solve the initial value problem

$$x' = 2t(x + 1), \quad x(0) = 0$$

with the method of successive approximations, take as a first approximation in the sequence the function $\varphi_0(t) = 0$, which satisfies the

Figure 2.5.6. Approximations and the exact solution of the equation in Example 6: $\varphi_0(t) = 1$, $\varphi_1(t) = 1 + t + t^2/2$, $\varphi_2(t) = 1 + t + t^2 + t^3/6$, $\varphi_3(t) = 1 + t + t^2 + t^3/3 + t^4/12$, and $\varphi(t) = 2e^t - t - 1$.

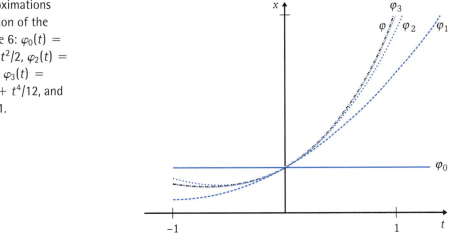

initial condition. Using equation (1), define the second term in the sequence as

$$\varphi_1(t) = \int_0^t 2s(1 + \varphi_0(s))\,ds = \int_0^t 2s\,ds = t^2.$$

Take further

$$\varphi_2(t) = \int_0^t 2s(1 + \varphi_1(s))\,ds = \int_0^t 2s(1 + s^2)\,ds = t^2 + \frac{t^4}{2} = \frac{t^2}{1!} + \frac{t^4}{2!}$$

and

$$\varphi_3(t) = \int_0^t 2s(1 + \varphi_2(s))\,ds$$

$$= \int_0^t 2s\left(1 + s^2 + \frac{s^4}{2}\right)ds$$

$$= t^2 + \frac{t^4}{2} + \frac{t^6}{6} = \frac{t^2}{1!} + \frac{t^4}{2!} + \frac{t^6}{3!}.$$

In general, obtain

$$\varphi_n(t) = \int_0^t 2s(1 + \varphi_{n-1}(s))\,ds$$

$$= \frac{t^2}{1!} + \frac{t^4}{2!} + \frac{t^6}{3!} + \cdots + \frac{t^{2n}}{n!},$$

so the solution φ of the initial value problem is given by the series

$$\varphi(t) = \sum_{k=1}^{\infty} \frac{t^{2k}}{k!}, \tag{2}$$

which, according to the ratio test (recall it from calculus, or see Section 7.1) is convergent for all t. The graphs of the first four approximations of the solution are drawn in Figure 2.5.7.

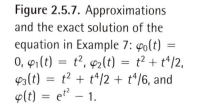

Figure 2.5.7. Approximations and the exact solution of the equation in Example 7: $\varphi_0(t) = 0$, $\varphi_1(t) = t^2$, $\varphi_2(t) = t^2 + t^4/2$, $\varphi_3(t) = t^2 + t^4/2 + t^4/6$, and $\varphi(t) = e^{t^2} - 1$.

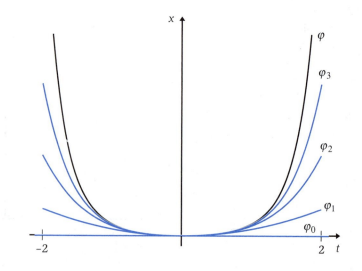

The exact solution can be obtained as in Section 2.2 by noticing that the given equation is separable. We thus obtain $\varphi(t) = e^{t^2} - 1$. Since, according to Cauchy's theorem, the solution of the above initial value problem is unique, it follows that $e^{t^2} = 1 + \sum_{n=1}^{\infty} t^{2k}/k!$. This shows another use of the uniqueness property: for identifying functions that look different but are in fact the same. Of course, students familiar with Taylor series expansions may see from the start that the sum of the series (2) is $e^{t^2} - 1$.

The method of successive approximations seems suited for solving any differential equation, but that impression is false. In most cases it is difficult, if not impossible, to find the limit function φ of the sequence of successive approximations. We can obtain better approximations by computing more and more terms of the sequence. However, it is hard to estimate how close these approximations are to the solution. In general, they are reasonably accurate only in a small interval around t_0.

The method of successive approximations can be justified by proving that the proper choice of the sequence always leads to the desired solution. This proof, however, is beyond our scope.

PROBLEMS

For the initial value problems in Problems 1 through 8, determine whether the theorems of Peano and Cauchy apply.

1. $x' = 2tx + \sin t, \; x(\pi) = 1$

2. $xx' - \frac{3}{2}xt + 4 = 0, \; x(1) = -\frac{2}{5}$

3. $3tx' = \dfrac{4t^2 + \sin x}{x}, \; x(0) = 0$

4. $x' = 4\tan(x^2), \; x\left(\dfrac{\pi}{4}\right) = \dfrac{1}{4}$

5. $x' = f(t, x), \; x(0) = 1,$
where $f(t, x) = \begin{cases} tx & \text{if } (t, x) \neq (0, 0) \\ 0 & \text{if } (t, x) = (0, 0) \end{cases}$

6. $x' = f(t, x), \; x(0) = 1,$
where $f(t, x) = \begin{cases} t^2 + x^2 & \text{if } (t, x) \neq (0, 0) \\ 0 & \text{if } (t, x) = (0, 0) \end{cases}$

7. $x' = f(t, x), \; x(0) = 0,$
where $f(t, x) = \begin{cases} t + x + 1 & \text{if } (t, x) \neq (0, 0) \\ 0 & \text{if } (t, x) = (0, 0) \end{cases}$

8. $x' = f(t, x), \; x(0) = 0,$
where $f(t, x) = \begin{cases} \sin t \sin x & \text{if } (t, x) \neq (0, 0) \\ 0 & \text{if } (t, x) = (0, 0) \end{cases}$

Show that each of the initial value problems in Problems 9 through 14 have the solution $\varphi_1(t) = 0$; then using methods studied in previous sections, find

another solution. Does this mean that the theorem of Cauchy is not applicable to those problems?

9. $x' = x^{1/3}, \; x(0) = 0$

10. $x' = 4x^{3/5}, \; x(0) = 0$

11. $x' = \pi x^{2/7}, \; x(0) = 0$

12. $x' = \sqrt{2}x^{1/9}, \; x(0) = 0$

13. $x' = \frac{2}{3}x^{3/7}, \; x(0) = 0$

14. $x' = \alpha x^{m/n}, \; x(0) = 0$, where $m, n = 1, 2, 3, \ldots, n$ is odd, $m < n$, and $\alpha \neq 0$.

15. For the initial value problem
$$x' = f(t, x), \; x(0) = 0,$$
where $f(t, x) = \begin{cases} \dfrac{1}{4t^2}x & \text{if } (t, x) \neq (0, 0) \\ 0 & \text{if } (t, x) = (0, 0), \end{cases}$

show that $\varphi_0(t) = 0$ is a solution, then find infinitely many solutions that are 0 for $t \leq 0$ and nonzero for $t > 0$.

16. Find the general solution of the equation $x' = 1/x$. Make sure that you establish the domain of the solution. Why does the domain of each particular solution depend on the integration constant?

For the initial value problems in Problems 17 through 20, given by linear equations, apply the *method of successive approximations and find φ_n for an arbitrary value of n. Then compute the solution using a different method and compare its graph with the graphs of the first four approximations.*

17. $x' = \frac{1}{2}x + t, \; x(0) = 0$

18. $x' = x - t + 1, \; x(0) = 0$

19. $x' = t^2x - 2t, \; x(0) = 0$

20. $x' = 2tx + t^2, \; x(0) = 0$

For the initial value problems in Problems 21 through 28, given by nonlinear equations, apply the method of successive approximations by taking as an

initial step the function $\varphi_0(t) = 0$ and then finding the next three iterations: $\varphi_1, \varphi_2,$ and φ_3.

21. $x' = \sin t - 1, \; x(0) = 0$

22. $x' = \cos t + 1, \; x(0) = 0$

23. $x' = x^2 + t^2, \; x(0) = 0$

24. $x' = 2x^2 + t^3, \; x(0) = 0$

25. $x' = t \ln(x + 1), \; x(0) = 0$

26. $x' = \ln x + t, \; x(1) = 0$

27. $x' = x^2 + x + 1, \; x(0) = 0$

28. $x' = x^3 + x^2 + x + 1, \; x(0) = 0$

2.6 Numerical Methods

Picard's method of successive approximations suggests the idea of finding functions as close as possible to the solution of an initial value problem. Especially in practice, an approximate solution is better than no solution at all. This raises the issue of designing methods that lead to approximate solutions and of controlling the error involved. In this section we will discuss first the Euler method, then the second-order Runge-Kutta method, and finally some rudiments of error theory.

With Picard's method we found a sequence of *functions* that locally approximate a solution near some initial value (t_0, x_0). The closer each function to the initial value, the better the approximation. Here we will do something different. The estimate we obtain will be a collection of points $(t_0, x_0), (t_1, x_1), \ldots, (t_n, x_n)$, which track a curve that approximates the graph of the solution passing through (t_0, x_0) (see Figure 2.6.1). So we don't end up with a function, but with a table of values. To do this we will use a self-feeding machine that works with numbers instead of functions. The prototype of all these numerical solvers is the *Euler method*, which we present below.

Figure 2.6.1. The goal of a numerical method for a differential equation is to obtain a collection of points $(t_0, x_0), (t_1, x_1), \ldots, (t_n, x_n)$ that track a curve that estimates a finite portion of the graph of the real solution passing through (t_0, x_0). The more dots we obtain and the closer they are to the real curve, the better the approximation.

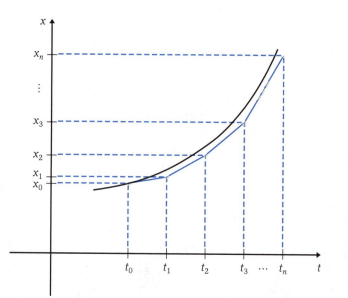

Euler's Method

Leonhard Euler.

Approximate methods for solving initial value problems first appeared in the 18th century. One of the notable ones was proposed in 1768 by the Swiss mathematician Leonhard Euler (1707–1783). Finding a numerical solution of the initial value problem

$$x' = f(t, x), \qquad x(t_0) = x_0, \tag{1}$$

means obtaining approximate values of the solution for a set of values of the independent variable t in some interval $[t_0, a]$ $(a > t_0)$, say, at $t_0 < t_1 < \cdots < t_n = a$.

The starting point of this method is (t_0, x_0) (see Figure 2.6.2). The slope of the tangent to the graph of the unknown solution φ is $\varphi'(t_0) = f(t_0, x_0)$, so the equation of the line through (t_0, x_0) and of slope $f(t_0, x_0)$ is

$$x = x_0 + f(t_0, x_0)(t - t_0).$$

Using this equation and moving along the tangent line from t_0 to t_1, we obtain an approximate value x_1 for $\varphi(t_1)$:

$$x_1 = x_0 + f(t_0, x_0)(t_1 - t_0).$$

Next we take $f(t_1, x_1)$ as an approximate value for the slope of the tangent to the solution's graph at $(t_1, \varphi(t_1))$. (We cannot obtain the exact value $f(t_1, \varphi(t_1))$ of the slope, since we do not know $\varphi(t_1)$—only its approximation x_1.) We further write the equation of the line through (t_1, x_1) and of slope $f(t_1, x_1)$ as

$$x = x_1 + f(t_1, x_1)(t - t_1).$$

Moving along this line to t_2, the approximation x_2 for $\varphi(t_2)$ is

$$x_2 = x_1 + f(t_1, x_1)(t_2 - t_1).$$

Continuing this procedure we obtain *Euler's numerical formula*:

$$x_{n+1} = x_n + f(t_n, x_n)(t_{n+1} - t_n). \tag{2}$$

Figure 2.6.2. Euler's method for solving initial value problems.

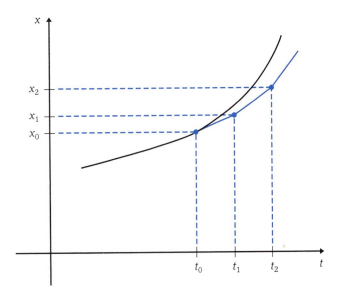

As in the method of successive approximations, formula (2) is also a self-feeding machine: we obtain x_1 using x_0, then x_2 using x_1, then x_3 using x_2, and so on.

If we divide the interval $[t_0, a]$ in equal parts, each of length h, formula (2) becomes

$$x_{n+1} = x_n + hf(t_n, x_n). \tag{3}$$

So this method computes the approximate values $x_1, x_2, \ldots, x_n$ instead of the exact values $\varphi(t_1), \varphi(t_2), \ldots, \varphi(t_n)$, respectively, for the solution φ. In other words, we have some idea of how the solution φ behaves in the interval $[t_0, a]$. The larger the value of n, the better. Let us now summarize the procedure.

The Algorithm of Euler's Method

Step 1. In the interval $[t_0, a]$ pick the points $t_0 < t_1 < \cdots < t_n = a$ at distance $h = (a - t_0)/n$ of each other.

Step 2. Compute the approximation x_1 of $\varphi(t_1)$ with the formula

$$x_1 = x_0 + hf(t_0, x_0).$$

Step 3. Compute the approximation x_2 of $\varphi(t_2)$ with the formula

$$x_2 = x_1 + hf(t_1, x_1).$$

Step 4. Continue the procedure and obtain the approximations $x_3, x_4, \ldots, x_n$ by successively using the formula

$$x_{n+1} = x_n + hf(t_n, x_n).$$

EXAMPLE 1   Let us apply Euler's method in the interval $[0, 1.5]$ with step size 0.25 to solve numerically the initial value problem

$$x' = 2tx + e^{t^2}, \qquad x(0) = 1$$

(see Example 4, Section 2.3). For comparison we will use its exact solution $\varphi(t) = (t + 1)e^{t^2}$.

Step 1. Notice first that the vector field is $f(t, x) = 2tx + e^{t^2}$, $t_0 = 0$, and $x_0 = 1$. We take as $[t_0, a]$ the interval $[0, 1.5]$ and divide it into six equal parts to have the step size $h = 0.25$. This makes $t_1 = 0.25$, $t_2 = 0.5$, $t_3 = 0.75$, $t_4 = 1$, $t_5 = 1.25$, and $t_6 = 1.5$.

Step 2. $x_1 = x_0 + hf(t_0, x_0) = 1 + 0.25 \cdot f(0, 1) = 1 + 0.25 \cdot 1 = 1.25$.

Step 3. $x_2 = x_1 + hf(t_1, x_1) = 1.25 + 0.25 \cdot f(0.25, 1.25) = 1.25 + 0.25 \cdot 1.6894945 = 1.6723736$.

Step 4. Analogously, we obtain $x_3 = 2.4114734$, $x_4 = 3.7545396$, $x_5 = 6.3113798$, and $x_6 = 11.4486755$.

These computations are summarized in Table 2.6.1 to seven decimal places. Compare the numerical values x_n with the seven-place approximation $\varphi(t_n)$ of the exact ones.

TABLE 2.6.1

n	t_n	x_n	$\varphi(t_n)$
0	0.0	1.0000000	1.0000000
1	0.25	1.2500000	1.3306181
2	0.50	1.6723736	1.9260381
3	0.75	2.4114734	3.0713457
4	1.00	3.7545396	5.4365637
5	1.25	6.3113798	10.7341497
6	1.50	11.4486755	23.7193396

TABLE 2.6.2

n	$\bar{t}_n$	$\bar{x}_n$	$\varphi(\bar{t}_n)$
0	0.0	1.0000000	1.0000000
1	0.1	1.1000000	1.1110552
2	0.2	1.2230050	1.2489729
3	0.3	1.3760062	1.4224266
4	0.4	1.5679841	1.6429152
5	0.5	1.8107739	1.9260381
6	0.6	2.1202539	2.2933271
7	0.7	2.5180173	2.7749376
8	0.8	3.0337713	3.4136656
9	0.9	3.7088228	4.2710252
10	1.0	4.6012017	5.4365637
11	1.1	5.7932702	7.0423178
12	1.2	7.4031381	9.2855307
13	1.3	9.6019608	12.4648056
14	1.4	12.6404187	17.0383840
15	1.5	16.8896687	23.7193396

What happens if we take a smaller step size (say, $\bar{h} = 0.1$) in the same interval $[0, 1.5]$? The points that divide the interval are $\bar{t}_0 = 0, \bar{t}_1 = 0.1, \bar{t}_2 = 0.2, \ldots, \bar{t}_{15} = 1.5$. The corresponding approximations $\bar{x}_1, \bar{x}_2, \ldots, \bar{x}_{15}$ as well as the exact values $\varphi(0)$, $\varphi(0.1)$, $\varphi(0.2)$, $\ldots, \varphi(1.5)$ are given in Table 2.6.2 to seven decimal places.

The aspects encountered in Tables 2.6.1 and 2.6.2 are illustrated in Figure 2.6.3. Notice that the smaller the step size, the better the approximations. There is a limit to that tendency, though. If the step is too small, round-off errors lead to less accurate results. A second observation is that the farther from the starting point, the worse the approximation gets. This happens because errors add up. In Table 2.6.2, at $n = 1$ the error is approximately 0.01, whereas at $n = 15$ it is almost 7. So we expect that the approximations are better on shorter intervals.

The fundamental question concerning any numerical method is that of the algorithm's convergence. This is, when h approaches 0, does x_k approach $\varphi(t_k)$ for every $k = 0, 1, 2, \ldots, n$? In other words, when the

Figure 2.6.3. In general, the approximation is better if the step size is smaller.

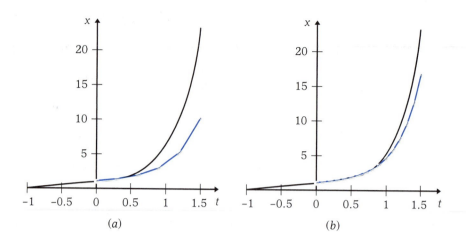

(a)   (b)

interval is divided in very many parts, is the theoretical approximation very close to the exact solution? If the answer is negative, the method is worthless. To make sense, every numerical method must be endorsed by a convergence theorem. It can be shown that Euler's method is convergent.

The Second-Order Runge-Kutta Method

The errors of Euler's method are usually large. Therefore, we introduce now the *second-order Runge-Kutta method*, also called the *improved Euler method*, which was developed in 1895 by the German mathematician and physicist Carl Runge (1856–1927) and improved to higher orders in 1901 by another German mathematician, Wilhelm Kutta (1867–1944). Instead of approximating $f(t, x)$ by the value at the left endpoint of the interval, as Euler's procedure does, the second-order Runge-Kutta method takes the average of the approximate values of $f(t, x)$ at both ends of the interval, so the formula is

$$x_{n+1} = x_n + (1/2)[f(t_n, x_n) + f(t_{n+1}, x_{n+1}^E)](t_{n+1} - t_n) \qquad (4)$$

(also called *Heun's formula* after the mathematician who derived it in 1900). This is a two-stage method. We first compute x_{n+1}^E using Euler's formula (E is not a power; it just stands for "Euler"), and then introduce it into the right-hand side of (4) to obtain a better value x_{n+1}. For intervals of equal step size h, Heun's formula becomes

$$x_{n+1} = x_n + (h/2)[f(t_n, x_n) + f(t_{n+1}, x_{n+1}^E)]. \qquad (5)$$

The algorithm of this method works as follows.

The Algorithm of the Second-Order Runge-Kutta Method

Step 1. In the interval $[t_0, a]$ pick the points $t_0 < t_1 < \cdots < t_n = a$ at distance $h = (a - t_0)/n$ of each other.

Step 2. Compute x_1^E with Euler's formula:

$$x_1^E = x_0 + hf(t_0, x_0).$$

Step 3. Compute x_1 with Heun's formula:

$$x_1 = x_0 + (h/2)[f(t_0, x_0) + f(t_1, x_1^E)].$$

Step 4. Compute x_2^E with Euler's formula:

$$x_2^E = x_1 + hf(t_1, x_1).$$

Step 5. Compute x_2 with Heun's formula:

$$x_2 = x_1 + (h/2)[f(t_1, x_1) + f(t_2, x_2^E)].$$

Step 6. Continue the procedure and obtain the approximation x_{n+1}^E, using successively the formula

$$x_{n+1}^E = x_n + hf(t_n, x_n),$$

and then x_{n+1} with the formula

$$x_{n+1} = x_n + (h/2)[f(t_n, x_n) + f(t_{n+1}, x_{n+1}^E)].$$

EXAMPLE 2 Let us use the second-order Runge-Kutta method to obtain a numerical solution for the equation

$$x' = 2tx + e^{t^2}, \quad x(0) = 1,$$

in the interval $[0, 1.5]$, with step size $h = 0.1$.

Step 1. Take $t_0 = 0, t_1 = 0.1, \ldots, t_{15} = a = 1.5, x_0 = 1$, and $h = 0.1$.

Step 2. $x_1 = x_0 + hf(t_0, x_0) = 1 + 0.1(0 + 1) = 1.1$.

Step 3. $x_1 = x_0 + (h/2)[f(t_0, x_0) + f(t_1, x_1)] = 1 + 0.05[f(0, 1) + f(0.1, 1.1)] = 1.1115025$.

Step 4. $x_2 = x_1 + hf(t_1, x_1)$, where x_1 is the one obtained at Step 3. So $x_2 = 1.1115025 + 0.1f(0.1, 1.1115025) = 1.2347376$.

Step 5. $x_2 = x_1 + (h/2)[f(t_1, x_1) + f(t_2, x_2)] = 1.1115025 + 0.05[f(0.1, 1.1115025) + f(0.2, 1.2347376)] = 1.2498553$.

Step 6. Continuing the procedure we obtain the values in Table 2.6.3 to seven decimal places.

TABLE 2.6.3

n	t_n	x_n	$\varphi(t_n)$
0	0.0	1.0000000	1.0000000
1	0.1	1.1115025	1.1110552
2	0.2	1.2498553	1.2489729
3	0.3	1.4237196	1.4224266
4	0.4	1.6445579	1.6429152
5	0.5	1.9278907	1.9260381
6	0.6	2.2950979	2.2933271
7	0.7	2.7760550	2.7749376
8	0.8	3.4130535	3.4136656
9	0.9	4.2667083	4.2710252
10	1.0	5.4249722	5.4365637
11	1.1	7.0170552	7.0423178
12	1.2	9.2351790	9.2855307
13	1.3	12.3689892	12.4648056
14	1.4	16.8606606	17.0383840
15	1.5	23.3942430	23.7193396

The values in Table 2.6.3 show that in this case, at the expense of doubling the amount of computations, the second-order Runge-Kutta method offers a better approximation of the solution. But are the errors of this method always smaller than in Euler's case, independent of the equation considered? To address this issue, let us do some rudimentary error analysis.

Errors   Evaluating the errors of numerical methods is an important but difficult task. Errors occur because both the formulas and the values we substitute into the formulas are approximate. Let us describe different types of errors.

The *global truncation error* E_n at step n is the absolute value of the difference between the exact solution and the numerical solution,

$$E_n = |\varphi(t_n) - x_n|,$$

assuming that the process of approximation has started at step 0 (i.e., the only exact value in the sequence $x_0, x_1, \ldots, x_n$ is x_0). In general, this error is difficult to determine; therefore, we usually deal with the so-called *local truncation error*.

The *local truncation error* e_n at step n is the absolute value of the difference between the exact value $\varphi(t_n)$ and the approximate value x_n^e,

$$e_n = |\varphi(t_n) - x_n^e|$$

(e stands for "exact"), but assuming that the process of approximation has started only at this step, i.e., $x_{n-1} = \varphi(t_{n-1})$. In other words, the data are considered exact, so the only source of error is the formula of the method.

The *round-off error* r_n at step n is the absolute value of the difference between the approximate value x_n, as given by the computer with a finite number of decimal places, and the theoretical approximation X_n, supposed to have infinitely many digits. So

$$r_n = |x_n - X_n|.$$

This error depends on computer, on the method of rounding off, on the sequence in which the computations are carried out, etc.

The *total error* T_n at step n (or the *theoretical error*) is given by the absolute value of the difference between the exact value $\varphi(t_n)$ and the theoretical approximation X_n,

$$T_n = |\varphi(t_n) - X_n|.$$

With the triangle inequality $|\alpha + \beta| \le |\alpha| + |\beta|$, we can estimate T_n with respect to E_n and r_n. Indeed,

$$T_n = |\varphi(t_n) - X_n| = |\varphi(t_n) - x_n + x_n - X_n|$$
$$\le |\varphi(t_n) - x_n| + |x_n - X_n| = E_n + r_n,$$

so the total error is never greater than the sum of the global truncation error and the round-off error.

Errors can be expressed with respect to the step size h. For Euler's method, e_n is of the order of h^2, whereas for the second-order Runge-Kutta method, e_n is of the order of h^3. This means that the error at every step is always less than αh^2 for Euler's method and less than βh^3 for the Runge-Kutta method, where $\alpha, \beta > 0$ are constants specific to each method and of order unity (i.e., numbers like 1.8, 2.53, or 3.1, but unlike 10, 100, or 10^{-2}). This means that for a step size of 0.1, the error at t_1 cannot exceed 0.01α in the first case and 0.001β in the second case. At t_5 the error is at most 0.05α and 0.005β, respectively, and so on. This explains why the Runge-Kutta method is better.

The equation of our example is linear. In general, linear equations raise fewer problems if treated numerically. Nonlinear equations are more troublesome, especially if the interval of approximation is large. After studying Chapter 5 we will understand the reason and nature of these difficulties.

To solve the problems below, you can use one of the programs of Section 2.7. Though Maple, *Mathematica*, and MATLAB have built-in numerical programs you are welcome to explore, those presented in Section 2.7 will allow you to control the parameters and have a better grasp of the methods.

PROBLEMS

For the initial value problems in Problems 1 through 8, use

- **(a)** *Euler's method with step size $h = 0.1$*
- **(b)** *Euler's method with step size $h = 0.05$*
- **(c)** *the second-order Runge-Kutta method with step size $h = 0.1$*
- **(d)** *the second-order Runge-Kutta method with step size $h = 0.05$*

to find approximate values of the solution in the interval $[0, 1]$. Then solve the equation using an exact method and compute the difference between the correct value and the approximate one at $t = 1$.

1. $x' = 2tx, \ x(0) = 1$

2. $x' = 3t^2 x, \ x(0) = 1$

3. $x' = -\dfrac{x}{1+t}, \ x(0) = 1$

4. $x' = \dfrac{2x}{1+t}, \ x(0) = 1$

5. $x' = e^{-x}, \ x(0) = 0$

6. $x' = x - t, \ x(0) = 2$

7. $x' = \sqrt{x + t}, \ x(0) = 3$

8. $x' = 2x - t - 1, \ x(0) = 1$

9. For the initial value problems (a) and (b) use Euler's method with step $h = 0.1$ in the interval $[0, 5]$, then solve them as separable equations. Compare the exact and the approximate solutions and explain the phenomenon you encounter.

- **(a)** $x' = -x^2 + 2x + 3, \ x(0) = 3,$
- **(b)** $x' = -x^2 + 2x + 3, \ x(0) = 0.$

10. For the initial value problem $x' = x^2 + t^2, \ x(0) = 0$, apply Euler's method and the second-order Runge-Kutta method in the interval $[0, 2]$ with step size

- **(a)** $h = 0.1,$
- **(b)** $h = 0.01.$

Compare the results at (a) and (b). For what reason are these results so different when you approach 2? (*Hint:* The solution has a vertical asymptote near $t = 2.00315$.)

Perform the calculations in Problems 11 through 14 using

- **(a)** *a two-digit round-off approximation*
- **(b)** *a three-digit round-off approximation*
- **(c)** *all the digits*

of all the numbers involved. Compare the results and draw a conclusion about round-off errors.

11. $10^7 \cdot (25.0009121 - 25.0009987)$

12. $2^{12} \cdot (49.0008212 - 49.0009763)$

13. $1024 \cdot (60 \cdot 60.00101 - 180.00401 \cdot 20.000401)$

14. $10^9 \cdot (5 \cdot 8.00112 - 10.00211 \cdot 4.00033)$

For the initial value problems in Problems 15 through 20, use

   (a) *Euler's method with step* $h = 0.1$
   (b) *Euler's method with step* $h = 0.05$
   (c) *second-order Runge-Kutta method with step* $h = 0.1$
   (d) *second-order Runge-Kutta method with step* $h = 0.05$

in the given intervals and graph each approximate solution in that interval. Draw an approximate graph of what you think would be the exact solution in that interval. Then solve the problem using an exact method and draw the graph of the solution. Compare it with the approximate graph.

15. $x' = 2x + 1,$ $x(0) = 1$ in $[0, 0.5]$

16. $x' = 3x + 2,$ $x(0) = 1$ in $[0, 1]$

17. $x' = tx - 2,$ $x(0) = 1$ in $[0, 0.5]$

18. $x' = 2tx - 7,$ $x(0) = 1$ in $[0, 1]$

19. $x' = x + 2\sin t,$ $x(0) = 1$ in $[0, 0.5]$

20. $x' = -x + \cos t,$ $x(0) = 1$ in $[0, 0.5]$

21. For the initial value problem $x' = \sqrt{x}$, $x(0) = 0$, apply Euler's method with step size $h = 0.01$ in the interval $[0, 1]$. Then

notice that an exact solution of the problem is $x(t) = t^2/4$. Compare the values obtained numerically with the corresponding values of the exact solution. Can you explain what happens?

22. For the initial value problem $x' = t \sin x$, $x(0) = 1$, use Euler's method with step size $h = 0.2$, $h = 0.1$, and $h = 0.05$ in the interval $[0, 0.6]$. Then compute the exact solution and determine the global error. Does the global error decrease linearly with h?

23. For the initial value problem $x' = x^2$, $x(0) = 1$, use Euler's method and then the second-order Runge-Kutta method in the interval $[0, 1]$. How small must the step size h be in each case such that the local truncation error at $t = 1$ is less than 0.5?

24. For the initial value problem $x' = x$, $x(1) = 2$, use Euler's method and the second-order Runge-Kutta method with step size 0.1 in the interval $[1, 2]$. In each case determine the sum of the local truncation errors made at each step of the computation. Then find the global error at $t = 1$ and compare it with the above sum. Can you draw any conclusion?

2.7 Computer Applications

Today's computers are an important tool in understanding differential equations. Users debate the quality of the existing packages, whose performances differ from problem to problem. We will use here three languages that have established themselves as basic during the last few years, and that complement each other well, are used in research, have been upgraded, and run under site licenses at most institutions of higher education: Maple, *Mathematica*, and MATLAB. We present them separately and in parallel. This will allow you to focus on any of them but also to compare their performance with respect to the types of problems we deal with. Students unfamiliar with these packages can use the on-line help and the built-in examples to understand the various meanings, to learn the syntax, and to apply the procedure techniques of each command. We encourage the more advanced students to improve our programs or come up with alternative ones. Our aim was to achieve our mathematical goals while keeping the programs simple.

Maple

Maple has been under development since 1980 by a group of professors and graduate students with the Faculty of Mathematics at the University of Waterloo in Waterloo, Ontario, Canada. We will first show through examples how Maple solves separable and linear differential equations and then present programs for finding numerical solutions and for drawing slope fields.

University of Waterloo, Ontario, Canada.

Exact solutions   In Maple the command dsolve is designed to solve differential equations and initial value problems.

EXAMPLE 1   To solve the separable equation $x' = 2t^2 x$, type

> dsolve(diff(x(t),t)=2*t^2*x(t), x(t));

and Maple will display the answer

$$x(t) = \_C1 e^{(\frac{2}{3}t^3)}$$

That is, the general solution is $x(t) = C_1 e^{(2/3)t^3}$, where C_1 is a constant.
    We can also proceed in two steps: first define the equation, say eq, and only then use the command dsolve.

EXAMPLE 2   To solve the linear nonhomogeneous equation $x' = 5x + t^3$, type

> eq:=diff(x(t),t)=5*x(t)+t^3;

and Maple will display the equation to solve:

$$eq := \frac{\partial}{\partial t} x(t) = 5x(t) + t^3$$

Then use the command dsolve as follows:

> dsolve(eq,x(t));

and obtain the result

$$x(t) = -\frac{1}{5}t^3 - \frac{3}{25}t^2 - \frac{6}{125}t - \frac{6}{625} + e^{(5t)}\_C1$$

That is, the general solution is $x(t) = -\frac{1}{5}t^3 - \frac{3}{25}t^2 - \frac{6}{125}t - \frac{6}{625} + C_1 e^{5t}$, where C_1 is a constant.

To solve an initial value problem for the above equation, specify an initial condition as

```
> dsolve({eq, x(1)=1},x(t));
```

and Maple will display the answer

$$x(t) = -\frac{1}{5}t^3 - \frac{3}{25}t^2 - \frac{6}{125}t - \frac{6}{625} + \frac{861}{625}\frac{e^{(5t)}}{e^5}$$

which is the solution of the initial value problem $x' = 5x + t^3$, $x(1) = 1$.

Numerical solutions   With Maple's help we will apply Euler's method to solve the initial value problem used in the examples of Section 2.6, $x' = 2tx + e^{t^2}$, $x(0) = 1$, in the interval $[0, 1.5]$, with step size $h = 0.1$. Let us first write a program and then explain how it works.

```
> f := (t,x) -> 2*t*x + exp(t^2);
                                  2
            f := (t,x) -> 2 t x + exp(t )
> t0:=0: x0:=1:
> h:=0.1:
> n:=15:
> t:=t0: x:=x0:
> for i from 1 to n do
  x:=x+h*f(t,x):
  t:=t+h:
  print(t,x);
  od:
.1, 1.1
.2, 1.223005017
.3, 1.376006295
.4, 1.567984101
.5, 1.810773916
.6, 2.120253849
.7, 2.518017252
.8, 3.033771289
.9, 3.708822783
1.0, 4.601201683
1.1, 5.793270202
1.2, 7.403138111
1.3, 9.601960840
1.4, 12.64041873
1.5, 16.88966868
```

We first defined the vector field f of the equation to solve (which Maple displays), the initial conditions $t_0 = 0$, $x_0 = x(t_0) = 1$, the step size $h = 0.1$, and the number of steps $n = 15$ necessary to cover the interval $[0, 1.5]$. We initialized the variables t and x at t_0 and x_0 respectively, and asked Maple to compute the variable x successively using Euler's formula and print all the values for t and x from $i = 1$ to $i = n$.

(od just marks the end of the do command, which makes all steps from 1 to n.)

To use the second-order Runge-Kutta method, we need to change the final part of the program used for the Euler method. According to Heun's formula, we use do as follows:

```
> for i from 1 to n do
  x:=x+h*(f(t,x)+f(t+h, x+h*f(t,x)))/2:
  t:=t+h:
  print(t,x);
  od:
```

This will produce the following numerical result:

```
.1, 1.111502508
.2, 1.249855332
.3, 1.423719610
.4, 1.644557865
.5, 1.927890673
.6, 2.295097885
.7, 2.776055020
.8, 3.413053474
.9, 4.266708306
1.0, 5.424972205
1.1, 7.017055181
1.2, 9.235178969
1.3, 12.36898917
1.4, 16.86066057
1.5, 23.39424293
```

Compare this with Table 2.6.3 and with the result obtained using Euler's method.

Slope fields To draw the slope field for the equation $x' = -x^2 + 2t$ in the rectangle $[-1, 9] \times [-5, 5]$, use the command DEplot. For this proceed as follows. First, load a package that contains programs related to differential equations:

```
> with(DEtools):
```

Then write the equation whose slope field you want to draw,

```
> eq:=diff(x(t),t)=-x^2+2*t;
```

Finally, give the DEplot command, also specifying the desired rectangle:

```
>  DEplot(eq,x(t),t=-1..9,x=-5..5);
```

This produces the picture in Figure 2.7.1.

Mathematica *Mathematica* was first released in 1988 by Wolfram Research, Inc., a company based in Champaign, Illinois, USA. We will first show through examples how *Mathematica* solves separable and linear differential equations and then present programs for finding numerical solutions and for drawing slope fields.

Figure 2.7.1. The slope field of the equation $x' = -x^2 + 2t$ drawn with Maple.

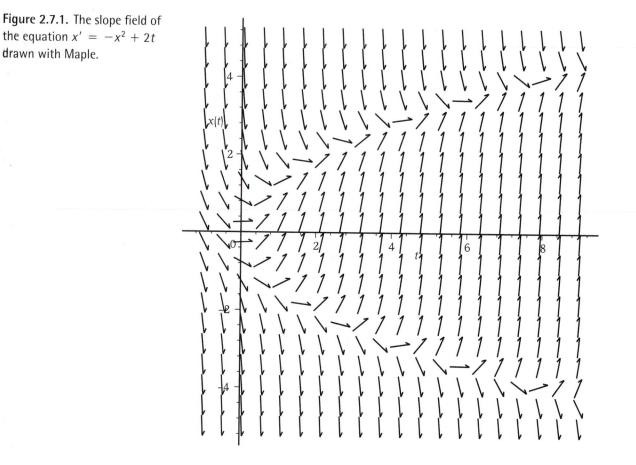

Exact solutions   In *Mathematica* the command DSolve is designed to solve differential equations and initial value problems.

EXAMPLE 1   To solve the separable equation $x' = 2t^2x$, type

    DSolve[x'[t]==2*t^2*x[t],x[t],t]

and *Mathematica* will display the answer

$$\{\{x[t]->E^{(2t^3)/3}C[1]\}\}$$

which is the general solution $x(t) = C_1 e^{(2/3)t^3}$, where C_1 is a constant.

We can also proceed in two steps: First define the equation, say eq, and only then use the command DSolve.

EXAMPLE 2   To solve the linear nonhomogeneous equation $x' = 5x + t^3$, type

    eq=x'[t]==5*x[t]+t^3

and *Mathematica* displays the equation to solve:

$$x'[t] == t^3 + 5x[t]$$

Then use the command `DSolve` as follows:

$$\{\{x[t]->-\left(\frac{6}{625}\right)-\frac{6t}{125}-\frac{3t^2}{25}-\frac{t^3}{5}+E^{5t}C[1]\}\}$$

That is, the general solution is $x(t) = -\frac{1}{5}t^3 - \frac{3}{25}t^2 - \frac{6}{125}t - \frac{6}{625} + C_1e^{5t}$, where C_1 is a constant.

To solve an initial value problem for the above equation, specify an initial condition as shown below:

```
DSolve[{eq,x[1]==1},x[t],t]
```

and *Mathematica* will display the answer:

$$\{\{x[t]->-\left(\frac{6}{625}\right)+\frac{861E^{-5+5t}}{625}-\frac{6t}{125}-\frac{3t^2}{25}-\frac{t^3}{5}\}\}$$

which is the solution of the initial value problem $x' = 5x + t^3$, $x(1) = 1$.

Numerical solutions   With *Mathematica*'s help we will apply Euler's method to solve the initial value problem used in the examples of Section 2.6, $x' = 2tx + e^{t^2}$, $x(0) = 1$, in the interval $[0, 1.5]$, with step size $h = 0.1$. Let us first write a program and then explain how it works.

```
f[t_,x_]:=2*t*x+Exp[t^2]
t0=0; x0=1;
h=0.1;
n=15;
t=t0; x=x0;
Do[x=x+h*f[t,x];
   t=t+h;
   Print[t," ",x],
   {i,1,n}]
```

This produces the following numerical result:

```
0.1    1.1
0.2    1.22301
0.3    1.37601
0.4    1.56798
0.5    1.81077
0.6    2.12025
0.7    2.51802
0.8    3.03377
0.9    3.70882
1.0    4.6012
```

1.1	5.79327
1.2	7.40314
1.3	9.60196
1.4	12.6404
1.5	16.8897

We first defined the vector field f of the equation to solve, the initial conditions $t_0 = 0$, $x_0 = x(t_0) = 1$, the step size $h = 0.1$, and the number of steps $n = 15$ necessary to cover the interval $[0, 1.5]$. We initialized the variables t and x at t_0 and x_0 respectively, and asked *Mathematica* to compute the variable x successively using Euler's formula and print all the values for t and x from $i = 1$ to $i = n$.

To use the second-order Runge-Kutta method, we need to change the final part of the program used for the Euler method. According to Heun's formula, we use Do as follows:

```
Do[x=x+h*(f[t,x]+f[t+h,x+h*f[t,x]])/2;
   t=t+h;
   Print[t," ",x],
   {i,1,n}]
```

This will produce the following numerical result:

0.1	1.1115
0.2	1.24986
0.3	1.42372
0.4	1.64456
0.5	1.92789
0.6	2.2951
0.7	2.77606
0.8	3.41305
0.9	4.26671
1.0	5.42497
1.1	7.01706
1.2	9.23518
1.3	12.369
1.4	16.8607
1.5	23.3942

Compare this with Table 2.6.3 and with the result obtained using Euler's method.

Slope fields   To draw the slope field for the equation $x' = -x^2 + 2t$ in the rectangle $[-1, 9] \times [-5, 5]$, proceed as follows. First load a certain graphics package by typing

```
<<Graphics'PlotField'
```

then define the vector field

```
f:=-x^2+2*t
```

and finally use the command PlotVectorField:

```
PlotVectorField[{Cos[ArcTan[f]],Sin[ArcTan[f]]},{t,-1,9},
   {x,-5,5}];
```

Figure 2.7.2. The slope field of the equation $x' = -x^2 + 2t$ drawn with *Mathematica*.

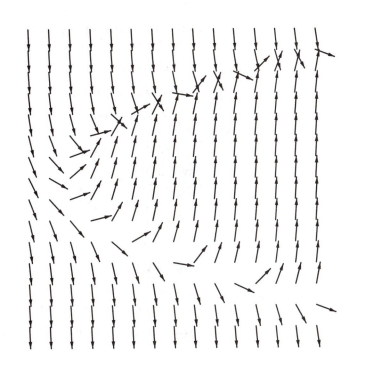

which displays the picture in Figure 2.7.2. The $\cos(\arctan(f))$ and $\sin(\arctan(f))$ are used to obtain vectors of unit length.

MATLAB MATLAB, which stands for "Matrix Laboratory," was developed in 1984 by The MathWorks, Inc., a company based in Natick, Massachusetts, USA. We will first show through examples how MATLAB solves separable and linear differential equations and then present programs for finding numerical solutions and for drawing slope fields.

Exact solutions

EXAMPLE 1 To solve the separable equation $x' = 2t^2 x$, type

```
>> dsolve('Dx=2*t^2*x')
```

and MATLAB will display the answer:

```
ans =
exp(2/3*t^3)*C1
```

That is, the general solution is $x(t) = C_1 e^{(2/3)t^3}$, where C_1 is a constant.

We can also proceed in two steps: define first the equation, say **eq**, and only then use the command **dsolve**.

EXAMPLE 2 To solve the linear nonhomogeneous equation $x' = 5x + t^3$, type

```
>> eq='Dx=5*x+t^3'
```

and MATLAB will display the equation to solve:

```
eq =
Dx=5*x+t^2
```

Then use the command `dsolve` as follows:

```
>> dsolve(eq)
ans =
-1/5*t^2-2/25*t-2/125+exp(5*t)*C1
```

which means that the general solution is $x(t) = -\frac{1}{5}t^3 - \frac{3}{25}t^2 - \frac{6}{125}t - \frac{6}{625} + C_1 e^{5t}$, where C_1 is a constant.

To solve an initial value problem for the above equation, specify an initial condition,

```
dsolve(eq2,'x(1)=1')
```

and MATLAB will display the answer

```
ans =
-1/5*t^2-2/25*t-2/125+162/125*exp(5*t)/exp(5)
```

which is the solution of the initial value problem $x' = 5x + t^3$, $x(1) = 1$.

Numerical solutions   We will apply Euler's method to solve with MATLAB the initial value problem used in the examples of Section 2.6, $x' = 2tx + e^{t^2}$, $x(0) = 1$, in the interval $[0, 1.5]$, with step size $h = 0.1$. Let us first write a program and then explain how it works.

First create in a text editor a file called f.m, and type into it

```
function xp=f(t,x)
xp=2.*t.*x+exp(t.^2);
```

(If the function is not f but g, then the file should be called g.m instead.) Then return to MATLAB and proceed as follows:

```
>> t0=0; x0=1;
>> h=0.1;
>> n=15;
>> t=t0; x=x0; T=t; X=x;
>> for i=1:n
     x=x+h*f(t,x);
     t=t+h;
     T=[T;t];
     X=[X;x];
   end
>> [T,X]
```

This produces the following numerical result:

```
ans =
        0      1.0000
   0.1000      1.1000
   0.2000      1.2230
   0.3000      1.3760
   0.4000      1.5680
```

0.5000	1.8108
0.6000	2.1203
0.7000	2.5180
0.8000	3.0338
0.9000	3.7088
1.0000	4.6012
1.1000	5.7933
1.2000	7.4031
1.3000	9.6020
1.4000	12.6404
1.5000	16.8897

We first defined the initial conditions $t_0 = 0$, $x_0 = x(t_0) = 1$, the step size $h = 0.1$, and the number of steps $n = 15$ necessary to cover the interval $[0, 1.5]$. We initialized the variables t and x at t_0 and x_0 respectively, asked MATLAB to compute the variable x successively using Euler's formula and print all the values for t and x from $i = 1$ to $i = n$.

To use the second-order Runge-Kutta method, we need to change the final part of the program used for the Euler method. According to Heun's formula we proceed as follows:

```
>> for i=1:n
   x=x+h*(f(t,x)+f(t+h,x+h*f(t,x)))/2;
   t=t+h;
   T=[T;t];
   X=[X;x];
   end
>> [T,X]
ans =
```

0	1.0000
0.1000	1.1115
0.2000	1.2499
0.3000	1.4237
0.4000	1.6446
0.5000	1.9279
0.6000	2.2951
0.7000	2.7761
0.8000	3.4131
0.9000	4.2667
1.0000	5.4250
1.1000	7.0171
1.2000	9.2352
1.3000	12.3690
1.4000	16.8607
1.5000	23.3942

Compare this with Table 2.6.3 and with the result obtained using Euler's method.

Slope fields To draw the slope field for the equation $x' = -x^2 + 2t$ in the rectangle $[-1, 9] \times [-5, 5]$, we will use the commands meshgrid and quiver. The program that displays the slope field in Figure 2.7.3 is

Figure 2.7.3. The slope field of the equation $x' = -x^2 + 2t$ drawn with MATLAB.

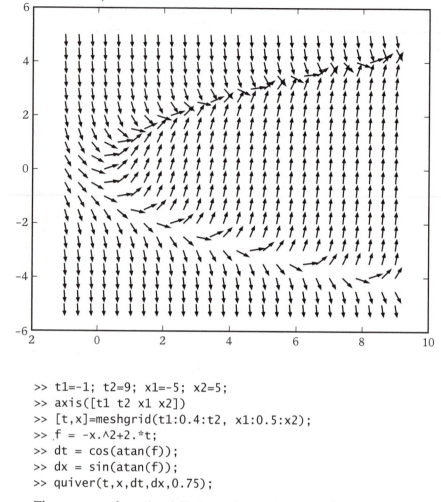

```
>> t1=-1; t2=9; x1=-5; x2=5;
>> axis([t1 t2 x1 x2])
>> [t,x]=meshgrid(t1:0.4:t2, x1:0.5:x2);
>> f = -x.^2+2.*t;
>> dt = cos(atan(f));
>> dx = sin(atan(f));
>> quiver(t,x,dt,dx,0.75);
```

The command meshgrid(t1:a:t2, x1:b:x2) defines a mesh of points (t, x) at distance a and b of each other relative to t and x, in the rectangle $[t1, t2] \times [x1, x2]$. The command quiver(t,x,dt,dx,s) draws vectors, of direction and magnitude determined by (dt, dx), at every point (t, x) defined by the mesh. The magnitude of the vectors can be scaled by the (optional) factor s. In our case, 0.75 makes the length of the vectors $\frac{3}{4}$ of that given by (dt, dx), which would be otherwise 1 due to the functions $\cos(\arctan(f))$ and $\sin(\arctan(f))$.

PROBLEMS

Use Maple, Mathematica, or MATLAB to find the general solution of the separable equations in Problems 1 through 8. If initial conditions are specified, solve the corresponding initial value problems also.

1. $x' = 3x^2 t^3$

2. $x' = x \cos t$

3. $x' = e^{x-t}$

4. $x' = \dfrac{x^2 - 9}{t}$

5. $x' = x \ln x, \ x(1) = e$

6. $x' = x \cot t, \ x\left(\dfrac{\pi}{2}\right) = 1$

7. $x' = 4x^2 - 9,$
$x(1) = 1$

8. $x' = (4x - 1)^2,$
$x(4) = 7$

Use Maple, Mathematica, or MATLAB to find the general solution of the linear equations in Problems 9 through 16. If initial conditions are specified, solve the corresponding initial value problems also.

9. $x' = 2x + t^2$

10. $x' = x + \sin t$

11. $x' = (\cot t)x + 4$

12. $x' = 2x - 3e^{27}$

13. $x' = (t + 1)x,$
$x(2) = 2$

14. $x' = tx + t^3,$
$x(1) = 1$

15. $x' = 2(\tan t)x + t,$
$x(0) = 1$

16. $x' = -x + t^5,$
$x(1) = 1$

Use Maple, Mathematica, or MATLAB to find the general solution of each of the equations in Problems 17 through 24 below. If you cannot obtain it directly, see whether you can combine analytical methods from previous sections with computer methods.

17. $x' = \dfrac{x}{t} + tx^{1/2}$

18. $x' = -\dfrac{x}{t} - \dfrac{x^3}{t^2}$

19. $x' = x + x^3$

20. $x' = (\cos t)x - x^2$

21. $x' = -x^2 + t^2 + 1$

22. $x' = x - \dfrac{x}{t} - \dfrac{1}{t}$

23. $x' = \dfrac{x^2 + t^2}{xt}$

24. $x' = \dfrac{t^2 x - x^3}{t^3 - tx^2}$

Use Maple, Mathematica, or MATLAB to find numerical solutions of Problems 25 through 32 by using Euler's method and the second-order Runge-Kutta method in the given interval and for the given step size.

25. $x' = tx + 1,$ $[1, 3],$
$h = 0.1,$
$x(1) = 1$

26. $x' = x - t + 6,$ $[0, 1],$
$h = 0.05,$
$x(0) = 2$

27. $x' = \sqrt{2x + t},$ $[1, 2],$
$h = 0.025,$
$x(1) = 0$

28. $x' = t \sin t,$ $[0, \pi/2],$
$h = 0.025,$
$x(0) = 1$

29. $x' = x + \ln t,$ $[1, 1.5],$
$h = 0.01,$
$x(1) = 3$

30. $x' = 2x + t^2,$ $[0, 4],$
$h = 0.1,$
$x(0) = 0$

31. $x' = x^4 + t,$ $[0, 1],$
$h = 0.05,$
$x(0) = 0$

32. $x' = x^6 - t,$ $[0, 1],$
$h = 0.025,$
$x(0) = 0$

Use Maple, Mathematica, or MATLAB to draw the slope field of the equations in Problems 33 through 40. In each case, choose several rectangles.

33. $x' = x \sin x$

34. $x' = x \cos x$

35. $x' = \dfrac{\ln x}{x - 1}$

36. $x' = t^2 \ln x$

37. $x' = t \sin x$

38. $x' = t \cos x$

39. $x' = x^2 + t^2$

40. $x' = x^2 - t^2$

41. Inside rectangles that contain the origin in the middle, draw the slope fields of the equations in Problems 34 through 39 in Section 2.2. What symmetry do you observe? How is this symmetry connected to the fact that the vector fields of homogeneous equations depend only on the ratio x/t?

2.8 Modeling Experiments

The problems posed in this section are very general. The goal of an experiment is unlike that of an end-of-the-section problem. The latter requires a precise result. The former is a research assignment that has a general goal but not a clear answer. Different students may obtain different outcomes and can draw different conclusions. You are supposed to use the knowledge acquired in this chapter, to think about the problem for several days, to ask new questions and to try to answer them, to stir your imagination, and to search for suitable references. In each case, discuss the model and see whether you can find an alternative or a better one. After ending your endeavors, outline your results and conclusions in an essay, whose content and originality are the prime aspects you must take into account.

Money Investments

Consider the money investments application in Section 2.2. Recall that if x is the amount of invested money and k the interest, then the equation describing the evolution in time of the invested amount with interest compounded continuously is given by the equation

$$x' = kx.$$

Experiment with several initial investments and different interest rates (small and large values) and determine the outcome. Then, find some reference (e.g., a textbook on the mathematics of finance) from which to learn how to compute simple interest and interest compounded annually, semiannually, monthly, weekly, and daily. Compute the outcome for each of these cases and then compare the results with the ones obtained for continuously compounded interest. Write the result first as a difference and then as a percentage of the compared amounts. If you wish, you can use one of the computer environments in Section 2.7 to save time with your computations. What conclusions can you draw from your analysis? Try to draw more than one conclusion. Can you state a general result (theorem) from any of your conclusions? Can you prove it? Are your conclusions consistent with existing investment strategies?

A Model for Memory

Like any mental process, memorization is difficult to describe by precise rules. Most laws formulated in psychology are therefore probabilistic. The following model is an approximate one, which you are asked to test through experiments and to improve if possible. We assume that the rate of memorization M' is proportional to the amount of information left to be memorized, i.e., M' changes like $k(1 - M)$, where 1 stands for the whole amount of information and k is a constant that depends on the individual.

To determine your constant k, consider the following two lists of 24 words each:

> tree, coffee, book, girl, lawn, picture, desk, leg, wall, bus, earth, computer, bed, coin, glass, carpet, dog, car, jacket, mug, chest, skirt, clock, wallet.

> idea, change, conclusion, love, interior, deception, strong, trauma, fantastic, law, boring, sweet, truth, happy, clue, reward, issue, free, complicated, chance, tender, state, pride, life.

Read the first list for one minute and try to memorize as many words as you can in the given order. Then ask a friend to test you and mark down the number of words correctly memorized. Then read the list for another minute and do another test. Continue this until you memorize the entire list. Use the results of the test to determine a value for k, say, k_1.

The next day, repeat the whole procedure with the second list and obtain a value for k, say k_2. Is there a big difference between the two values? This should not surprise you, because the first list represents concrete objects, while the second consists of abstract notions. Which number is larger, k_1 or k_2? Then take the average $k = (k_1 + k_2)/2$.

Use any method you like to find the solutions of the equation for the above value of k. What relationship do you see between these solutions and your experiments? What results did your fellow students obtain? Are they comparable? Is this model realistic? Can you improve it? If so, in what sense? Take the time into consideration? What other factors would matter? Do you think that a test with numbers should also be included? Can you draw any conclusions? Are your conclusions related with the ones drawn by psychologists?

The Landing of Apollo 11

Recall that in the model describing the landing of Apollo 11 in Section 2.4, the equation that gives the change of velocity v in time is

$$v' = g - \frac{k}{m}v^2,$$

where g is the gravitational acceleration, m is the mass of the module, and k is a constant of the order of 10^1, which depends on the size and shape of the module and on air resistance. Also recall that the terminal velocity $v = \sqrt{mg/k}$ is an equilibrium solution.

Use one of the computer programs in Section 2.7 for the second-order Runge-Kutta method to determine the approximate values of the velocity v during the fall. Try several values for the initial velocity (close to 11,000 m/s), k (between 7 and 25), and m (between 3000 and 6000 kg). When is v varying faster: at the beginning of the fall or toward the end? Graph the numerical results. What other conclusions can you draw from these experiments? Can you state some general theorem? Can you prove it? What do physics textbooks conclude about falling objects? Are your conclusions consistent with those results? How does the equation change if you include the braking effect of a parachute? What equations do the NASA specialists use to model this motion? How can you approach their equations? What is the difference between the previous conclusions and the ones you can draw now?

Population Dynamics

Consider a problem similar to Problem 33 in Section 2.4, which models the dynamics of a fish population. This time, however, assume that the rate of change also depends on a periodic function $A \cos Bt$, where A and B are positive constants. This function takes into account the way fishing varies every year. The equation is of the form

$$P' = k\left(\frac{P}{n_1} - 1\right)\left(1 - \frac{P}{n_2}\right) - A \cos Bt.$$

First, fix two large numbers n_1 and n_2 and use them throughout. Then, with the help of one of the computer programs in Section 2.7, find numerical solutions of the equation for various values of A and B and for several initial conditions. Also consider various lengths of the intervals of integration. Graph your solutions. Do certain values of A and B and certain initial conditions lead to solutions that appear to be periodic functions of t? What do most solutions look like? What conclusions can you draw about this model? Is it realistic? Can it give us some hints about how much to fish? What happens if A is large? Can you state some general theorem regarding this model? Can you prove it?

Second-Order Equations

The miracle of the appropriateness of the language of mathematics for the formulation of the laws of physics is a wonderful gift which we neither understand nor deserve.

EUGENE PAUL WIGNER
Nobel Prize for Physics, 1963

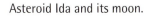

3.1 Homogeneous Equations

The goal of this chapter is to study linear second-order equations, which are important for their wide range of applications. In this section we will first present an existence and uniqueness theorem, then introduce and characterize the notion of linear independence, and finally describe the structure of the general solution for linear homogeneous second-order equations. This fundamental result will later help us understand several phenomena, among them the Tacoma Narrows Bridge collapse and the origin of the Kirkwood gaps in the asteroid belt of our solar system.

Asteroid Ida and its moon.

Second-Order Equations

Second-order differential equations involve an unknown function x and its derivatives x' and x''. Except for x'', each of the other functions may be absent from the equation. We will use the *normal form*, defined as follows:

DEFINITION 3.1.1

The *normal form* of a second-order differential equation is

$$x'' = f(t, x, x'), \qquad (1)$$

where f is a function of three variables. (What would be the *general form*?)

EXAMPLE 1   The second-order equation

$$x'' = 2tx'x - 5e^{-t} \qquad (2)$$

is in normal form because $f(t, x, x') = 2tx'x - 5e^{-t}$.

EXAMPLE 2   The second-order equation

$$x'' = -\frac{1}{3}x^2 + 3t^2x \qquad (3)$$

is also in normal form because $f(t, x) = -\frac{1}{3}x^2 + 3t^2x$. Notice that f does not depend on x' in this case.

A solution of a second-order equation is a function that satisfies the equation. We can formally write this as follows.

DEFINITION 3.1.2

A *solution* of equation (1) is a function φ defined on an open interval, at least twice differentiable, that satisfies the equation

$$\varphi''(t) = f(t, \varphi(t), \varphi'(t)).$$

EXAMPLE 3   The function $\varphi(t) = e^t$ is a solution of the equation

$$x'' = \frac{1}{2}(x' + x). \qquad (4)$$

Indeed, φ is infinitely many times differentiable and satisfies the identity

$$(e^t)'' = \frac{1}{2}[(e^t)' + e^t].$$

EXAMPLE 4   The function $\varphi(t) = t^3$ defined in $(0, \infty)$ is a solution of the equation

$$x'' = \frac{1}{9t^3}(x')^2 + \frac{5}{t^2}x. \qquad (5)$$

Indeed, φ is infinitely many times differentiable, and a direct computation shows that it also satisfies the identity

$$(t^3)'' = \frac{1}{9t^3}[(t^3)']^2 + \frac{5}{t^2}t^3.$$

In contrast to first-order equations, initial-value problems for second-order equations must satisfy two initial conditions: one for the function x and the other for the derivative x'.

DEFINITION 3.1.3

An *initial value problem* for equation (1) has the form

$$x'' = f(t, x, x'), \qquad x(t_0) = x_0, \ x'(t_0) = x_0',$$

where t_0, x_0, and x_0' are given numbers. A solution of the initial value problem is a solution φ of (1) that also satisfies the initial conditions, i.e., $\varphi(t_0) = x_0$ and $\varphi'(t_0) = x_0'$.

EXAMPLE 5 For the initial value problem

$$x'' = -5x' - 6x, \qquad x(0) = 2, \ x'(0) = 3, \tag{6}$$

the function $\varphi(t) = 9e^{-2t} - 7e^{-3t}$ is a solution. Indeed, φ is infinitely many times differentiable, and a straightforward computation shows that

$$\varphi''(t) = -5\varphi'(t) - 6\varphi(t), \qquad \varphi(0) = 2, \quad \varphi'(0) = 3.$$

Linear Equations Nonlinear second-order equations are in general complicated, and most existing results refer to the autonomous case. In Section 3.6 we will numerically study an initial value problem that leads to a complicated solution. In Chapter 5 we will understand that many equations like this are characterized by so-called chaotic behavior.

  Therefore we first restrict our study to that of linear second-order equations. Our immediate goal is to describe the structure of their general solutions. We now consider the homogeneous case; we will discuss the nonhomogeneous case in Section 3.3. Notice that equations (2), (3), and (5) are nonlinear, whereas (4) and (6) are linear. We formally define linear equations as follows.

DEFINITION 3.1.4

The *normal form* of a linear second-order equation is

$$x'' = p(t)x' + q(t)x + r(t), \tag{7}$$

where p, q, r are known functions. Equation (7) is called *nonhomogeneous* if r is not the zero function, and *homogeneous* if $r(t) = 0$ for all t.

  As with first-order equations, we can provide existence and uniqueness results. The following is the analogue of Cauchy's theorem. Its proof is beyond our scope.

Existence and Uniqueness Theorem

The initial value problem

$$x'' = p(t)x' + q(t)x + r(t), \qquad x(t_0) = x_0, \ x'(t_0) = x_0',$$

where p, q, r are continuous in some interval I containing t_0, has a unique solution φ, defined in the whole interval I.

Notice that unlike Cauchy's theorem, this result is global; i.e., the solution is defined not only in some interval around t_0 but in the whole domain I of the functions $p, q,$ and r. The reason for the global existence and uniqueness is the simple (linear) structure of the vector field with respect to x and x'.

Linear Independence An important concept in linear algebra is that of linear dependence, which in the case of two functions means that one equals the other times a nonzero constant. Their graphs can be similar or similar up to a symmetry, as for $f(t) = \sin t$ and $g(t) = \frac{1}{2}\sin t$ or $f(t) = t^3$ and $g(t) = -2t^3$ (see Figures 3.1.1(a) and (b)). The graphs of two linearly independent functions, however, can be quite different (see Figures 3.1.1(c) and (d)).

Figure 3.1.1. The graphs of two linearly dependent functions are similar, *(a)* $f(t) = \sin t$ and $g(t) = \frac{1}{2}\sin t$, or similar up to a symmetry, *(b)* $f(t) = t^3$ and $g(t) = -2t^3$. Two linearly independent functions can have quite different graphs: *(c)* $f(t) = t^3$ and $g(t) = t^2$, *(d)* $f(t) = \sin t$ and $g(t) = \frac{1}{3}t$.

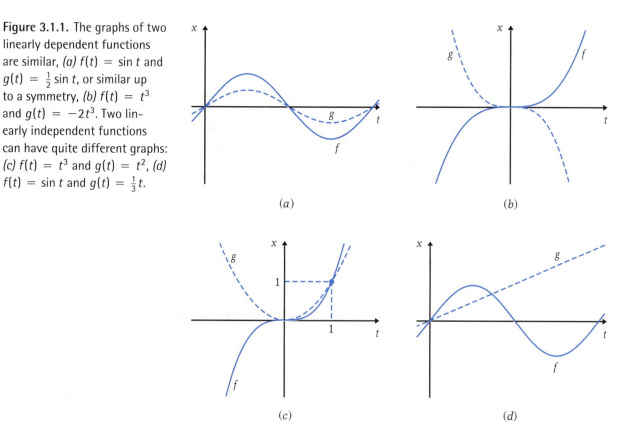

DEFINITION 3.1.5

The nonzero functions f and g are called *linearly dependent* on the interval I if there exist two constants c_1 and c_2, not both zero, such that

$$c_1 f(t) + c_2 g(t) = 0, \tag{8}$$

for all t in I. The functions f and g are called *linearly independent* if they are not linearly dependent, i.e., if relation (8) is satisfied only for $c_1 = c_2 = 0$.

EXAMPLE 6 The functions $f(t) = \sqrt{t}$ and $g(t) = 5\sqrt{t}$ are linearly dependent on $[0, \infty)$. Indeed, we can choose $c_1 = -5, c_2 = 1$. Is this choice unique?

EXAMPLE 7 The functions $f(t) = t^2$ and $g(t) = t^3$ are linearly independent on $(-\infty, \infty)$. Indeed, the equality $c_1 t^2 + c_2 t^3 = 0$ must be satisfied for all t, in particular for $t = t_1 \neq 0$ and for $t = t_2 \neq 0$, with $t_1 \neq t_2$. This leads to $t_1^2(c_1 + c_2 t_1) = 0$ and $t_2^2(c_1 + c_2 t_2) = 0$. If $c_2 \neq 0$, then $t_1 = -c_1/c_2$ and $t_2 = -c_1/c_2$, so $t_1 = t_2$, which contradicts the hypothesis. So $c_2 = 0$, and consequently $c_1 = 0$.

EXAMPLE 8 The functions

$$f(t) = \begin{cases} t^2, & \text{if } t < 0 \\ t^3, & \text{if } t \geq 0 \end{cases} \quad \text{and} \quad g(t) = \begin{cases} 2t^2, & \text{if } t < 0 \\ 4t^4, & \text{if } t \geq 0 \end{cases}$$

are linearly dependent on $(-\infty, 0)$ but linearly independent on $(-\infty, \infty)$. Indeed, linear dependence on $(-\infty, 0)$ can be proved as in Example 6 and linear independence on $(0, \infty)$ as in Example 7. But since only $c_1 = c_2 = 0$ satisfy relation (8) on $(0, \infty)$, no other constants can satisfy (8) on an interval containing $(0, \infty)$. Therefore f and g are linearly independent on $(-\infty, \infty)$.

As we can see from the above examples, linear independence is harder to prove than linear dependence. A more convenient way of checking linear independence is given by the following result, valid only in the class of differentiable functions.

Theorem 3.1.1

If f, g are differentiable functions on an open interval I and if there is a t_0 in I such that $f(t_0)g'(t_0) \neq f'(t_0)g(t_0)$, then f and g are linearly independent on I.

PROOF

Assume that f and g are linearly dependent on I. Then there exist constants c_1 and c_2, at least one nonzero, such that $c_1 f(t) + c_2 g(t) = 0$ for all

t in I. Therefore $c_1 f'(t) + c_2 g'(t) = 0$ for all t in I. These two equations form the algebraic system in unknowns c_1, c_2

$$\begin{cases} c_1 f(t_0) + c_2 g(t_0) = 0 \\ c_1 f'(t_0) + c_2 g'(t_0) = 0. \end{cases}$$

Multiplying the first equation by $f'(t_0)$ and the second by $-f(t_0)$ and adding the two new equations, obtain

$$c_2[f'(t_0)g(t_0) - f(t_0)g'(t_0)] = 0.$$

Since $f(t_0)g'(t_0) \neq f'(t_0)g(t_0)$, it follows that $c_2 = 0$. A similar trick (can you do it?) leads to the conclusion that $c_1 = 0$. This proves the linear independence of f and g.•

Does Theorem 3.1.1 contradict the conclusion in Example 8?

EXAMPLE 9   The linear independence of the functions $f(t) = t^2$ and $g(t) = t^3$, proved in Example 7, follows more easily with the above theorem: $f'(t) = 2t$, $g'(t) = 3t^2$, so $f(t)g'(t) - f'(t)g(t) = t^4$, which is nonzero for every $t \neq 0$.

EXAMPLE 10   The linear independence of $f(t) = \cos t$ and $g(t) = \sin t$ on $(-\infty, \infty)$ also follows without difficulty since $f(t)g'(t) - f'(t)g(t) = \cos^2 t + \sin^2 t = 1$. Can you prove this directly from the definition?

Remark. The converse of Theorem 3.1.1 (which is equivalent to saying that if $f(t)g'(t) = f'(t)g(t)$ for all t, then f and g are linearly dependent) is false. The next two examples show that f and g are linearly dependent in some cases and linearly independent in others.

EXAMPLE 11   The functions $f(t) = t$ and $g(t) = 2t$ are differentiable, satisfy the condition $f(t)g'(t) = f'(t)g(t)$ for all t, and are linearly dependent on $(-\infty, \infty)$.

EXAMPLE 12   The functions $f(t) = t^3$ and $g(t) = |t^3|$ are differentiable and satisfy the condition $f(t)g'(t) = f'(t)g(t)$ for all t, but are linearly independent on $(-\infty, \infty)$. Indeed, g is differentiable everywhere (including 0, since $\lim_{t \to 0, t<0} g'(t) = \lim_{t \to 0, t>0} g'(t) = 0$), and a direct computation shows that the above relation is satisfied for all t. To prove linear independence on $(-\infty, \infty)$, we must show that the equations

$$c_1 t^3 + c_2(-t^3) = 0, \quad \text{for } t < 0,$$

$$c_1 t^3 + c_2 t^3 = 0, \quad \text{for } t \geq 0$$

are simultaneously satisfied for all t only if $c_1 = c_2 = 0$. But this is obviously true, because the first relation implies $c_1 = c_2$ and the second yields $c_1 = -c_2$, so $c_1 = c_2 = 0$.

The General Solution for Homogeneous Equations

The connection between two linearly independent solutions and the condition in Theorem 3.1.1 is given by the following result, proved in 1827 by the Norwegian mathematician Niels Henrik Abel (1802–1829).

Theorem 3.1.2

If the functions x_1 and x_2, defined on the open interval I, are solutions of the equation

$$x'' = p(t)x' + q(t)x, \tag{9}$$

where p and q are continuous on I, then x_1 and x_2 are linearly independent on I if and only if $x_1(t)x_2'(t) \neq x_1'(t)x_2(t)$ for all t in I.

Theorems 3.1.1 and 3.1.2 show that if x_1 and x_2 are solutions of equation (9), then the following statements are equivalent:

1. x_1 and x_2 are linearly independent on I.
2. There is a t_0 in I such that $x_1(t_0)x_2'(t_0) \neq x_1'(t_0)x_2(t_0)$.
3. $x_1(t)x_2'(t) \neq x_1'(t)x_2(t)$ for all t in I.

The structure of the general solution for a linear second-order homogeneous equation follows now from the following result.

Fundamental Theorem for Homogeneous Equations

Any solution φ of the linear homogeneous second-order equation (9) is of the form

$$\varphi(t) = c_1 x_1(t) + c_2 x_2(t),$$

where c_1, c_2 are constants and x_1, x_2 are linearly independent solutions of equation (9).

PROOF

Let x_1 and x_2 be two solutions of equation (9), linearly independent on some interval I. Solutions of this kind always exist. Indeed, according to the existence theorem it is enough to assume that x_1 is a solution of the initial value problem $x'' = p(t)x' + q(t)x$, $x(t_0) = 0$, $x'(t_0) = 1$, where t_0 belongs to I, and x_2 is a solution of the initial value problem $x'' = p(t)x' + q(t)x$, $x(t_0) = 1$, $x'(t_0) = 0$. To prove their linear independence, notice that $0 = x_1(t_0)x_2'(t_0) \neq x_1'(t_0)x_2(t_0) = 1$.

Let φ be another solution of equation (9), different from x_1 and x_2, also defined in I. Denote $x_0 = \varphi(t_0)$ and $x_0' = \varphi'(t_0)$. Consider the initial value problem

$$x'' = p(t)x' + q(t)x, \qquad x(t_0) = x_0, \ x'(t_0) = x_0'. \tag{10}$$

Obviously φ is a solution of (10). Let us now show that there exist constants c_1, c_2, at least one nonzero, such that the function $\theta(t) = c_1 x_1(t) + c_2 x_2(t)$ is also a solution of (10). Indeed, appropriate constants c_1 and c_2

can be found by solving the linear nonhomogeneous algebraic system in unknowns c_1, c_2

$$\begin{cases} c_1 x_1(t_0) + c_2 x_2(t_0) = x_0 \\ c_1 x_1'(t_0) + c_2 x_2'(t_0) = x_0'. \end{cases}$$

This system has a nonzero solution because x_1, x_2 are linearly independent, so $x_1(t)x_2'(t) - x_1'(t)x_2(t) \neq 0$ for all t in I, in particular for t_0. From the uniqueness of the solution of (10), $\varphi = \theta$, so $\varphi(t) = c_1 x_1(t) + c_2 x_2(t)$ for all t in I. This completes the proof. ∎

Finding the general solution of a linear homogeneous second-order equation is thus equivalent to obtaining two linearly independent solutions. This is a strong property. All solutions (infinitely many) can be described in terms of only two solutions. Therefore, any set of two linearly independent solutions is called a *fundamental set of solutions*.

EXAMPLE 13   The functions $x_1(t) = e^t$ and $x_2(t) = te^t$ form a fundamental set of solutions for the equation

$$x'' = 2x' - x.$$

Indeed, a direct computation shows that x_1 and x_2 are solutions, and since $x_1(t)x_2'(t) - x_1'(t)x_2(t) = e^{2t}$, they are also linearly independent. (Can you find another fundamental set of solutions?)

EXAMPLE 14   The functions $x_1(t) = t$ and $x_2(t) = t \ln t$ defined on $(0, \infty)$ form a fundamental set of solutions for the equation

$$x'' = \frac{1}{t} x' - \frac{1}{t^2} x.$$

Indeed, a direct computation shows that x_1 and x_2 are solutions. Also $x_1(t)x_2'(t) - x_1'(t)x_2(t) = t$, which is nonzero for all t except $t = 0$, so x_1 and x_2 are linearly independent. (Can you find another fundamental set of solutions?)

PROBLEMS

In Problems 1 to 6, write the given second-order equations in normal form.

1. $5(t + 1)x'' + 2x' - x^2 + \frac{2}{5}x = 4t$

2. $\frac{2}{t}u'' + 2u'' - 4(u')^2 - 1 = 0$

3. $v'' + 3x^2v'' - 2v' + e^{-x}v = 0$

4. $\dfrac{w'' + w'}{w'' - w} = 2$

5. $(7ty - 1)y'' - 3y' + 5y - 4t^5 = -6$

6. $5(24x'' + 2x' - x^2) = 2\pi t$

In Problems 7 to 12, check whether the given functions are solutions of the corresponding second-order equations.

7. $\varphi(t) = 2t^2 \ln t$ for $x'' = \frac{3}{t}x' - \frac{4}{t^2}x$

8. $\psi(x) = \frac{1}{2}x^{-1/2} \sin x$ for

$$x^2 w'' = -xw' + \left(\frac{1}{4} - x^2\right)w$$

9. $r(z) = 4z \sin(\ln z)$ for $u'' = \frac{1}{z}u' - \frac{2}{z^2}u$

10. $q(\theta) = -2(\theta^2 + \theta^3)$ for $\theta^2 r'' = 4\theta r' - 6r$

11. $s(p) = 2\pi p^{10}$ for $p^2 q'' = 7pq' + 20q$

12. $u(v) = -e^v$ for $u'' = \left(1 + \dfrac{1}{v}\right)u' - \dfrac{u}{v}$

In Problems 13 to 18, check whether the given functions are solutions of the corresponding initial value problems.

13. $\varphi(t) = t^4$ for $x'' = \dfrac{20}{t^2}x,$

$x(0) = 0, \quad x'(0) = 0$

14. $u(x) = 1 + x$ for $(1 - 2x - x^2)u'' = -2(1 + x)u' + 2x, \quad u(0) = 1, \quad u'(0) = 1$

15. $r(\theta) = e^{3\theta}$ for $r'' = \dfrac{9\theta + 6}{3\theta + 1}r' -$

$\dfrac{9}{3\theta + 1}r, \quad r(0) = 1, \quad r'(0) = 3$

16. $v(s) = 2\ln s$ for $\phi'' = -\dfrac{1}{s}\phi',$

$\phi(e) = 1, \quad \phi'(e) = \dfrac{2}{e}$

17. $v(t) = 2\sqrt{t}\ln t$ for $w'' = -\dfrac{1}{4t^2}w,$

$w(e) = 2\sqrt{e}, \quad w'(e) = \dfrac{3}{5e}$

18. $y(x) = \cos x$ for $(1 + x)y'' = (5 - x)y' + x^2 y, \quad y(1) = 0, \quad y'(1) = -\sin 1$

For the linear second-order equations in Problems 19 to 24, determine the domain of the coefficient functions $p, q,$ and r (see Definition 3.1.4) and find the largest intervals I in which the solution of each initial value problem exists and is unique.

19. $x'' = 7tx' - t^2 x, \quad x(0) = 0, \quad x'(0) = 2$

20. $y'' - 3(\sin x)y' = 3y + 1, \quad y(1) = \dfrac{\pi}{3},$

$y'(1) = 0, \quad y'(1) = 0$

21. $z'' = \dfrac{1}{t}z' + \dfrac{3}{t^2}z, \quad z(2) = \dfrac{2}{3}, \quad z'(2) = 5$

22. $tp'' - 2p' - 5p = 0, \quad p(1) = 2e, \quad p'(1) = 0$

23. $(1 - \theta)r'' - \theta r' = 6r, \quad r(-2) = 1, \quad r'(-2) = 2$

24. $(1 - x^2)w'' = 4w, \quad w(0) = 0, \quad w'(0) = 0$

Using the definition of linear dependence/independence, determine whether the functions in Problems 25 through 32 are linearly dependent or independent on the given intervals.

25. $f(t) = t, \quad g(t) = 3t$ on $(-\infty, \infty);$ on $(-\infty, 0);$ on $(0, \infty).$

26. $f(t) = t^2, \quad g(t) = -t^2$ on $(-\infty, \infty);$ on $(-1, 1);$ on $[-1, 1].$

27. $\tan t, \quad \cot t$ on $(-\pi/2, 0);$ on $(0, \pi/4);$ on $(0, \pi/2).$

28. $\sec t, \quad \csc t$ on $(-\pi/2, 0);$ on $(0, \pi/4);$ on $(0, \pi/2).$

29. $f(t) = \begin{cases} t, & \text{if } t < 0 \\ 0, & \text{if } t \geq 0, \end{cases}$

$g(t) = \begin{cases} 0, & \text{if } t < 0 \\ -t, & \text{if } t \geq 0, \end{cases}$

on $(-\infty, \infty);$ on $(-\infty, 0);$ on $(0, \infty).$

30. $f(t) = \begin{cases} t^2, & \text{if } t < 0 \\ t^3, & \text{if } t \geq 0, \end{cases}$

$g(t) = \begin{cases} 2t^3, & \text{if } t < 0 \\ 4t^2, & \text{if } t \geq 0, \end{cases}$

on $(-\infty, \infty);$ on $(-\infty, 0);$ on $(0, \infty).$

31. $f(t) = t^4|t|, \quad g(t) = t^5$ on $(-\infty, \infty);$ on $(-\infty, 0);$ on $(0, \infty).$

32. $f(t) = |t|, \quad g(t) = t$ on $(-\infty, \infty);$ on $(-\infty, 0);$ on $(0, \infty).$

In Problems 33 through 36 we give the form of the function $W(t) = f(t)g'(t) - f'(t)g(t)$ for some functions f and g. Are f and g linearly dependent or independent in each of the given intervals?

33. $W(t) = t\sin t$ on $(-\infty, \infty);$ on $(0, \infty).$

34. $W(t) = t\cos t$ on $(-\infty, \infty);$ on $(0, \pi).$

35. $W(t) = t^2|t|$ on $(-\infty, 0);$ on $(0, \infty).$

36. $W(t) = \begin{cases} 0, & \text{on } (-\infty, 0] \\ t^2, & \text{on } (0, \infty), \end{cases}$

on $(-\infty, \infty);$ on $(-\infty, 0);$ on $(0, \infty).$

In Problems 37 through 40 compute the function $fg' - f'g$. Show then that f and g are linearly independent on $(-\infty, \infty)$. Do these results contradict Theorem 3.1.1? Why? In each case, can these functions be solutions of some equation of the form (9) with p and q continuous? Why?

37. $f(t) = |t|t^8, \quad g(t) = 2t^9$

38. $f(t) = \frac{1}{5}|t^7|, \quad g(t) = -2t^7$

39. $f(t) = \sin(t^2), \quad g(t) = \cos(t^2)$

40. $f(t) = t^2\sin(t^2), \quad g(t) = t^2\cos(t^2)$

41. Abel's formula states that the function $W(t) = x_1(t)x_2'(t) - x_1'(t)x_2(t)$ of equation (9) is given by

$$W(t) = c \cdot e^{\int p(t)\, dt},$$

where c is a constant that depends on x_1 and x_2 ($c = 0$ if x_1 and x_2 are linearly dependent and $c \neq 0$ if they are linearly independent). To prove this formula, show that if x_1 and x_2 are two distinct solutions of (9), then $x_1 x_2'' - x_2 x_1'' = p(t)(x_1 x_2' - x_2 x_1')$. Further notice that $W' = x_1 x_2'' - x_2 x_1''$, so W satisfies the differential equation $W' = p(t)W$. Then solve this equation.

In Problems 42 to 45 determine whether the given functions form a fundamental set of solutions for the corresponding equations.

42. $x_1(t) = 2e^{2t}$, $x_2(t) = 3te^{2t}$
for $x'' - 4x' + 4x = 0$

43. $y_1(t) = -e^{2t}$, $y_2(t) = -\frac{1}{2}e^{3t}$
for $y'' - 5y' + 6y = 0$

44. $u_1(t) = 3e^{-t/2}\cos\left(\frac{\sqrt{3}}{2}t\right)$,

$u_2(t) = 5e^{-t/2}\sin\left(\frac{\sqrt{3}}{2}t\right)$ for $u'' + u' + u = 0$

45. $z_1(t) = e^{-t/2}\cos\left(\frac{\sqrt{3}}{2}t\right) + e^{-t/2}\sin\left(\frac{\sqrt{3}}{2}t\right)$,

$z_2(t) = e^{-t/2}\cos\left(\frac{\sqrt{3}}{2}t\right) - e^{-t/2}\sin\left(\frac{\sqrt{3}}{2}t\right)$

for $z'' + z' + z = 0$

3.2 Integrable Cases

Equations that can be completely solved are called *integrable*. As we learned in Section 3.1, solving (or *integrating*) a second-order linear homogeneous equation means finding two linearly independent solutions. We lack methods that apply to the general case, but we can always deal with certain classes of equations. In this section we will introduce the *characteristic-equation method*, which provides the general solution of any homogeneous equation with constant coefficients, and the *reduction-of-order method*, suitable for some homogeneous equations with variable coefficients. In the end we will apply these methods to understand the motion of Galileo's pendulum for small amplitudes and to study the oscillations of a water segment in a pipe.

Method for Homogeneous Equations with Constant Coefficients

Linear homogeneous second-order equations with constant coefficients b and c,

$$x'' + bx' + cx = 0, \tag{1}$$

can be solved using the *characteristic-equation method*, which has its roots in a paper published by Leonhard Euler in 1739. The idea is to seek solutions of the form $x(t) = e^{rt}$, where r is a constant to be determined. The substitution of $x'(t) = re^{rt}$ and $x''(t) = r^2 e^{rt}$ into equation (1) reduces the problem to finding r from the equation $e^{rt}(r^2 + br + c) = 0$, which is equivalent to

$$r^2 + br + c = 0, \tag{2}$$

called the *characteristic equation* of (1). This is remarkable: If the solution is of exponential type, then the differential equation (1) reduces to the quadratic equation (2), which we can solve. The method works as follows.

Step 1. Attach to equation (1) the characteristic equation (2) and find its roots r_1 and r_2,

$$r_1 = \frac{-b - \sqrt{\Delta}}{2}, \qquad r_2 = \frac{-b + \sqrt{\Delta}}{2}, \qquad \text{where } \Delta = b^2 - 4c.$$

If $\Delta > 0$, $\Delta = 0$, or $\Delta < 0$, go to Step 2, 3, or 4, respectively.

Step 2. If $\Delta > 0$, then r_1 and r_2 are real, $r_1 \neq r_2$, and two linearly independent solutions of (1) are

$$x_1(t) = e^{r_1 t}, \qquad x_2(t) = e^{r_2 t}. \tag{3}$$

Step 3. If $\Delta = 0$, then $r_1 = r_2 = b/2$, and two linearly independent solutions of (1) are

$$x_1(t) = e^{r_1 t}, \qquad x_2(t) = te^{r_1 t}. \tag{4}$$

Step 4. If $\Delta < 0$, then $r_1 = \alpha + \beta i$, $r_2 = \alpha - \beta i$ (i.e., complex conjugates), where α and β are constants that depend on b and c, and two linearly independent solutions of (1) are

$$x_1(t) = e^{\alpha t} \cos(\beta t), \qquad x_2(t) = e^{\alpha t} \sin(\beta t). \tag{5}$$

The method can be justified by showing that in each case discussed above, the functions x_1 and x_2 are linearly independent solutions of equation (1). (Can you check this?)

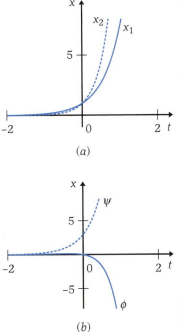

(a)

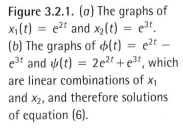

(b)

Figure 3.2.1. (a) The graphs of $x_1(t) = e^{2t}$ and $x_2(t) = e^{3t}$. (b) The graphs of $\phi(t) = e^{2t} - e^{3t}$ and $\psi(t) = 2e^{2t} + e^{3t}$, which are linear combinations of x_1 and x_2, and therefore solutions of equation (6).

EXAMPLE 1 To solve the equation

$$x'' - 5x' + 6x = 0, \tag{6}$$

notice that for $r^2 - 5r + 6 = 0$ we have $\Delta > 0$, $r_1 = 2$ and $r_2 = 3$, so according to Step 2 two linearly independent solutions of (6) are $x_1(t) = e^{2t}$ and $x_2(t) = e^{3t}$ (see Figure 3.2.1).

EXAMPLE 2 To solve the equation

$$x'' - 4x' + 4x = 0, \tag{7}$$

notice that for $r^2 - 4r + 4 = 0$ we have $\Delta = 0$ and $r_1 = r_2 = 2$, so according to Step 3 two linearly independent solutions of (7) are $x_1(t) = e^{2t}$ and $x_2(t) = te^{2t}$ (see Figure 3.2.2).

EXAMPLE 3 To solve the equation

$$x'' + x' + x = 0, \tag{8}$$

notice that for $r^2 + r + 1 = 0$ we have $\Delta < 0$ and $r_{1,2} = -\frac{1}{2}(1 \pm i\sqrt{3})$, so according to Step 4 two linearly independent solutions of (8) are $x_1(t) = e^{-t/2} \cos(\sqrt{3}/2)t$ and $x_2(t) = e^{-t/2} \sin(\sqrt{3}/2)t$ (see Figure 3.2.3).

Reduction of Order   The reduction of order applies to two different classes of homogeneous equations with variable coefficients: those in which the coefficient of x is 0 and those in which one solution is known. Let us study each of them.

Equations without an x-term

Consider equations of the form

$$x'' = p(t)x', \qquad (9)$$

where p is continuous. They can be solved in two steps.

Step 1. Make the change of variable $y = x'$, which reduces (9) to the first-order separable equation

$$y' = p(t)y, \qquad (10)$$

whose general solution y depends on a constant.

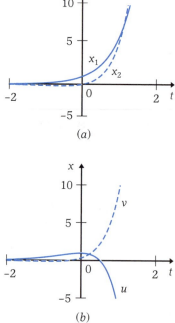

(a)

Step 2. Integrate the equation $x' = y$ with respect to t and obtain the general solution, which depends on two constants.

EXAMPLE 4   To solve the equation

$$x'' = \frac{2}{t}x',$$

proceed as follows.

Step 1. Make the change of variable $y = x'$, which leads to the first-order separable equation

$$y' = \frac{2}{t}y,$$

already solved in Example 5 of Section 2.2. Its solution is $y(t) = kt^2$, where k is a constant.

Step 2. Integrating the equation $x'(t) = kt^2$, obtain the general solution

$$x(t) = c_1 t^3 + c_2,$$

where c_1 and c_2 are real constants ($c_1 = k/3$).

Notice that since $x_1(t) = t^3$ and $x_2(t) = 1$ are two linearly independent solutions of the equation $x'' = (2/t)x'$, the above formula indeed gives the general solution.

Figure 3.2.2. (a) The graphs of $x_1(t) = e^{2t}$ and $x_2(t) = te^{2t}$. (b) The graphs of $u(t) = e^{2t} - 2te^{2t}$ and $v(t) = \frac{1}{3}e^{2t} + te^{2t}$, which are linear combinations of x_1 and x_2, and therefore solutions of equation (7).

EXAMPLE 5   To solve the equation

$$x'' = \frac{2t}{1 + t^2}x',$$

make the change of variable $y = x'$, which leads to the first-order separable equation

$$y' = \frac{2t}{1 + t^2}y.$$

Its general solution is $y(t) = k(1 + t^2)$, where k is a constant. By integrating the equation $x'(t) = k(1 + t^2)$, obtain the general solution

$$x(t) = c_1(t^3 + 3t) + c_2,$$

where c_1 and c_2 are constants.

Figure 3.2.3. (*a*) The graphs of $x_1(t) = e^{-t/2}\cos(\sqrt{3}/2)t$ and $x_2(t) = e^{-t/2}\sin(\sqrt{3}/2)t$.
(*b*) The graphs of $y(t) = \frac{2}{3}e^{-t/2}\cos(\sqrt{3}/2)t - e^{-t/2}\sin(\sqrt{3}/2)t$ and $z(t) = -e^{-t/2}\cos(\sqrt{3}/2)t + 2e^{-t/2}\sin(\sqrt{3}/2)t$, which are linear combinations of x_1 and x_2, and therefore solutions of equation (8).

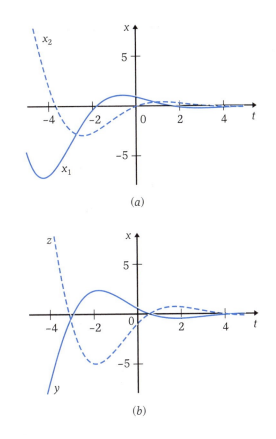

(*a*)

(*b*)

Equations for which a solution is known

Now consider the general case, i.e., equations of the form

$$x'' = p(t)x' + q(t)x, \tag{11}$$

with p, q continuous, but assume that x_1 is a known (not identically zero) solution. Further restrict this solution to an interval I in which it never takes the value 0. The method of finding a second linearly independent solution in I works as follows.

Step 1. Seek a second solution of the form

$$x_2(t) = u(t)x_1(t), \tag{12}$$

where u is a function to be determined.

Step 2. Substitute x_2, $x_2' = u'x_1 + ux_1'$, and $x_2'' = u''x_1 + 2u'x_1' + ux_1''$ into equation (11) and, after performing the computations, obtain the equation

$$u'' = \frac{px_1 - 2x_1'}{x_1}u', \tag{13}$$

which is of the same type as (9).

Step 3. Solve equation (13) using the previous method. Choose a nonconstant solution u and obtain a solution of the form (12) for the equation (11).

Remark. Since u is not a constant, $x_1 x_2' - x_1' x_2 = u' x_1^2$ is nonzero for at least one value of t. Therefore the solutions x_1 and x_2 are linearly independent on their common interval of existence.

EXAMPLE 6   Knowing that $x_1(t) = 1/t$, defined on $(0, \infty)$, is a solution of the equation

$$x'' = -\frac{3}{2t} x' + \frac{1}{2t^2} x,$$

let us find a second solution x_2, linearly independent with x_1.

Step 1. Seek a solution of the form

$$x_2(t) = \frac{u(t)}{t},$$

where u is to be determined.

Step 2. Substitute

$$x_2 = \frac{u(t)}{t}, \qquad x_2' = \frac{u'}{t} - \frac{u}{t^2}$$

and

$$x_2'' = \frac{u''}{t} - \frac{2u'}{t^2} + \frac{2u}{t^3}$$

into the given second-order equation. This leads to the equation

$$u'' = \frac{1}{2t} u'.$$

Step 3. Using the substitution $u' = y$, obtain the equation

$$y' = \frac{1}{2t} y,$$

which has the general solution $y(t) = kt^{1/2}$, where k is a constant. Integration of the equation $u' = kt^{1/2}$ yields $u(t) = ct^{3/2}$, where c is a constant. Taking, say, $c = 1$, u becomes $t^{3/2}$. Thus a second solution, linearly independent with x_1 on $(0, \infty)$, is $x_2(t) = t^{3/2}/t = t^{1/2}$.

EXAMPLE 7   Knowing that $x_1(t) = t$, defined on $(0, \infty)$, is a solution of the equation

$$x'' = -\frac{2}{t} x' + \frac{2}{t^2} x,$$

let us find another solution, x_2, such that x_1 and x_2 are linearly independent on $(0, \infty)$. We seek a solution of the form $x_2(t) = tu(t)$, which means that $x_2'(t) = u(t) + tu'(t)$ and $x_2''(t) = 2u'(t) + tu''(t)$. Substituting x_2, x_2', and x_2'' into the given equation, we obtain

$$u'' = -\frac{4}{t} u',$$

which with the substitution $v = u'$ reduces to the separable equation

$$v' = -\frac{4}{t}v.$$

After solving this equation, choose a solution $v(t) = t^{-4}$, so $u'(t) = t^{-4}$, which if integrated yields $u(t) = ct^{-3}$. The choice $c = 1$ leads to $u(t) = t^{-3}$, so $x(t) = t^{-3}t = t^{-2}$. Therefore a second solution, linearly independent with x_1 on $(0, \infty)$, is $x_2(t) = t^{-2}$.

Applications

Galileo's pendulum

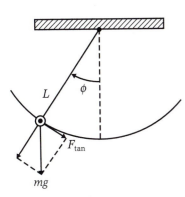

Figure 3.2.4. The forces acting during the motion of a simple pendulum.

Galileo contemplating the oscillations.

Legend has it that in 1583, in his first year of study at the University of Pisa, Galileo Galilei (1564–1642) discovered the law of the pendulum by comparing his pulse rate with the swings of a lamp in the city's cathedral. His later experiments showed that, for small oscillations, the period of a simple pendulum is proportional to the square root of its length but is independent of mass. Let us determine and solve the equation of a pendulum of mass m and length L near the downward equilibrium position for small swings and negligible frictions, as in Galileo's case.

Solution. Let m be the mass of the bob, and assume that the rod is rigid and weightless. The gravitational force acting downward on the bob can be decomposed into a radial component and a tangential component (see Figure 3.2.4). Since the rod is rigid, the radial component does not influence the motion. If ϕ denotes the angular position of the upward-moving pendulum ($\phi > 0$), with $\phi = 0$ at the downward equilibrium, then the tangential component of the gravitational force is $F_{tan} = -mg \sin \phi$, where g is the gravitational acceleration (the minus sign appears because F_{tan} always has the sign of $-\phi$). By Newton's second law,

Figure 3.2.5. The graphs of the functions t and $\sin t$ (a) in the interval $[-5, 5]$ and (b) in the interval $[0, 0.6]$. Notice that for $t > 0$ small, $\sin t$ and t are almost identical.

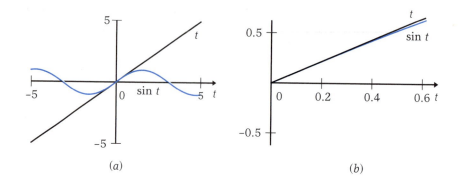

(a)   (b)

the force acting on the bob is $F = ma$. The acceleration of the bob is defined as the product between the length of the rod and the angular acceleration, i.e., $a = L\phi''$. Since we ignore frictions, $F = F_{\text{tan}}$. Substituting all these into $F = ma$, we obtain the equation of motion

$$\phi'' = -\frac{g}{L} \sin \phi.$$

For small oscillations, however, we can approximate $\sin \phi$ with ϕ (see Figure 3.2.5), so the above equation becomes

$$\phi'' = -\frac{g}{L} \phi, \tag{14}$$

which is a model for the so-called linear pendulum. The characteristic equation,

$$r^2 + \frac{g}{L} = 0,$$

has the solutions $r_{1,2} = \pm i \sqrt{g/L}$. The general solution,

$$\phi(t) = c_1 \cos \omega t + c_2 \sin \omega t,$$

is periodic in t, where $\omega = \sqrt{g/L}$, called *angular frequency*, shows how fast the pendulum swings (see Figure 3.2.6). Since the *period* of $\sin \omega t$ and $\cos \omega t$ is $T = 2\pi/\omega$, it means that the period of the pendulum, $T = 2\pi \sqrt{L/g}$, is independent of m and proportional to $\sqrt{L}$. (If, for example, $L = 1$ m and $g = 9.8$ m/s^2, then the period of the pendulum is $T \approx 2$ seconds.) This proves the correctness of Galileo's experimental conclusions.

Figure 3.2.6. The graph of the function $\sin 2\pi t + \cos 2\pi t$, which is a particular solution of equation (14) for certain values of the constants involved. This function is typical among those describing the motion of the pendulum for small oscillations, in the hypothesis that no friction forces are present.

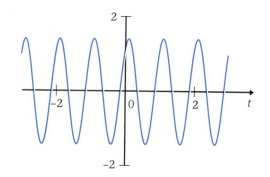

Water oscillating in a pipe

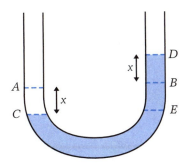

Figure 3.2.7. A water segment oscillating in a pipe.

A pipe of cross-sectional area S (see Figure 3.2.7) contains water. The length of the water segment is L. Assuming that in spite of frictions the water segment moves as a whole and that the force of friction is proportional to the velocity, describe the oscillatory motion if the segment is initially out of equilibrium.

Solution. Take the water segment out of the AB equilibrium to the CD position and denote $x = BD$. If we remove the portion $ED = 2x$, the remaining portion CE stays in equilibrium. Therefore the water segment CD is pushed down by the weight of the ED segment. By Newton's law,

$$mx'' = F_{\text{weight}} + F_{\text{friction}},$$

where m is the mass of the water in CD; x'' is the acceleration; $F_{\text{weight}} = -\rho Vg$, where ρ is the density of the liquid ($\rho_{\text{water}} = 1$); V is the volume of the water segment ($V = 2xS$ in our case); and g is the gravitational acceleration. Since the friction is proportional to the velocity of motion, $F_{\text{friction}} = kx'$, where $k > 0$ is a small constant that depends on the pipe material. Substituting $m = \rho LS$ and $F_{\text{weight}} = -\rho 2xSg$ into Newton's law, we obtain the equation of motion

$$x'' = -\frac{k}{m}x' - \frac{2g}{L}x. \qquad (15)$$

To solve it, consider the characteristic equation

$$r^2 + \frac{k}{m}r + \frac{2g}{L} = 0,$$

which has the solutions

$$r_{1,2} = -\frac{k}{2m} \pm \sqrt{\frac{k^2}{4m^2} - \frac{2g}{L}}.$$

Since k is small,

$$\frac{k^2}{4m^2} - \frac{2g}{L} < 0,$$

so the solutions are complex conjugates:

$$r_{1,2} = -\frac{k}{2m} \pm i\sqrt{\frac{2g}{L} - \frac{k^2}{4m^2}}.$$

This corresponds to Step 4 of the method, so taking $\alpha = -(k/2m)$ and $\beta = \sqrt{2g/L - (k^2/4m^2)}$, we obtain the general solution

$$x(t) = e^{-kt/2m}\left[c_1 \cos\left(t\sqrt{\frac{2g}{L} - \frac{k^2}{4m^2}}\right) + c_2 \sin\left(t\sqrt{\frac{2g}{L} - \frac{k^2}{4m^2}}\right)\right]$$

(see Figure 3.2.8). The factor $e^{-(kt/2m)}$ in front of the periodic expression in the brackets makes the oscillations decrease in amplitude. The oscillations tend to 0 when $t \to \infty$.

Figure 3.2.8. The graph of the function $e^{-t}(\cos 3t + \sin 3t)$, which is a particular solution of equation (15) for certain values of the constants involved. This function is typical among those describing the oscillations of a water segment in a pipe, assuming that frictions are taken into account.

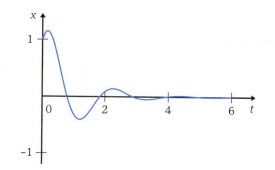

PROBLEMS

Solve the linear homogeneous equations and initial value problems in Problems 1 through 12.

1. $x'' = 7x' - 12x$

2. $x'' = 3x' - 2x$

3. $x'' = 6x' - 9x$

4. $x'' = x' - \frac{1}{4}x$

5. $x'' = -3x' - 5x$

6. $x'' = -\frac{1}{4}x' - 8x$

7. $x'' = 7x' - 10x,\ x(0) = 1,\ x'(0) = 2$

8. $x'' = 11x' - 24x,\ x(0) = -1,\ x'(0) = 1$

9. $x'' = 10x' - 25x,\ x(1) = x'(1) = e$

10. $x'' = 16x' - 64x,\ x(0) = x'(0) = 0$

11. $x'' = -7x' - 7x,\ x(0) = 1,\ x'(0) = -1$

12. $x'' = -\pi^2 x' - \pi x,\ x(0) = x'(0) = \pi$

Solve by reduction of order the equations and the initial value problems in Problems 13 through 20.

13. $x'' = 2\dfrac{x'}{t}$

14. $x'' = \dfrac{3t^2}{2 + t^3}x'$

15. $x'' = \dfrac{5 + \cos t}{5t + \sin t}x'$

16. $x'' = \dfrac{1 + e^t}{t + e^t}x'$

17. $x'' = (\tan t)x',\ x(0) = x'(0) = 1$

18. $x'' = (\cot t)x',\ x(\pi/2) = x'(\pi/2) = 2$

19. $x'' = t(x')^2,\ x(0) = 1,\ x'(0) = -2$

20. $x'' = -\frac{1}{2}(x')^3 + tx',\ x(0) = x'(0) = 1$

For each of Problems 21 through 28, check that the given function is a solution; then solve the equation or the initial value problem by reduction of order.

21. $x'' = \dfrac{t + 2}{t}x' - \dfrac{t + 2}{t^2}x,\ x_1(t) = t$

22. $x'' = \dfrac{t}{t - 1}x' - \dfrac{x}{t - 1},\ x_1(t) = e^t$

23. $x'' = -\dfrac{x'}{t} + \dfrac{x}{t^2},\ x_1(t) = t,\ t > 0$

24. $x'' = -\dfrac{1}{t}x' - \dfrac{4t^2 - 1}{4t^2}x,\ x_1(t) = \dfrac{\sin t}{\sqrt{t}}$

(the *Bessel equation* of order $\frac{1}{2}$)

25. $x'' = \dfrac{1}{t}x' - 4t^2 x,\ x_1(t) = \sin(t^2),\ t > 0$

26. $x'' = \dfrac{2}{t}x' - \dfrac{t^2 + 2}{t^2}x,\ x_1(t) = t \sin t$

27. $x'' = \dfrac{4}{t}x' - \dfrac{6}{t}x,\ x_1(t) = t^2,\ t > 0$

28. $x'' = -\dfrac{3}{t}x' - \dfrac{x}{t^2},\ x_1(t) = \dfrac{1}{t},\ t > 0$

29. Euler equations. An equation of the form

$$x'' = \frac{b}{t}x' + \frac{c}{t^2}x,$$

where b and c are constants and the vector field is defined on $(0, \infty)$, is called a *homogeneous Euler equation*. To solve it, check a solution of the form $x(t) = t^\lambda$, which reduces the above differential equation to the algebraic equation

$$\lambda^2 - (1 + b)\lambda - c = 0.$$

Denote its solutions by λ_1, λ_2. Prove the following:

(a) If λ_1, λ_2 are real and distinct, the general solution of the Euler equation is

$$x(t) = c_1 t^{\lambda_1} + c_2 t^{\lambda_2}.$$

(b) If λ_1, λ_2 are real and equal, the general solution of the Euler equation is

$$x(t) = t^{\lambda_1}(c_1 + c_2 \ln t).$$

(c) If λ_1, λ_2 are complex conjugates ($\lambda_1 = \alpha + i\beta$, $\lambda_2 = \alpha - i\beta$), the general solution of the Euler equation is

$$x(t) = t^{\alpha}[c_1 \cos(\beta \ln t) + c_2 \sin(\beta \ln t)].$$

Using the method described in Problem 29, solve Problems 30 through 35.

30. $x'' = -\dfrac{2}{t}x' + \dfrac{12}{t^2}x$

31. $x'' = \dfrac{1}{t}x' - \dfrac{3}{4t^2}x$

32. $x'' = \dfrac{5}{t}x' - \dfrac{9}{t^2}x$

33. $x'' = \dfrac{3}{t}x' - \dfrac{4}{t^2}x$

34. $x'' = -\dfrac{4}{t}x' - \dfrac{7}{t^2}x$

35. $x'' = -\dfrac{3}{2t}x' - \dfrac{5}{2t^2}$

36. A pipe of cross-sectional area S (see Figure 3.2.7) contains a liquid of density ρ. The length of the liquid-occupied portion is L. Assuming that the friction can be neglected and that all molecules move at the same speed, determine the up-and-down motion of the liquid if it is taken out of the equilibrium position. Compare the result with that obtained for Galileo's pendulum.

37. The linear damped pendulum. For a simple pendulum of mass m and length L, describe the motion near the downward equilibrium position, assuming that the friction force is proportional to the velocity of motion, $F_{\text{friction}} = kv$. Discuss all possible cases of the proportionality constant k.

38. A simple electric circuit. Assume that an electric circuit has a resistor, a capacitor, and an inductor connected in series. The resistance of the resistor is $R = u_r/i$, where i is the intensity of the current and u_R is the potential through the resistor. The capacitance of the capacitor is $C = q/u_C$, where q is the total charge on the plates and u_C is the potential through the capacitor. The inductance of an inductor is $L = u_L/i'$, where $i' = di/dt$ is the rate of change of the intensity and u_L is the potential through the inductor. For an electric circuit of this type $i = q'$ and $u_R + u_C + u_L = 0$ (Kirchhoff's law). Assuming that $L = 1$ henry (H), $C = \frac{2}{1405}$ farads (F), and $R = 1$ ohm (Ω), determine the variation in time of the charge q, knowing that $q(0) = 0.1$ coulomb (C) and $q'(0) = 0$.

39. A cubic box that can float is dropped into a lake from a helicopter. Once in the water, the box oscillates periodically (up and down). Neglecting all frictions and recalling Archimedes' law according to which an object submerged in a fluid is pushed upward with a force that equals the weight of the displaced fluid, determine the period of motion with respect to the density ρ of the box and its size L.

40. The physical pendulum. The physical pendulum describes the motion of a rigid body of mass m that oscillates about an axis that does not pierce its center of mass (see Figure 3.2.9). The distance from the axis to the center of mass is r and the angular

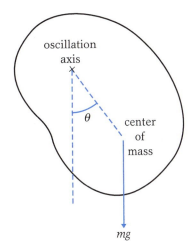

Figure 3.2.9. A cross section through the center of mass of the physical pendulum, perpendicular to the oscillation axis.

position is θ. The law of motion (which assumes that there are no frictions) is described by the equation $I\theta'' = M$, where I is the moment of inertia and $M = mgr \sin\theta$ is the torque, g being the gravitational acceleration. Describe the motion of the body if $m = 1$ kg, $r = 1$ m, and $I = 5$ kg·m^2.

41. Assume that Galileo's pendulum is in an elevator that moves upward, decelerating at 0.8 m/s^2, starting at rest at $t = 0$. Describe the future motion of the pendulum near the downward equilibrium position if the initial angle is 2 degrees of arc and the initial angular velocity is 0.

3.3
Nonhomogeneous Equations

Once we determine two linearly independent solutions of a linear homogeneous second-order equation, we can obtain the general solution of any nonhomogeneous equation derived from the homogeneous one. In this section we will present two methods for finding the general solution of linear nonhomogeneous second-order equations: the *method of variation of parameters* and the *method of undetermined coefficients*. Then, together with the reduction of order treated in Section 3.2, we will apply these methods to describe the motion of a maglev transportation device.

The General Solution for Nonhomogeneous Equations

In this section we will deal with equations of the form

$$x'' = p(t)x' + q(t)x + r(t), \tag{1}$$

where p, q, r are continuous functions. We will first show that obtaining the general solution of a linear nonhomogeneous second-order equation is equivalent to finding one particular solution of it and two linearly independent solutions of the corresponding linear homogeneous equation.

Fundamental Theorem for Nonhomogeneous Equations

Any solution x of equation (1) is of the form

$$x(t) = c_1 x_1(t) + c_2 x_2(t) + \varphi(t),$$

where c_1 and c_2 are constants, x_1, x_2 are linearly independent solutions of the corresponding homogeneous equation, and φ is some particular solution of equation (1).

PROOF

If x and φ are two particular solutions of equation (1), then, since $x'' = p(t)x' + q(t)x + r(t)$ and $\varphi'' = p(t)\varphi' + q(t)\varphi + r(t)$, it follows that

$$(x - \varphi)'' = p(t)(x - \varphi)' + q(t)(x - \varphi),$$

which means that $x - \varphi$ is a solution of the corresponding linear homogeneous equation. So there exist constants c_1, c_2 such that $x - \varphi = c_1 x_1 + c_2 x_2$, where x_1, x_2 are linearly independent solutions of the linear homogeneous equation. This is true for every solution x of (1), however, so the conclusion of the theorem follows. This completes the proof.

As we have seen in Sections 3.1 and 3.2, there is no general method for finding two linearly independent solutions of the homogeneous equation unless the coefficients p and q are constant or the reduction of order applies. Now we will show that the method of variation of parameters yields a particular solution of the nonhomogeneous equation if the general solution of the corresponding homogeneous equation is known.

Variation of Parameters The method of variation of parameters for second-order equations is a generalization of the one used in Section 2.3 for first-order equations and is also due to the French mathematician Joseph Louis Lagrange (1736–1813). It works as follows.

Step 1. After having obtained two linearly independent solutions x_1, x_2 of the corresponding linear homogeneous equation $X'' = p(t)X' + q(t)X$, check for equation (1) a solution of the form

$$\varphi(t) = s_1(t)x_1(t) + s_2(t)x_2(t) \tag{2}$$

that satisfies the supplementary condition[1]

$$s_1'(t)x_1(t) + s_2'(t)x_2(t) = 0. \tag{3}$$

Step 2. Substituting φ into (1) and using (3), obtain the equation in unknowns s_1' and s_2',

$$s_1'(t)x_1'(t) + s_2'(t)x_2'(t) = r(t). \tag{4}$$

Step 3. Solve system (3)–(4) to find s_1' and s_2', and then integrate to obtain

$$s_1(t) = -\int \frac{r(t)x_2(t)}{x_1(t)x_2'(t) - x_1'(t)x_2(t)}dt, \qquad s_2(t) = \int \frac{r(t)x_1(t)}{x_1(t)x_2'(t) - x_1'(t)x_2(t)}dt.$$

Step 4. Evaluate the integrals and, omitting the integration constants, write a particular solution

$$\varphi(t) = s_1(t)x_1(t) + s_2(t)x_2(t).$$

Step 5. Write the general solution of equation (1) as

$$x(t) = c_1x_1(t) + c_2x_2(t) + \varphi(t),$$

where c_1 and c_2 can be any real constants.

EXAMPLE 1   To find a particular solution of the equation

$$x'' = \frac{1+t}{t}x' - \frac{1}{t}x + te^{2t}, \tag{5}$$

---

[1]When substituting equation (2) into equation (1), we obtain one equation in two unknowns, s_1' and s_2'. (Can you do this computation?) This equation, which is usually very complicated, has infinitely many solutions. If we impose a supplementary linear relation between s_1' and s_2', we form a linear algebraic system that is likely to have only one solution. Computations prove easier if we impose condition (3). This leads to the linear nonhomogeneous algebraic system (3)–(4), which has a unique solution, indeed, because $x_1(t)x_2'(t) \neq x_1'(t)x_2(t)$.

where the vector field is defined on $(0, \infty)$, notice first that the functions $x_1(t) = 1 + t$ and $x_2(t) = e^t$ are linearly independent solutions of the corresponding homogeneous equation $X'' = [(1 + t)/t]X' - (1/t)X$.

Step 1. For equation (5) check a solution of the form

$$\varphi(t) = s_1(t)(1 + t) + s_2(t)e^t$$

that satisfies the supplementary condition

$$s_1'(t)(1 + t) + s_2'(t)e^t = 0.$$

Step 2. Substituting φ into (5) and using the above condition, obtain the equation

$$s_1'(t) + e^t s_2'(t) = te^{2t}.$$

Step 3. These two equations form the linear algebraic system of unknowns s_1', s_2',

$$\begin{cases} s_1'(t) + e^t s_2'(t) = te^{2t} \\ (1 + t)s_1'(t) + e^t s_2'(t) = 0, \end{cases}$$

whose solution is $s_1'(t) = -e^{2t}$, $s_2'(t) = (1 + t)e^t$.

Step 4. Integrating, and ignoring the integration constants, obtain $s_1(t) = -\frac{1}{2}e^{2t}$ and $s_2(t) = te^t$, so a particular solution of equation (5) is

$$\varphi(t) = \frac{1}{2}(t - 1)e^{2t}.$$

Step 5. The general solution of equation (5) is therefore

$$x(t) = c_1(1 + t) + c_2 e^t + \frac{1}{2}(t - 1)e^{2t},$$

where c_1 and c_2 can be any real constants.

EXAMPLE 2   To find a particular solution of the equation

$$x'' = -9x + \frac{3}{\cos 3t}, \tag{6}$$

first use the method of Section 3.2 to see that two linearly independent solutions of the equation $X'' + 9X = 0$ are $x_1(t) = \sin 3t$ and $x_2(t) = \cos 3t$. Then check for equation (6) a solution of the form

$$\varphi(t) = s_1(t)\sin 3t + s_2(t)\cos 3t.$$

Using the condition

$$s_1' \sin 3t + s_2' \cos 3t = 0, \tag{7}$$

compute that $\varphi' = 3s_1 \cos 3t - 3s_2 \sin 3t$ and $\varphi'' = (3s_1' - 9s_2)\cos 3t - (9s_1 + 3s_2')\sin 3t$. Substituting φ and φ'' into (6), obtain the equation

$s_1' \cos 3t - s_2' \sin 3t = 1/\cos 3t$. This together with condition (7) forms the algebraic system of unknowns s_1', s_2',

$$\begin{cases} s_1' \cos 3t - s_2' \sin 3t = \dfrac{1}{\cos 3t} \\ s_1' \sin 3t + s_2' \cos 3t = 0, \end{cases}$$

which has the solution $s_1'(t) = 1$ and $s_2'(t) = -\tan 3t$. Integrating and choosing one element of each antiderivative, obtain $s_1(t) = t$ and $s_2(t) = \frac{1}{3} \ln(\cos 3t)$. So a particular solution of equation (6) is

$$\varphi(t) = t \sin 3t + \frac{1}{3} \ln(\cos 3t) \cos 3t,$$

and therefore the general solution has the form

$$x(t) = c_1 \sin 3t + c_2 \cos 3t + t \sin 3t + \frac{1}{3} \ln(\cos 3t) \cos 3t,$$

where c_1 and c_2 can be any real constants.

Undetermined Coefficients

The method of undetermined coefficients also assumes that we have found in some way two linearly independent solutions of the corresponding homogeneous equation. Unlike the variation of parameters, however, the method of undetermined coefficients works only for those variable coefficients r of equation (1) that are constants, polynomials, exponentials, sines, cosines, or any sums or products of these functions. The idea of the method is to check into the equation a solution of a form that can be determined from Table 3.3.1. This can be done according to the following two rules.

RULE 1 If $r(t)$ is of the form in the left column of the table or is the sum or product of such functions, then check a particular solution of the corresponding form as indicated in the right column of the table.

For example, if $r(t) = 3e^{2t}$, then seek a particular solution of (1) of the form $\varphi(t) = ae^{2t}$, where a is a coefficient to be determined; if $r(t) = -\frac{1}{5}t^2$, seek a solution of the form $\varphi(t) = a + bt + ct^2$ (i.e., a complete polynomial of the same degree as r), where a, b, c are the coefficients to be determined; if $r(t) = 5 \sin 3t$, then check a solution of the

TABLE 3.3.1

$r(t)$	Attempted solution
$\alpha e^{\beta t}$	$a e^{\beta t}$
$\alpha \cos \omega t + \beta \sin \omega t$	$a \cos \omega t + b \sin \omega t$
$\alpha \ (\neq 0)$	a
$\alpha + \beta t$	$a + bt$
$\alpha + \beta t + \gamma t^2$	$a + bt + ct^2$
$\alpha + \beta t + \gamma t^2 + \delta t^3$	$a + bt + ct^2 + dt^3$
$\vdots$	$\vdots$

form $\varphi(t) = a\cos 3t + b\sin 3t$, where a and b are coefficients to be determined.[2] In case of sums or products of the functions in the table, also seek the corresponding sums or products of the same type. For example, if $r(t) = 4t^2 e^t$, check a solution of the form $\varphi(t) = e^t(a + bt + ct^2)$. Or, if $r(t) = -7t\sin 5t$, then seek a solution of the form $\varphi(t) = a\cos 5t + b\sin 5t + ct\cos 5t + dt\sin 5t$, where a, b, c, d must be determined. There is, however, an exception to Rule 1.

RULE 2 If $r(t)$ contains terms that duplicate any solution of the homogeneous equation, then each such term must be multiplied by t^n, where n is the smallest natural number that eliminates the duplication.

For example, if the solution of the homogeneous equation has the general form $c_1\sin 2t + c_2\cos 2t$ and if $r(t) = 3\sin 2t$, then instead of checking a particular solution of the form $a\cos 2t + b\sin 2t$ as Rule 1 suggests, seek one of the form $\varphi(t) = at\cos 2t + bt\sin 2t$, where a and b must be determined. Or, if the general solution of the homogeneous equation is $c_1 e^{-2t} + c_2 t e^{-2t}$ and if $r(t) = t + 10e^{-2t}$, then check a solution of the form $\varphi(t) = a + bt + ct^2 e^{-2t}$.

EXAMPLE 3   To find a particular solution of the equation

$$x'' = 3x' + 4x + 12e^{2t},$$

first solve the corresponding homogeneous equation $X'' - 3X' - 4X = 0$ with the method of Section 3.2 and obtain the general solution $c_1 e^{-t} + c_2 e^{4t}$, where c_1, c_2 are constants. Neither e^{-t} nor e^{4t} is involved in the coefficient $r(t) = 12e^{2t}$, so check, for the nonhomogeneous equation, a particular solution of the form

$$\varphi(t) = ae^{2t},$$

where a is a constant to be determined. For this substitute $\varphi(t)$, $\varphi'(t) = 2ae^{2t}$, and $\varphi''(t) = 4ae^{2t}$ into the nonhomogeneous equation and obtain $-6ae^{2t} = 12e^{2t}$, which implies that $a = -2$. So a particular solution of the given equation is $\varphi(t) = -2e^{2t}$. Thus, by the fundamental theorem, the general solution of the nonhomogeneous equation is

$$x(t) = c_1 e^{-t} + c_2 e^{4t} - 2e^{2t}.$$

EXAMPLE 4   To find a particular solution of the equation

$$x'' = 2x' - x + 8e^t,$$

first solve the homogeneous equation $X'' - 2X' + X = 0$ and obtain the general solution $c_1 e^t + c_2 t e^t$, where c_1, c_2 are constants. Since $r(t) = 8e^t$, and since e^t is already a solution of the homogeneous equation, you

---

[2]Even if only sin or cos appears alone in the expression of r, still seek a complete expression of the type $\varphi(t) = a\cos\omega t + b\sin\omega t$, because sin and cos switch roles if differentiated.

could attempt a solution of the form ate^t. But te^t is also a solution of the homogeneous equation, so go one step further and check a solution of the form

$$\varphi(t) = at^2 e^t.$$

Substituting $\varphi(t)$, $\varphi'(t) = a(2t + t^2)e^t$, and $\varphi''(t) = a(2 + 4t + t^2)e^t$ into the nonhomogeneous equation obtain $2ae^t = 8e^t$, so $a = 4$. Therefore $\varphi(t) = 4t^2 e^t$. Thus the general solution of the nonhomogeneous equation is

$$x(t) = c_1 e^t + c_2 t e^t + 4t^2 e^t.$$

EXAMPLE 5 To solve the initial value problem

$$x'' = -x + 2t + 6\cos t, \qquad x(0) = x'(0) = 1,$$

determine first the general solution of the given equation. For this solve the corresponding homogeneous equation $X'' + X = 0$ and find its general solution to be $c_1 \cos t + c_2 \sin t$. Since $r(t) = 2t + 6\cos t$, take a look at lines 2 and 4 in Table 3.3.1. But $\cos t$ is already a solution of the homogeneous equation, so according to Rule 2 applied to line 2 in Table 3.3.1, check a particular solution of the form

$$\varphi(t) = at\cos t + bt\sin t + c + dt,$$

where a, b, c, d must be determined. Substituting $\varphi(t)$ and $\varphi''(t) = (2b - at)\cos t - (2a + bt)\sin t$ into the given equation, obtain

$$c + (d - 2)t + 2(b - 3)\cos t - 2a\sin t = 0$$

which implies that $a = 0, b = 3, c = 0$, and $d = 2$. So a particular solution of the nonhomogeneous equation is $\varphi(t) = 2t + 3t\sin t$, and its general solution is

$$x(t) = c_1 \cos t + c_2 \sin t + 2t + 3t\sin t.$$

To find the solution ψ of the initial value problem, use the initial conditions and compute that $1 = x(0) = c_1$ and $1 = x'(0) = c_2 + 2$, so $c_1 = 1$ and $c_2 = -1$, and therefore

$$\psi(t) = \cos t + (3t - 1)\sin t + 2t.$$

Applications

Maglev transportation

While designing a maglev transportation device, which can float in a magnetic field, a team of scientists at the Jet Propulsion Laboratory in Pasadena, California, performed the following experiment. A horizontal iron bar of mass $m = 1$ kg slides (with negligible friction) along a frame consisting of two vertical posts connected at the bottom through a variable resistor (as shown in Figure 3.3.1). Let x denote the distance from the bar to the bottom of the frame. The system is in a uniform horizontal magnetic field, which creates electricity. This produces a resistance force $-(1/t)x'$, $t > 1$, opposed to the fall (the coefficient $-1/t$ appears due to the variance of the resistor at the bottom). How did the

Modern train moving by
magnetic levitation.

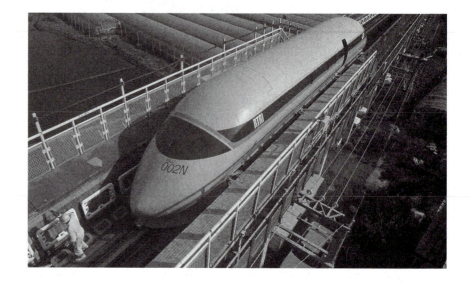

scientists determine the formula that describes the motion of the bar,
and what is its physical interpretation?

Solution. By Newton's second law, the motion of the bar is given by

$$mx'' = F_{\text{weight}} + F_{\text{resistance}},$$

where $m = 1$, $F_{\text{weight}} = mg$, g is the gravitational acceleration, and
$F_{\text{resistance}} = -(1/t)x'$ is due to the magnetic field. So the equation of
motion is

$$x'' = -\frac{1}{t}x' + g. \tag{8}$$

Using reduction of order, solve first the homogeneous equation $X'' = -(1/t)X'$. For this make the change of variable $y = X'$ and obtain the
separable equation $y' = -(1/t)y$, whose general solution is $y(t) = ct^{-1}$.
Returning to the change of variable, obtain $X'(t) = ct^{-1}$, which if inte-
grated gives the general solution of the homogeneous equation, $X(t) = c_1 \ln t + c_2$, where c_1, c_2 are constants.

   To obtain the solution of the nonhomogeneous equation, apply
the method of the variation of parameters by checking a solution of
the type $\varphi(t) = s_1(t)\ln t + s_2(t)$, with s_1 and s_2 satisfying the condition $s_1'(t)\ln t + s_2'(t) = 0$. This leads to $\varphi'(t) = (1/t)s_1(t)$ and $\varphi''(t) =$

Figure 3.3.1. An iron bar
sliding down a frame inside
a horizontal magnetic field.

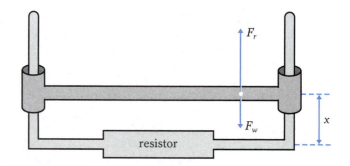

$-(1/t^2)s_1(t) + (1/t)s_2(t)$, which if substituted into (8) yield $s_1'(t) = gt$; a choice for s_1 is then $s_1(t) = gt^2/2$. From the above condition obtain $s_2'(t) = -gt \ln t$, and integration by parts leads to the choice

$$s_2(t) = -\frac{gt^2 \ln t}{2} + \frac{gt^2}{4}$$

A particular solution of (8) is therefore $\varphi(t) = gt^2/4$, so the general solution of the nonhomogeneous equation has the form

$$x(t) = \frac{gt^2}{4} + c_1 \ln t + c_2.$$

Because $x(t) \to \infty$ when $t \to \infty$, this formula shows that after a while the resistance force is stronger than the gravitational one, so the bar is finally pushed upward. Of course, the magnetic field is not infinite in space, so the bar will never go too high.•

PROBLEMS

In Problems 1 to 8, find two linearly independent solutions of the corresponding homogeneous equation and then solve the given nonhomogeneous equation using variation of parameters.

1. $x'' = -x + \tan t$

2. $x'' = x + 1 + e^t$

3. $x'' = -3x' - 2x + e^t$

4. $x'' = -2x' - x + t$

5. $x'' = 2x' - x + e^t \arctan t$

6. $x'' = x' - \frac{1}{4}x + \frac{1}{4}e^{t/2}\sqrt{1 - t^2}$

7. $x'' = x + \dfrac{2(t - 1)e^t}{t^2}$

8. $x'' = -4x + \sec 2t$

For the linear nonhomogeneous equations in Problems 9 to 12, find a particular solution using the method of variation of parameters and write the general solution. In each case check first that the given functions are linearly independent solutions of the corresponding homogeneous equation.

9. $x'' = \dfrac{2}{t^2}x - \dfrac{1}{t^2} + 3$ on $(0, \infty)$;

$x_1(t) = \dfrac{1}{t}, \ x_2(t) = t^2$

10. $x'' = \dfrac{t + 2}{t}x' - \dfrac{t + 2}{t^2}x + 2t^3$ on $(0, \infty)$;

$x_1(t) = te^t, \ x_2(t) = t$

11. $x'' = \dfrac{t}{t - 1}x' + \dfrac{x}{1 - t} + 2(t - 1)^2 e^{-t}$ on $(0, 1)$;

$x_1(t) = t, \ x_2(t) = e^t$

12. $x'' = \dfrac{t + 1}{t}x' - \dfrac{x}{t} + t^2 e^{2t}$ on $(0, \infty)$;

$x_1(t) = e^t, \ x_2(t) = t + 1$

Solve the initial value problems in Problems 13 to 16 using variation of parameters and knowing that in each case the given function is the general solution x_h of the corresponding homogeneous equation.

13. $x'' = -\dfrac{1}{t}x' + \dfrac{4}{t^2}x + t^2, \ x(1) = 1,$

$x'(1) = 2, \ x_h(t) = c_1 t^2 + c_2 t^{-2}$

14. $x'' = \dfrac{2}{t}x' - \dfrac{t^2 + 2}{t^2}x - t, \ x(1) = 1,$

$x'(1) = 3, \ x_h(t) = c_1 t \sin t + c_2 t \cos t$

15. $x'' = \dfrac{1 - t}{t}x' + \dfrac{3(4t - 1)}{t}x + t,$

$x(2) = x'(2) = 1, \ x_h(t) = c_1 e^{3t} + c_2(7t + 1)e^{-4t}$

16. $x'' = \dfrac{t + 3}{t}x' - \dfrac{3}{t}x + \dfrac{1}{t}, \ x(1) = x'(1) = 0,$

$x_h(t) = c_1 e^t + c_2(t^3 + 3t^2 + 6t + 6)$

In Problems 17 to 24, find two linearly independent solutions of the corresponding homogeneous equation and then solve the given nonhomogeneous equation using the method of undetermined coefficients.

17. $x'' = 9x + 7$

18. $x'' = -2x' - x + 3t - 1$

19. $x'' = -x' - x + 2e^{3t}$

20. $x'' = -3x' - 2x + \frac{2}{3}t + 7e^{-2t}$

21. $x'' = 2x' - x + 2\sin t - 1$

22. $x'' = -3x' + 10x - 3t(e^t + 1)$

23. $x'' = -9x + 13\cos 3t$

24. $x'' = -x' - x + e^t$

In Problems 25 to 32, solve the given initial value problems using the method of undetermined coefficients.

25. $x'' = 16x + 4, \quad x(0) = 1, \; x'(0) = 0$

26. $x'' = -x' - 2t, \quad x(0) = 1, \; x'(0) = 0$

27. $x'' = 9x' + 3t - 1, \quad x(0) = 0, \; x'(0) = 3$

28. $x'' = -5x' + 6x + 4e^{2t}, \quad x(0) = x'(0) = 1$

29. $x'' = -x + \sin t, \quad x(\pi/2) = 0,$
    $x'(\pi/2) = -1$

30. $x'' = -25x + 4\sin 5t, \quad x(0) = x'(0) = 0$

31. $x'' = x - te^t, \quad x(0) = 0, \; x'(0) = 1$

32. $x''' = x'' + t + e^t, \quad x(0) = x'(0) = x''(0) = 0$

33. Using the method of undetermined coefficients, solve the maglev transportation problem. Do you obtain the same answer as in the text?

34. Imagine that in the maglev transportation problem the resistor is varied such that the equation describing the motion of the bar is

$$x'' = -\frac{x'}{t} + \frac{x}{t^2} + g, \; t > 1.$$

Find the formula that describes the motion of the bar and give the physical interpretation.

35. Recall from Section 3.2 that for small oscillations the motion of Galileo's undamped pendulum is given by the equation

$$\phi'' = -\frac{g}{L}\phi,$$

where g is the gravitational constant and L the length of the pendulum. If the pendulum is shaken periodically by a vertical force of the type $\cos 2\pi t$, the equation describing the motion is

$$\phi'' = -\frac{g}{L}\phi + \cos 2\pi t.$$

Solve this equation for $g/L = 1 \text{ s}^{-2}$ and initial conditions $\phi(0) = 1$, $\phi'(0) = 0$ and describe the motion of the pendulum in this case.

36. Recall from Section 3.2 that the motion of water in a pipe is given by the equation

$$x'' = \frac{k}{m}x' - \frac{2g}{L}x,$$

where x is the height of the column of displaced water, k the friction constant, g the gravitational acceleration, and L the length of the water-occupied portion of the pipe. Imagine that during an earthquake the pipe is shaken periodically by a force of the type $\frac{1}{10}\cos 8\pi t$. This means that the equation describing the motion of water in the pipe takes the form

$$x'' = \frac{k}{m}x' - \frac{2g}{L}x + \frac{1}{10}\cos 8\pi t.$$

Solve this equation for $k/m = 1 \text{ s}^{-2}$, $g/L = 1 \text{ s}^{-2}$, and initial conditions $x(0) = x'(0) = 0$. Then give the physical interpretation of the solution.

3.4 Harmonic Oscillators

In this section we will study an important class of linear second-order equations that model certain oscillatory phenomena, a typical representative being the spring-mass system in Figure 3.4.1. We will first consider motions *damped* by some resistance force and then deal with the case of *undamped* motions perturbed by periodic external forces. We will discover surprising phenomena such as *beats* and *resonance* and finally shed some light on the collapse of the Tacoma Narrows Bridge mentioned in the Introduction.

Figure 3.4.1. A spring-mass system. (*a*) The equilibrium position without weight attached. (*b*) The equilibrium position with weight attached. (*c*) An out–of–equilibrium position.

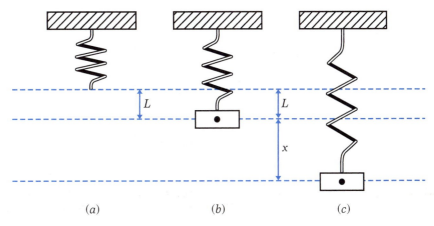

(*a*)   (*b*)   (*c*)

Damped Oscillations

The difference L between the lengths of the spring at rest with and without the mass m attached is called *elongation*. Two opposite forces act on the mass at rest: the weight, $F_w = mg$, directed downward, where g is the gravitational acceleration, and the restoring force, reliably modeled by *Hooke's law* as proportional to the elongation, i.e., of the form $-kL$, $k > 0$. (The minus sign shows that the force acts upward, opposed to the gravitational one.) Since the mass is at rest, the two forces cancel each other, so $mg - kL = 0$. This allows us to determine the *spring constant*

$$k = \frac{mg}{L}.$$

If we pull the mass down from equilibrium to a position[1] x, the restoring force becomes $F_r = -k(L+x)$. In addition, we assume the resistance force due to internal friction to be proportional to the velocity of motion, i.e., $F_f = -\nu x'$, where $\nu > 0$ is the *damping coefficient*. By Newton's second law, $F = mx''$, where $F = F_w + F_r + F_f$, so the equation of motion of the *damped oscillator* is

$$x'' = -\frac{\nu}{m}x' - \frac{k}{m}x.$$

The roots of the characteristic equation are

$$r_{1,2} = \frac{\nu}{2m}(-1 \pm \sqrt{\Delta}),$$

where $\Delta = 1 - 4km/\nu^2$. There are three cases to discuss, depending on the sign of Δ.

The case $\Delta > 0$

Since $\nu^2 > 4km$, the damping coefficient is large, and the motion is said to be *overdamped*. The general solution is

$$x(t) = c_1 e^{r_1 t} + c_2 e^{r_2 t},$$

---

[1]We will consider $x > 0$ below the equilibrium and $x < 0$ above the equilibrium.

Figure 3.4.2. The graph of the function $\alpha(t) = \frac{1}{5}e^{-t/2} + e^{-t/3}$, which is a typical overdamped solution.

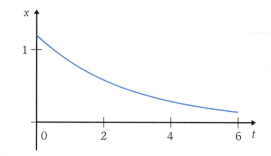

and since $-1 \pm \sqrt{\Delta} < 0$, r_1 and r_2 are negative (can you show this?), so $x(t) \to 0$ when $t \to \infty$. A typical graph of such a function is given in Figure 3.4.2.

The physical interpretation of the solution is that the weight is initially at some distance x_0 below the equilibrium position; when released, it slowly moves up without oscillating, tending to reach the equilibrium when $t \to \infty$. This is the case of a very rigid spring with high internal frictions.

The case $\Delta = 0$ Since $\nu^2 = 4km$, we say that the motion is *critically damped*. We have $r_1 = r_2 = -\nu/(2m)$, and the general solution is

$$x(t) = (c_1 + c_2 t)e^{-\nu t/2m}.$$

This is a nonperiodic function that tends to 0 when $t \to \infty$ and reaches the equilibrium without oscillating. A typical graph is represented in Figure 3.4.3.

The physical interpretation of such a solution is that the weight, which is initially at some distance x_0 below the equilibrium position, is pushed even lower; then it slowly moves up without oscillating, tending to reach the equilibrium when $t \to \infty$. This is still the case of a rigid spring with high internal frictions.

The case $\Delta < 0$ Since $\nu^2 < 4km$, we say that the motion is *underdamped*. We have $r_{1,2} = (-\nu \pm i\sqrt{4km - \nu^2})/(2m)$, and the general solution is

$$x(t) = e^{-\nu t/(2m)}\left(c_1 \sin \frac{\sqrt{4km - \nu^2}}{2m}t + c_2 \cos \frac{\sqrt{4km - \nu^2}}{2m}t\right).$$

The motion is not periodic, and $x(t) \to 0$ when $t \to \infty$, but now oscillations occur. They are similar to those we encountered when studying the

Figure 3.4.3. The graph of the function $\beta(t) = (0.1 + t)e^{-t}$, which is a typical critically damped solution.

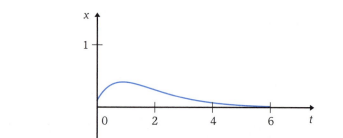

Figure 3.4.4. The graph of the function $\gamma(t) = e^{-t/8}(\sin 2t + \cos 2t)$, which is a typical under-damped solution.

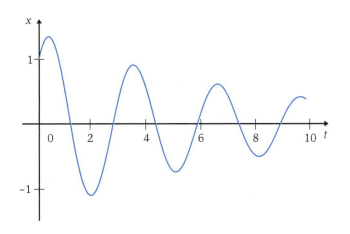

motion of a water segment in a pipe (see Section 3.2). A typical graph is drawn in Figure 3.4.4.

The physical interpretation of the solution is that the weight, initially at some distance x_0 below the equilibrium position, moves and down, passing through the equilibrium. The oscillations decrease exponentially in amplitude, and the motion tends to rest when $t \to \infty$. This is the case of an elastic spring with small internal frictions. (Can you characterize the motion when friction is negligible? What is the form of the solution in that case?)

Forced Undamped Oscillations

Assume now that the internal frictions are negligible but that a periodic external force, say $\alpha \cos \omega t$ (where ω is the *frequency* of the external force and $\alpha > 0$), acts vertically on the spring-mass system. This happens if, for example, the system is in a frame that is periodically shaken up and down. Then the equation of motion of the *forced undamped oscillator* is

$$x'' = -\omega_0^2 x + a \cos \omega t, \qquad (1)$$

where $\omega_0 = \sqrt{k/m}$ is the *natural frequency* of the spring, and $a = \alpha/m$. Using the method of variation of parameters, we can obtain the general solution. There are two cases to consider: $\omega \neq \omega_0$ and $\omega = \omega_0$.

The case $\omega \neq \omega_0$: Beats

In this case the general solution is

$$x(t) = c_1 \sin \omega_0 t + c_2 \cos \omega_0 t + \frac{a \cos \omega t}{\omega_0^2 - \omega^2},$$

where c_1, c_2 are constants (see Problem 25). The solution is the sum of two functions that represent motions of different periods and amplitudes. The simplest motion occurs when the spring-mass system is at rest and only the external force acts on it. This means that the initial conditions are $x(0) = x'(0) = 0$, which provide the constants $c_1 = 0$, $c_2 = -a/(\omega_0^2 - \omega^2)$ and the particular solution

$$\varphi(t) = \frac{a}{\omega_0^2 - \omega^2}(\cos \omega t - \cos \omega_0 t).$$

Using the formula

$$\cos a - \cos b = 2 \sin \frac{a+b}{2} \sin \frac{b-a}{2},$$

this solution takes the form

$$\varphi(t) = A(t) \sin \frac{(\omega_0 + \omega)t}{2}, \tag{2}$$

where

$$A(t) = \frac{2a}{\omega_0^2 - \omega^2} \sin \frac{(\omega_0 - \omega)t}{2}$$

can be interpreted as a variable amplitude that ranges in value between $-2a/(\omega_0^2 - \omega^2)$ and $2a/(\omega_0^2 - \omega^2)$ (see Problem 25). When ω and ω_0 are close but unequal, a solution as in (2) is called a *beat*. The typical graph of a beat is drawn in Figure 3.4.5.

EXAMPLE 1   For the initial value problem

$$x'' = -4.81x + 0.81 \cos 2t, \qquad x(0) = x'(0) = 0, \tag{3}$$

the coefficients defining the equation, as compared to equation (1), are $\omega_0 = 2.2$, $\omega = 2$, and $a = 0.81$. This means that the solution, as given by equation (2), takes the form

$$\varphi(t) = 2 \sin \frac{t}{10} \sin \frac{21t}{10}.$$

Its graph, which is typical for a beat, is represented in Figure 3.4.5. Notice the two types of oscillations characteristic for beats: a fast one, which crosses the t-axis frequently, and a slow one, which crosses the t-axis after each period.

Beats are also encountered in acoustics. If two vibrating tuning forks of close but unequal frequency are placed upward on a wooden table, we hear an up-and-down beatlike periodic sound. Sometimes the same effect occurs when a radio is tuned.

Figure 3.4.5. The graph of the function $\varphi(t) = 2 \sin(t/10) \sin(21t/10)$, which is typical for a beat, drawn in the interval $[0, 65]$.

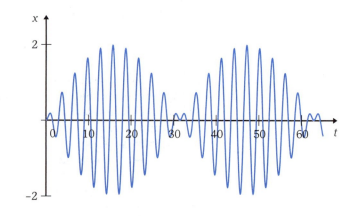

The case $\omega = \omega_0$: Resonance   In this case the period of the spring-mass system is equal to that of the external force. The general solution of equation (1) is

$$x(t) = \left(c_1 + \frac{at}{2\omega_0}\right)\sin \omega_0 t + c_2 \cos \omega_0 t \tag{4}$$

(see Problem 26). This means that when $t \to \infty$, x becomes unbounded and the amplitude, defined as

$$A(t) = \sqrt{\left(c_1 + \frac{at}{2\omega_0}\right)^2 + c_2^2},$$

tends to infinity (see Figure 3.4.6). This phenomenon is called *resonance*. Physically, the spring never reaches too large amplitudes, because it breaks.

EXAMPLE 2 Consider the initial value problem

$$x'' = -4.81x + 0.81 \cos 2.2t, \qquad x(0) = x'(0) = 0. \tag{5}$$

Notice that this equation slightly differs from equation (3) in Example 1. The only change is the frequency ω of the external force, 2.2 instead of 2, but this small change leads to a totally different solution. Indeed, according to formula (4), the initial conditions $x(0) = x'(0) = 0$ imply that $c_1 = c_2 = 0$, and therefore the solution of our initial value problem is

$$\psi(t) = \frac{0.81t}{4.4} \sin 2.2t.$$

The graph of ψ is shown in Figure 3.4.6 and is typical for a resonant solution. Notice that the amplitude increases linearly in time. Compare this graph to the one of the beat in Figure 3.4.5 and notice how the initial value problems (3) and (5), though almost identical, yield such different solutions.

Resonance has important applications. The existence of some of the *Kirkwood gaps* (zones of low-density population in the belt of asteroids that gravitate around the sun between the orbits of Mars and Jupiter) can be explained through resonance. This belt, formed by thousands of small planets, is similar to the rings of Saturn, except that due to the large-scale distribution, the gaps between rings can be "seen" only

Figure 3.4.6. The graph of the function $\psi(t) = (0.81\,t/4.4)\sin 2.2t$, which is a typical resonant solution, drawn in the interval $[1, 100]$.

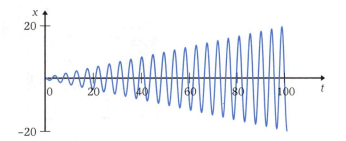

statistically. They were named after Daniel Kirkwood, an American mathematics teacher and amateur astronomer, who discovered them in 1860. The gravitational force of the sun acts on the asteroid in the same way as the spring does on the weight, while Jupiter plays the role of the external periodic force. When the corresponding frequencies coincide, resonance expels the asteroid out of the belt. The proof of this fact, however, requires special training in celestial mechanics and is therefore beyond our scope.

Another notable application concerns suspension bridges, which may encounter resonance if an external periodic force with suitable frequency acts on them. A British newspaper, the *Manchester Guardian* of April 16, 1831, described the collapse of the Broughton Bridge near Manchester when the soldiers of the 60th regiment marched four abreast upon it. The explanation of the accident was that the natural frequency ω_0 of the bridge, regarded as a spring, had been the same as the frequency ω of the external force due to marching. This is the reason today's military regulations forbid soldiers to march in step on bridges.

A more complicated situation appears when bridges collapse for different reasons—the action of the wind, for example. It is then unreasonable to think that the external force is periodic. Let us further discuss the case of the short-lived Tacoma Narrows Bridge, built in 1940 across Puget Sound in the northwest region of Washington State.

The Tacoma Narrows Bridge Disaster

Soon after its opening on July 1, 1940, the Tacoma Narrows Bridge near Seattle, a suspension structure more than a mile long, began twisting and oscillating in the wind, to break and collapse on November 7. For many years the accident was attributed solely to resonance, considering the vertical cables connecting the upper cable and the roadbed as springs and the wind as an external periodic force. Today we know that the answer

Old Tacoma Narrows Bridge.

is more complicated and can be given within the theory of partial differential equations (which also explains other large-scale motions and traveling waves in suspension structures, such as the Bronx-Whitestone Bridge in New York and the Golden Gate Bridge in San Francisco), but these aspects are beyond the scope of our textbook. However, we can look into a simpler model (a version of the one proposed in 1989 by the American mathematicians J. Glover, A. C. Lazer, and P. J. McKenna) that gives a reasonable explanation of the 1940 event.

Our model considers that the vertical cables act indeed like springs when the roadbed is below the equilibrium position, but above the equilibrium the roadbed behaves differently, because the slack suspension cables do not push down as a spring does. The equation is therefore

$$x'' = -bx' - cx + w(t), \tag{6}$$

where x is the distance of the roadbed from its undisturbed position ($x > 0$ if the roadbed is below equilibrium and $x < 0$ if it is above it), w is a function that describes the force of the wind, $b > 0$ is a constant that measures the damping,

$$c = \begin{cases} \delta + g, & \text{if } x < 0 \\ g, & \text{if } x \geq 0 \end{cases}$$

is introduced to model the motion of the roadbed, $\delta > 0$ is a constant, and g is the gravitational acceleration.

If no wind blows, equation (6) becomes

$$x'' = -bx' - cx, \tag{7}$$

with b and c as above. This is not exactly the damped harmonic oscillator, for c is not constant but piecewise constant. However, the model further assumes that $b^2 - 4c < 0$. This suggests that the general solution is

$$x(t) = e^{-bt/2}(c_1 \sin \beta t + c_2 \cos \beta t), \tag{8}$$

where

$$\beta = \begin{cases} \rho, & \text{if } x < 0 \\ \sigma, & \text{if } x \geq 0, \end{cases}$$

$\rho = \sqrt{|b^2 - 4(\delta + g)|}/2$, and $\sigma = \sqrt{|b^2 - 4g|}/2$. We seem to have overcome the difficulties. A closer look, however, reveals a flaw. In formula (8) the solution x is a function of β, which depends on the sign of x. Consequently, x is defined in terms of x. This does not make much sense.

Therefore we need to express β in terms of t. In other words, we have to discuss the sign of x with respect to t. For this, first notice that

$$x(t) = Ae^{-bt/2} \cos(\beta t - \gamma), \tag{9}$$

where $A = \sqrt{c_1^2 + c_2^2}$ and $\tan \gamma = c_1/c_2$ (see Problems 13–18). So the sign of x depends on the sign of $\cos(\beta t - \gamma)$, which can be determined in terms of t. Let us start with the case where $-\pi/2 \leq \beta t - \gamma \leq 3\pi/2$, which leads to two subcases: $-\pi/2 \leq \beta t - \gamma \leq \pi/2$ and $\pi/2 < \beta t - \gamma < 3\pi/2$.

1. $-\pi/2 \leq \beta t - \gamma \leq \pi/2$. This implies that $\cos(\beta t - \gamma) \geq 0$; therefore $\beta = \sigma$. Solving the inequalities we find that t

Figure 3.4.7. The graph of x in the interval $[a_1, b_2)$ before and after the approximation. The dotted line represents the approximation.

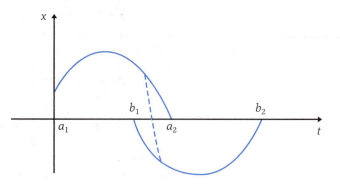

belongs to the interval $[a_1, a_2]$, where $a_1 = [\gamma - (\pi/2)]/\sigma$ and $a_2 = [\gamma + (\pi/2)]/\sigma$. Thus, for any t in $[a_1, a_2]$, $x(t) \geq 0$.

2. $\pi/2 < \beta t - \gamma < 3\pi/2$. This implies that $\cos(\beta t - \gamma) < 0$; therefore $\beta = \rho$. Solving the inequalities we find that t belongs to the interval (b_1, b_2), where $b_1 = [\gamma + (\pi/2)]/\rho$ and $b_2 = [\gamma + (3\pi/2)]/\rho$. Thus, for any t in (b_1, b_2), $x(t) < 0$.

Since $\rho \neq \sigma$, the ends of the corresponding intervals don't match; i.e., $a_2 \neq b_1$. More precisely, $\sigma < \rho$, which means that $b_1 < a_2$, so the intervals $[a_1, a_2]$ and (b_1, b_2) overlap. The graph of the solution depicted in Figure 3.4.7 in the interval (a_1, b_2) does not represent a function, because for every t in (b_1, a_2) there are two possible choices for x. We have obviously run into a new problem.

To overcome this new problem, we can approximate the solution as the dotted line shows. This gives a more realistic (though not fully satisfactory) idea of the bridge's behavior in the interval $[a_1, b_2]$. From formula (9), we see that the factor $e^{-bt/2}$ makes the oscillations' amplitude decrease in time. Therefore, if no wind acts, a bridge taken out of equilibrium would oscillate somewhat according to the graph in Figure 3.4.8. So, in the absence of exterior forces, a bridge taken out of equilibrium would qualitatively behave as the damped oscillator studied at the beginning of this section and would therefore tend to reach the equilibrium position.

Let us now take into consideration the action of the wind. It is tempting but unrealistic to assume that w is a periodic function, say of the form

Figure 3.4.8. The approximate behavior of a suspension bridge taken out of equilibrium but against which no other force acts. Recall that $x > 0$ describes the motion below the equilibrium, while $x < 0$ shows what happens above the equilibrium.

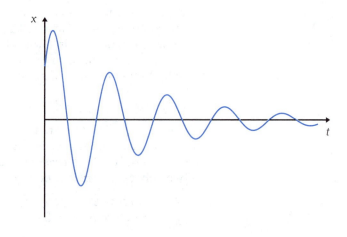

$w(t) = a \cos \omega t$, and then apply the theory of resonance to justify the amplitude increase of the oscillations. On the other hand, we don't know how the wind acts, so it seems hard even to estimate what the function w looks like. Therefore, it seems more realistic to imagine that every gust of wind acts by changing the values of the coefficients of equation (7). So instead of working with equation (6), which we cannot understand (even qualitatively) because we do not know the function w, we assume that between two wind gusts, i.e., on a finite (possibly small) interval of time, the motion is described by equation (7) for fixed values of the coefficients. Every new wind gust changes the coeffcients $a, b,$ and c. The initial conditions of the new problem are inherited from the old one to ensure the continuity of the solution. This leads to the study of a sequence of differential equations on consecutive intervals of time. Though we still do not know how to change the values of the new coefficients, we can at least say that on each interval the solution has a similar qualitative behavior to the one in Figure 3.4.8. Therefore the motion of the roadbed must be a function that qualitatively behaves according to the graph in Figure 3.4.9. Of course, we do not know how much the amplitude changes under the action of a wind gust, but we know that the oscillations are likely to increase and then tend to ameliorate until a new gust hits the bridge.

Though this is only a qualitative result, it explains why, after repeated oscillations that weaken the entire structure, a suspension bridge can collapse. As we see, this has nothing to do with resonance, which is unlikely to appear. (Can you explain why resonance is unlikely?) Therefore the collapse of the Tacoma Narrows Bridge was rather the consequence of irregular oscillations that occurred due to the repeated action of the wind and to the frailty of an ill-designed construction. The replacement bridge, which cost 250% more, is heavier, wider, more rigid, and more robust.

A recent test shows that engineers have learned from the mistakes of the past and can now build strong and resistant bridges. On June 5, 1998, at the inauguration of the 18-km-long Bit-Belt Bridge in Denmark, 30,000 people ran along it during a marathon event. No significant oscillations were detected, and the bridge is as sound today as it was then.

Figure 3.4.9. The qualitative behavior of an ill-designed suspension bridge under the action of the wind.

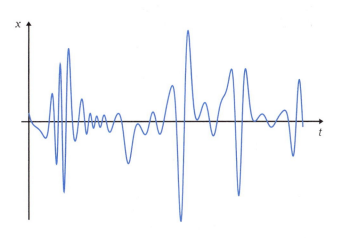

PROBLEMS

1. A spring has an elongation of 32 cm for a mass of 0.5 kg. Determine the constant k specific to this spring-mass system.

2. A mass of 1 kg stretches a spring 25 cm. If the mass is pulled down another 50 cm and then released, determine the motion of the mass if all resistance forces are negligible and given $g = 9.8$ m/s^2.

3. A mass of 0.5 kg that would stretch a spring by 50 cm is pushed up 25 cm and then released. Given that $g = 9.8$ m/s^2 and assuming that all resistance forces are negligible, determine the motion at every instant.

4. Write the general equation of the *undamped oscillator*, i.e., a spring-mass system without resistance forces, and then find the general solution. What is the physical interpretation of the solution? Compare the result with Galileo's pendulum in Section 3.2.

5. In the case of an *undamped oscillator* (see Problem 4), the *angular frequency* is defined as $\sqrt{k/m}$, where k is the constant of the spring and m is the mass. The period is defined as $2\pi\sqrt{m/k}$. Can you determine the angular frequency of an undamped harmonic oscillator knowing only the elongation and the value of the gravitational acceleration at the given location of the system? What about the period?

6. A mass of 1 kg stretches a spring 0.49 m. The damping coefficient is 80, and $g = 9.8$ m/s^2. Knowing that the spring is pulled down, out of the equilibrium position, with an initial velocity of 2 m/s, determine the motion of the mass at every moment. Draw the graph of the solution.

7. A mass of 0.2 kg stretches a spring 0.2 m. Given that $g = 9.8$ m/s^2, what damping coefficient must the resistance forces have to make the system critically damped?

8. Assume that a sensitive spring-mass system of mass 20 g and elongation $L = 20$ cm, in which friction can be neglected, is accelerated periodically up and down by a force of the type $2\cos(t/4)$. If $g = 9.8$ m/s^2 and the system is initially at rest, find the solution of the corresponding initial value problem and draw its graph.

9. Derive the solution of the spring-mass system in the resonant case using the method of variation of parameters.

10. The motion of a spring-mass system without damping is given by the initial value problem

$$x'' = -cx, \qquad x(0) = 1, \ x'(0) = \epsilon.$$

The measured period is $2\pi/3$ s and the measured amplitude is 2 m. Determine c and ϵ.

11. Find the general solution of the forced undamped oscillator if the external force is of the type $a\sin\omega t$ instead of $a\cos\omega t$. Compare the formula for beats with the one obtained in the text.

12. A spring-mass system in which friction can be neglected is accelerated periodically up and down by the force $\cos(\pi t)$. Assuming that the system is initially at rest and knowing that the spring breaks at amplitude 2.5 m, determine the time when this happens.

The functions in Problems 13 through 18 are of the type $c_1\sin\omega_0 t + c_2\cos\omega_0 t$. Put each of them in the form $A\cos(\beta t - \gamma)$. (Hint: Use the formula $\cos(a - b) = \sin a \sin b + \cos a \cos b$ to express $\cos(\beta t - \gamma)$; then compare the expression you obtain with $c_1\sin\omega_0 t + c_2\cos\omega_0 t$ and determine A, β, and γ with respect to c_1, c_2, and ω_0.)

13. $\sin 3t + 2\cos 3t$

14. $\frac{1}{2}\sin t + \frac{3}{2}\cos t$

15. $5\sin\dfrac{t}{2} - \cos\dfrac{t}{2}$

16. $-\pi\sin\pi t - 2\pi\cos\pi t$

17. $13\sin 6.28t - 15\cos 6.28t$

18. $-\frac{3}{7}\sin\dfrac{5\pi}{2}t + 2\cos\dfrac{5\pi}{2}t$

The functions in Problems 19 through 24 are of the form

$$\frac{a}{\omega_0^2 - \omega^2}(\cos\omega t - \cos\omega_0 t).$$

Put each of them in the form $A(t)\sin(\omega_0 + \omega)t/2$.

19. $5(\cos 2t - \cos 3t)$

20. $10\left(\cos 2t - \cos\dfrac{t}{2}\right)$

21. $\dfrac{\pi}{2}(\cos t - \cos 5\pi t)$

22. $\dfrac{2}{7}\left(\cos\dfrac{t}{3} - \cos\dfrac{t}{4}\right)$

23. $\cos 6t - \cos\frac{17}{2}t$

24. $3\left(\cos t - \cos\dfrac{t}{\pi}\right)$

25. Show that in case of beats, the general solution is indeed of the form

$$x(t) = c_1 \sin \omega_0 t + c_2 \cos \omega_0 t + \frac{a\cos\omega t}{\omega_0^2 - \omega^2},$$

and then, using the formula

$$\cos a - \cos b = 2\sin\frac{a+b}{2}\sin\frac{b-a}{2},$$

show that the initial conditions $x(0) = x'(0) = 0$ lead to the particular solution

$$x(t) = A(t)\sin\frac{(\omega_0 + \omega)t}{2},$$

where

$$A(t) = \frac{2a}{\omega_0^2 - \omega^2}\sin\frac{(\omega_0 - \omega)t}{2}.$$

26. Using the method of undetermined coefficients, show that in case of resonance the general solution is indeed

$$x(t) = \left(c_1 + \frac{at}{2\omega_0}\right)\sin\omega_0 t + c_2\cos\omega_0 t.$$

27. The motion of a spring-mass system is given by the initial value problem

$$x'' = -x + f(t), \qquad x(0) = 5,\ x'(0) = 0,$$

where

$$f(t) = \begin{cases} t, & \text{on } \left[0, \dfrac{\pi}{2}\right) \\[2mm] \pi - t, & \text{on } \left[\dfrac{\pi}{2}, \pi\right] \\[2mm] 0, & \text{on } (\pi, \infty). \end{cases}$$

To obtain an approximate solution, solve the problem on each interval independently, and then draw an approximate graph. What is the difference between this system and the one describing the motion of the Tacoma Narrows Bridge studied in this section?

28. If, in the spring-mass system with a periodic external vertical force, resistance forces are taken into consideration, the equation of motion becomes

$$x'' = -\frac{\nu}{m}x' - \frac{k}{m}x + a\cos\omega_0 t.$$

Find its general solution. Then consider the values $m = 1, \mu = \frac{1}{5}, k = a = \omega_0 = 1$, and the initial values $x(0) = 10, x'(0) = 0$. Solve the corresponding initial value problem and draw the graph of the solution.

3.5 Qualitative Methods

As in Section 2.4, we restrict our study to *autonomous* equations, which in this case have the form

$$x'' = f(x, x'). \tag{1}$$

We will introduce the *phase plane*, draw vector fields and phase-plane portraits (also called *flows*), classify equilibrium solutions, and define the notions of *structural stability* and *bifurcation*, which point out a new way of looking at differential equations. In the end we will apply these results to simple electric circuits and to some previous models.

Phase Plane

Let us split equation (1) into a system of two first-order equations. With the new variable $y = x'$, equation (1) becomes

$$\begin{cases} x' = y \\ y' = f(x, y). \end{cases} \tag{2}$$

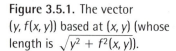

Figure 3.5.1. The vector $(y, f(x, y))$ based at (x, y) (whose length is $\sqrt{y^2 + f^2(x, y)}$).

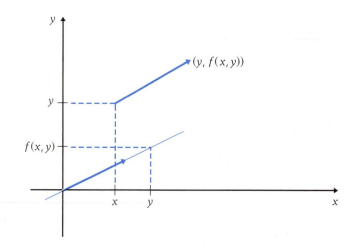

A *solution* of (2) is a pair of functions (φ, ψ) defined on an open interval, differentiable, and that satisfy the system; here this means

$$\begin{cases} \varphi'(t) = \psi(t) \\ \psi'(t) = f(\varphi(t), \psi(t)). \end{cases}$$

This may lead to some confusion. Equation (1) and system (2) are equivalent, but their solutions apparently are not, because φ is the solution of (1) and (φ, ψ) is the solution of (2). However, once we know φ, we can determine ψ by differentiation, so ψ depends entirely on φ. Therefore, though expressed differently, the solutions of (1) and (2) are still equivalent. As we will see below, working with the pair (φ, ψ) instead of φ alone has certain advantages.

The set of variables (x, y) is called the *phase plane*, and it is the natural extension of the *phase line* studied in Section 2.4. The phase line has two directions, whereas the phase plane has infinitely many. The *vector field* of (2) consists of all vectors $(y, f(x, y))$ based at (x, y) (see Figure 3.5.1). To draw the vector $(y, f(x, y))$ at (x, y), we sketch the vector $(y, f(x, y))$ at $(0, 0)$ and then translate it (parallel with itself) to (x, y). Alternatively, the *direction field* is similar to the vector field, except that all vectors have the same length. To draw the vector field or the direction field, we can use one of the programs in Section 3.7.

EXAMPLE 1   For the chosen mesh of points, the vector field of the second-order equation $x'' = -x$, or the system

$$\begin{cases} x' = y \\ y' = -x, \end{cases} \tag{3}$$

looks as in Figure 3.5.2(*a*). Indeed, the vector based at $(2, 0)$ is obtained by translating the vector $(y, f(x, y)) = (y, -x) = (0, -2)$ (which has length $\sqrt{y^2 + f^2(x, y)} = \sqrt{0^2 + (-2)^2} = 2$) to $(2, 0)$. Analogously, the vector based at $(-1, -1)$ is obtained by translating the vector $(-1, 1)$ to $(-1, -1)$. To avoid overlaps and have a clearer picture, we have proportionally rescaled all vectors.

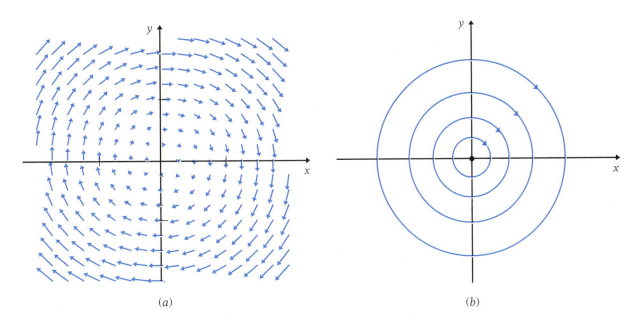

Figure 3.5.2. (*a*) The vector field and (*b*) the flow of the equation $x'' = -x$.

Remark. Notice the difference between vector or direction fields for second-order equations (which involve the dependent variables x and y) and slope fields for first-order equations (which involve the dependent variable x and the independent variable t). The slope field of a second-order equation is three-dimensional: It inhabits the space of the variables (t, x, y). We will not use it here.

The Flow Imagine a raft floating on the Amazon. The river is huge and the raft is small, so an abstract analogue of this picture is a point (x, y) moving in the phase plane. If the raft is initially released from the point with

The Amazon River.

coordinates (x_0, y_0), it will describe a path given by a curve $(x(t), y(t))$. The raft's velocity at every time can be represented by the vectors $(x'(t), y'(t))$, tangent to the curve.

Now suppose that only the velocity vectors are given, i.e., that we know the expression of x' and y' for every point (x, y) of the curve. Solving the initial value problem

$$\begin{cases} x' = y \\ y' = f(x, y), \end{cases} \qquad x(t_0) = x_0, \quad y(t_0) = y_0,$$

when t_0 is given, means determining the curve that is tangent to the given vectors. More generally, solving system (2) is equivalent to finding all possible curves that are tangent to the vectors of the vector field. These curves form the *flow*.

For example, some curves belonging to the flow of system (3) are depicted in Figure 3.5.2(b). Comparing this flow with the vector field in Figure 3.5.2(a), we can see that every circle is tangent to the corresponding vectors.

To understand what these circles represent, let us also solve system (3) with the method of Section 3.2. This leads us to the general solution

$$x(t) = c_1 \cos t + c_2 \sin t,$$

where c_1 and c_2 are real constants. Differentiating, we obtain

$$y(t) = x'(t) = -c_1 \sin t + c_2 \cos t.$$

Squaring each equality and then adding them yields

$$x^2 + y^2 = c_1^2 + c_2^2.$$

For each choice of c_1 and c_2, this equation represents a circle, except for $c_1 = c_2 = 0$, where it is a point. But these are exactly the curves in Figure 3.5.2(b), which form the flow of system (3). So now we can see that the flow of a second-order equation depicts the parametric plot of x versus x' (or x versus y in system notation), for all possible values of the constants c_1 and c_2. In many cases we cannot obtain the general solution, but we can draw the flow and thus gain qualitative information about the solution.

We have drawn the arrowheads along each curve in Figure 3.5.2(b) to show how x and y change when t varies. The direction of the arrow is that of increasing t. The independent variable t, however, is not represented in the picture. For example, the particular solution

$$x(t) = \cos t + \sin t, \qquad y(t) = x'(t) = -\sin t + \cos t,$$

which corresponds to the circle $x^2 + y^2 = 2$, is for $t = 0$ at the point $(x(0), y(0)) = (1, 1)$ and for $t = \pi/2$ at the point $(x(\pi/2), y(\pi/2)) = (1, -1)$ (see Figure 3.5.3).

Also, notice that any point on the curve can be taken as an initial condition at some t_0. For example, for $t_0 = 0$ we have $(x(0), y(0)) = (1, 1)$, whereas for $t_0 = \pi/2$ we obtain $(x(\pi/2), y(\pi/2)) = (1, -1)$. Indeed, solving the initial value problems

$$\begin{cases} x' = y \\ y' = -x, \end{cases} \qquad x(0) = 1, \ y(0) = 1$$

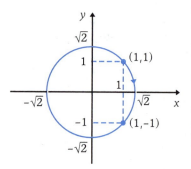

Figure 3.5.3. The arrow along a curve of the flow shows the direction of increasing t. The independent variable t, however, is not represented in the phase plane as a separate dimension. Therefore the flow must be imagined as "flowing" in time.

or

$$\begin{cases} x' = y \\ y' = -x, \end{cases} \quad x\left(\frac{\pi}{2}\right) = 1, \ y\left(\frac{\pi}{2}\right) = -1,$$

we obtain the same solution,

$$x(t) = \cos t + \sin t, \qquad y(t) = -\sin t + \cos t.$$

Equilibrium Solutions   In the phase plane, the equilibrium solutions of second-order equations are represented by fixed, unmoving points. They are curves that reduce to a point. In terms of the Amazon metaphor, the center of a whirlpool, for example, is a fixed point. For system (3) the point $(0, 0)$ is the equilibrium solution $x(t) = 0$, $y(t) = 0$ (see Figure 3.5.2(b)).

DEFINITION 3.5.1

An *equilibrium solution* of equation (1) is a constant function x that cancels the vector field f.

Remark 1. In terms of system (2), every equilibrium solution is of the form

$$(x(t), y(t)) = (\text{constant}, 0).$$

This is indeed so since $x'(t) = 0$ and $y(t) = x'(t)$, so $f(x(t), y(t)) = 0$.

Remark 2. Since $y = 0$ for every equilibrium solution of (2), equilibria of second-order autonomous equations belong to the x-axis.

Figure 3.5.4. The equilibrium solutions of the equation $x'' = x' + x^2 - x$.

EXAMPLE 2   To find the equilibrium solutions of the equation

$$x'' = x' + x^2 - x, \tag{4}$$

solve the algebraic equation $x^2 - x = 0$, which yields the solutions $x_1 = 0$ and $x_2 = 1$. The equilibrium solutions of (4) are thus the constant functions $\varphi_1(t) = 0$ and $\varphi_2(t) = 1$. In the phase plane they correspond to the points $(0, 0)$ and $(1, 0)$ (see Figure 3.5.4).

Linear Equations   We will study the equilibria of equation (1) when f is nonlinear in the more general context of Chapter 5. We now restrict our analysis to the linear case and thus consider only the autonomous second-order equation of the form

$$x'' + bx' + cx = 0, \tag{5}$$

where b and c are real constants, which is equivalent to the system

$$\begin{cases} x' = y \\ y' = -by - cx. \end{cases} \tag{6}$$

Definition 3.5.1 implies that if $c \neq 0$, equation (6) has exactly one equilibrium: the zero function, which corresponds to the origin of the coordinate system in the phase plane (see also Problems 17–20). It turns out that the behavior of the solutions near this equilibrium depends on the sign of the solutions r_1 and r_2 of the characteristic equation

$$r^2 + br + c = 0. \tag{7}$$

Indeed, the solutions x_1, x_2 of (5) can take the form (3), (4), or (5) in Section 3.2. If $r_1, r_2 < 0$ (or if their real part, α, is negative), then $x_1(t), x_2(t) \to 0$ when $t \to \infty$. Therefore $x(t) = c_1 x_1(t) + c_2 x_2(t)$ and $y(t) = c_1 x_1'(t) + c_2 x_2'(t)$ will also tend to 0 when $t \to \infty$. Examples of such flows appear in Figures 3.5.5(a) and 3.5.5(b).

If $r_1, r_2 > 0$ (or if their real part, α, is positive), then $x_1(t), x_2(t) \to \infty$ when $t \to \infty$. Then $x(t) = c_1 x_1(t) + c_2 x_2(t)$ and $y(t) = c_1 x_1'(t) + c_2 x_2'(t)$ will also tend to infinity. Examples of such flows appear in Figures 3.5.5(c) and 3.5.5(d).

If r_1, r_2 are real and $r_1 < 0$, $r_2 > 0$ (or vice versa), then $x_1(t) \to 0$ and $x_2(t) \to \infty$ (or vice versa). In this case the behavior of $x(t) = c_1 x_1(t) + c_2 x_2(t)$ and $y(t) = c_1 x_1'(t) + c_2 x_2'(t)$ depends on the values of c_1 and c_2. Depending on this choice, x and y can tend to zero or to infinity. An example of such a flow is given in Figure 3.5.5(e).

If r_1, r_2 are complex and the real part is zero (i.e., $r_1 = \beta i, r_2 = -\beta i$), then $x_1(t) = \cos \beta t$, $x_2(t) = \sin \beta t$ are periodic functions. Then $x(t) = c_1 \cos \beta t + c_2 \sin \beta t$ and $y(t) = -c_1 \beta \sin \beta t + c_2 \beta \cos \beta t$ are also periodic. An example of such a flow is given in Figure 3.5.5(f). See also Example 1 and the flow in Figure 3.5.2(b), which is of the same type.

The case when r_1 or r_2 is 0 is left as an exercise (see Problem 31). The above discussion suggests the following definition.

DEFINITION 3.5.2

Consider equation (5) with $c \neq 0$ and the corresponding characteristic equation (7), with solutions r_1 and r_2. Then the equilibrium 0 is called a

(i)   *source*, if r_1, r_2 are real and $r_1, r_2 > 0$ (*spiral source* if $r_{1,2} = \alpha \pm \beta i$ and $\alpha > 0$),

(ii)   *sink*, if r_1, r_2 are real and $r_1, r_2 < 0$ (*spiral sink* if $r_{1,2} = \alpha \pm \beta i$ and $\alpha < 0$),

(iii) *saddle*, if r_1, r_2 are real and $r_1 < 0, r_2 > 0$ or vice versa,

(iv)   *center*, if $r_{1,2} = \pm \beta i$.

The flow of system (6) near the equilibrium can be also expressed in terms of the coefficients b and c, as the next result shows. The proof follows directly from the Definition 3.5.2 and is therefore left as an exercise (see Problem 40).

Theorem 3.5.1

Assume $c \neq 0$. Then the equilibrium $(0, 0)$ of the system

$$\begin{cases} x' = y \\ y' = -by - cx \end{cases}$$

is a

  (i) *source* if and only if $b < 0$ and $c > 0$; if additionally $b^2 - 4c < 0$, then the equilibrium is a *spiral source*,
  (ii) *sink* if and only if $b > 0$ and $c > 0$; if additionally $b^2 - 4c < 0$, then the equilibrium is a *spiral sink*,
  (iii) *saddle* if and only if $c < 0$,
  (iv) *center* if and only if $b = 0$ and $c > 0$.

EXAMPLE 3   The direction fields and the flows of the systems

(a) $\begin{cases} x' = y \\ y' = -3y - x, \end{cases}$ (b) $\begin{cases} x' = y \\ y' = -2y - 2x, \end{cases}$ (c) $\begin{cases} x' = y \\ y' = 4y - x, \end{cases}$

(d) $\begin{cases} x' = y \\ y' = y - 2x, \end{cases}$ (e) $\begin{cases} x' = y \\ y' = 2y + x, \end{cases}$ (f) $\begin{cases} x' = y \\ y' = -x \end{cases}$

are drawn in Figures 3.5.5(a) through (f), respectively.

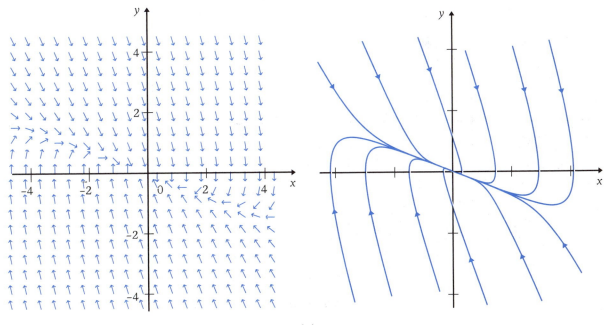

(a)

Figure 3.5.5. The direction fields of the cases (a) through (f) in Example 3 and the corresponding flows near the equilibria: (a) sink, (b) spiral sink, (c) source, (d) spiral source, (e) saddle, (f) center.

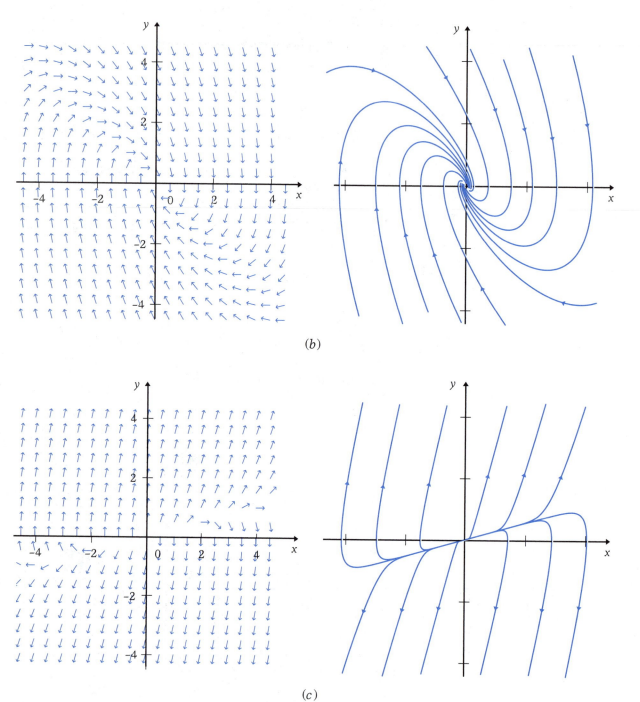

(b)

(c)

Figure 3.5.5. (*continued*)

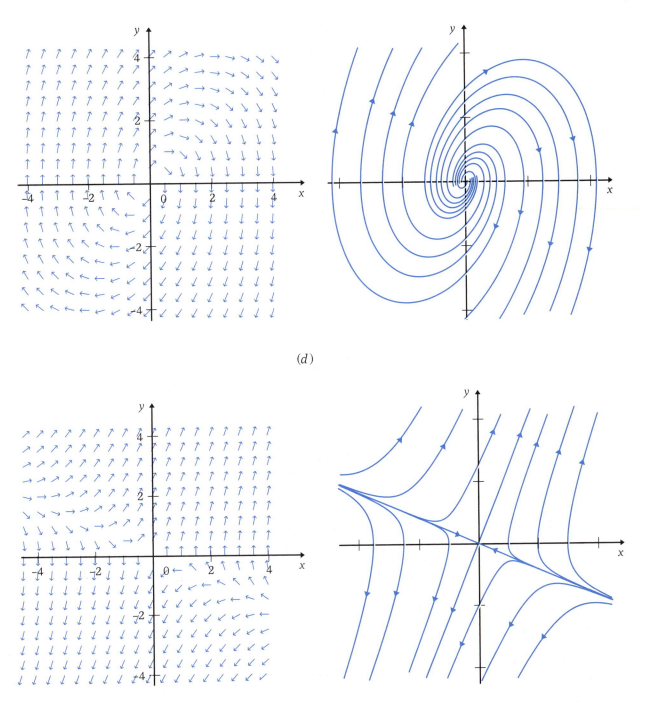

(d)

(e)

Figure 3.5.5. (continued)

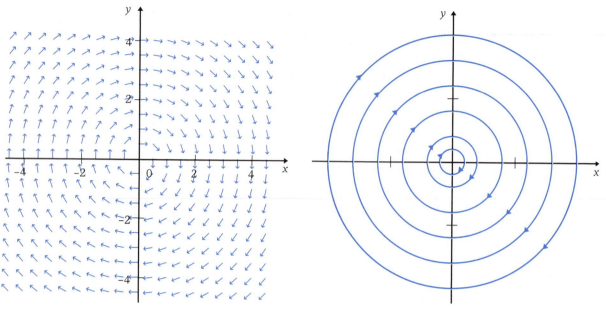

(f)

Figure 3.5.5. (*continued*)

The diagram in Figure 3.5.6 shows the distribution of the above types of equilibria in the plane of coordinates (b, c), except for the case $c = 0$, which is left as an exercise (see Problem 31). This is called the *parameter plane* and is different from the phase plane. Every point of the parameter plane corresponds to an equation of the form (5) (or a system of the form (6)). The correspondence is one to one.

Structural Stability and Bifurcation

Mathematical models are only approximations of reality. Solving a differential equation may not suffice for understanding the phenomenon it describes; we might need additional information. For example, if observations or experiments suggest as a model of a physical phenomenon a

Figure 3.5.6. The distribution of the types of equilibria with respect to b and c represented in the parameter plane. The parabola is given by the equation $c = b^2/4$.

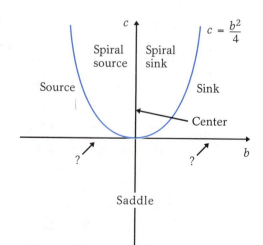

certain equation, say

$$x'' + 2.1x' + 0.76x = 0, \tag{8}$$

then we might need to determine the behavior of the solutions for all equations of the type (5), $x'' + bx' + cx = 0$, where b and c belong to some intervals that contain the coefficients in (8), say b in $(1.9, 2.3)$ and c in $(0.6, 0.9)$. Since all measurements are approximate, the value 2.1 could in truth be 2.05 or 2.17. If the solutions of all equations given by (5), for b and c in the chosen intervals, behave like the solution of (8), then the results are trustworthy because the qualitative structure of the phase-plane portrait remains the same. We then say that (5) is *structurally stable*. More precisely:

DEFINITION 3.5.3

System (6) is *structurally stable* around the values b_0 and c_0 if the nature of the equilibrium remains unchanged (i.e., a source, a sink, a saddle, or a center) for all b and c near b_0 and c_0, respectively. At the values b_0 or c_0 at which the nature of the equilibrium changes, we say that the solutions encounter a *bifurcation*.

For example, in the regions of the parameter plane (b, c) in which the equilibrium is a sink, the equation is structurally stable. At the points (b, c) at the border between sinks and saddles, we have a bifurcation. Figure 3.5.6 now suggests the following conclusion.

Theorem 3.5.2

System (6) is structurally stable in the whole parameter plane except for the b-axis and the positive part of the c-axis.

This means that models involving linear second-order equations that have coefficients belonging to the interior of structurally stable regions of the parameter plane are robust and can be trusted. Models whose coefficients come close to the bifurcation lines must be regarded more carefully. Depending on the nature of the phenomena they describe, such models may be misleading.

Applications

Electric circuits   Consider an electric circuit (called an *RLC* circuit) as in Figure 3.5.7, which has a resistor, an inductor, and a capacitor all connected in series. The circuit is characterized by q, the total charge in the plates of the capacitor and by i, the intensity of the current, which are connected by the relation $q' = i$. According to *Ohm's law*, the resistance of the resistor is $R = u_R/i$, where u_R is the potential difference across the resistor. The capacitance is defined as $C = q/u_C$, where u_C is the potential difference across the capacitor. A consequence of *Faraday's law* is that $L = u_L/i'$,

Figure 3.5.7. A simple series *RLC* electric circuit.

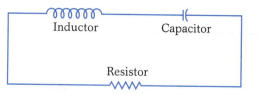

where L denotes the inductance of the inductor and u_L is the potential difference across the inductor. By *Kirchhoff's law*,

$$u_C + u_R + u_L = 0.$$

Let us find out whether or not the initial charge in the plates remains constant in time.

Solution. From Kirchhoff's law and the above expressions for the potential differences, we have $0 = u_C + u_R + u_L = Li' + Ri + q/C = Lq'' + Rq' + (1/C)q$. Thus the equation that describes the charge in plates is

$$q'' + \frac{R}{L}q' + \frac{1}{LC}q = 0.$$

If compared to (5), this equation gives $b = R/L > 0$ and $c = 1/(LC) > 0$, so according to Theorem 3.5.1 the equilibrium solution 0 is a sink. This means that no matter what initial positive value it has, the charge in plates will eventually tend to 0, so it cannot be constant in time.●

Remark. There is a close resemblance between *RLC* circuits and spring-mass systems. The capacitor is the analogue of the spring, the inductor that of the mass, and the resistor that of the resistance force. Moreover, if we add a *generator* to the circuit, Kirchhoff's law becomes $u_G = u_C + u_R + u_L$, where u_G is the potential of the generator, which in general is a periodic function of time. Therefore all the features of the spring-mass system, including beats and resonance, appear for electric circuits too.

Some models revisited

Recall the model describing the motion of water in a pipe studied in Section 3.2. If all frictions are neglected, the equation of motion is

$$x'' = -\frac{2g}{L}x,$$

where g and L are positive constants. In the phase plane the equilibrium corresponds to a center, so the solutions around it are periodic.

Viewing this model through the light of structural stability, we see that it corresponds to the point $(b, c) = (0, 2g/L)$ of the parameter plane. This point is on the bifurcation line, so this model may be deceptive. In Section 3.2 we took friction into consideration and obtained a model that corresponds to a point (b, c) of the parameter plane with $b > 0$ and $c = 2g/L > 0$. The equilibrium is therefore a sink to which all solutions tend. Physically, every motion in the pipe comes eventually to rest. So the model in which frictions are taken into consideration is structurally stable, whereas the one in which friction is neglected is sensitive to changes in the parameters.

The linear pendulum is in the same situation. The model without resistance forces leads to periodic solutions, the equilibrium is a center, and the coefficients correspond to the bifurcation line in the parameter plane. The model with damping corresponds to a sink. There is, however, a difference between this phenomenon and the motion of water in the pipe. The model of the pendulum is essentially nonlinear; we have considered linearity as an approximation for small oscillations only. So the pendulum fits better in the context of structural stability in the class of nonlinear equations and will be studied in detail in Chapter 5.

PROBLEMS

For the equations in Problems 1 through 8, sketch the vector field by drawing a few direction vectors around the origin. Use, if you wish, one of the programs in Section 3.7.

1. $x'' = 2x' + x$
2. $y'' = 3y' - 5y$
3. $u'' = -u' + \frac{1}{2}u$
4. $v'' = -v' - \sqrt{2}v$
5. $w'' = 2w' - 3w^2$
6. $z'' = (z')^3 + \frac{2}{3}z^2$
7. $\theta'' = \sin(\theta') + 2\theta$
8. $\phi'' = \phi' - \cos\phi$

Find the equilibrium solutions of the equations in Problems 9 through 16.

9. $x'' = x' + x^3 - x$
10. $\rho'' = 3\rho' + \rho^4 - 2\rho^3 + \rho^2$
11. $z'' = -z' + \sin z - 1$
12. $r'' = -r' + \cos r + r$
13. $y'' = \pi y' - 32y^5 - 1$
14. $u'' = (u')^3 + \frac{1}{3}$
15. $v'' = \sin(v') + b - \cos v$
16. $p'' = 2(p')^2 - ap^2 + bp + c$

The change of variable $X = x + d/c$, which is a shift in the plane, reduces the linear second-order nonhomogeneous equation $x'' + bx' + cx + d = 0$ to the linear second-order homogeneous equation $X' + bX' + cX = 0$. For the equations in Problems 17 through 20, determine the equilibria and then transform the equations into linear homogeneous ones. What are the equilibria of the new equations?

17. $x'' = x' - 3x + 5$
18. $x'' = 2x' + x - 4.5$

19. $x'' = -x' + 2x + \dfrac{\sqrt{13} - 1}{2}$
20. $x'' = 4x' - 3x - \frac{4}{3}$

For the linear equations in Problems 21 through 30, find the equilibria and determine their nature. Draw a few vectors of the vector or direction field and then sketch the phase-plane portrait. For drawing the vector or direction field, use, if you wish, one of the programs in Section 3.7.

21. $x'' = -2x' + x$
22. $y'' = -y' - 2y$
23. $z'' = 3z' - z$
24. $X'' = X' - 3X$
25. $u'' = u' + \frac{1}{2}u$
26. $\theta'' = 4\theta$
27. $\rho'' = -3\rho$
28. $\sigma'' = 2\sigma' + \sigma + 1$
29. $Y'' = Y' - 3Y - 2$
30. $r'' = -r' + \frac{1}{2}r - \frac{2}{3}$

31. Find the general solutions and then draw and discuss the phase-plane portraits of the family of equations $x'' + bx' = 0$, where b can take any real value.

32. Is it true that linear models for which the equilibrium is a center are sensitive to the change of parameters? Explain your answer. What about a saddle? Do you have a unique answer in this case?

33. Sketch the phase-plane portraits for the equations describing the motion of a spring-mass system in the case of undamped and then of damped oscillations. Discuss all possibilities and compare your results with those obtained for the *RLC* circuit.

34. Sketch the phase-plane portrait for the equation describing the motion (with friction) of water in a pipe. Discuss all possibilities.

35. Sketch the phase-plane portraits for the equations describing the motion without friction and then with friction for the linear pendulum. Discuss all possibilities.

36. Consider the problem of the bar sliding in a magnetic field (Section 3.3), but assuming that the resistance force is constant throughout the motion. Find the equation of motion and sketch the phase-plane portrait. Discuss all possibilities.

37. Write the equation that describes the motion of a cubic box floating in water (Problem

39 in Section 3.2). Determine the nature of the equilibrium. Is this model structurally stable?

38. Write the equation that describes the motion of the physical pendulum (Problem 40 in Section 3.2). Determine the nature of the equilibrium. Discuss all possibilities. Is this model structurally stable?

39. Draw the direction field for the model of the Tacoma Narrows Bridge (equation (7) in Section 3.4). What happens with the direction vectors at the x-axis? Why does this happen? Is this model structurally stable?

40. Prove Theorem 3.5.1.

3.6 Numerical Methods

As we saw for the Tacoma Narrows Bridge example, even a slight change away from linearity leads to theoretical difficulties. To get some understanding of the equation of motion, Glover, Lazer, and McKenna used numerical methods. In this section we will see how such methods apply to initial value problems of the form

$$x'' = f(t, x, x'), \qquad x(t_0) = x_0, \ x'(t_0) = x_0' \qquad (1)$$

and then use them to understand the motion of a bungee jumper and that of a cantilever beam with forced vibrations.

As in the previous section, we transform the second-order equation in (1) into a system of two first-order equations. With the new variable $y = x'$, the initial value problem (1) becomes

$$\begin{cases} x' = y \\ y' = f(t, x, y), \end{cases} \qquad x(t_0) = x_0, \ y(t_0) = y_0, \qquad (2)$$

where $y_0 = x_0'$. It will turn out that this trick, which helped us implement the qualitative methods of Section 3.5, allows a straightforward generalization of the numerical solvers introduced in Section 2.6. Let us see how this works.

Euler's Method

We can apply the method of Section 2.6 to each of the equations in (2). Taking in the interval $[t_0, a]$ the points $t_0 < t_1 < \cdots < t_n = a$ at distance $h = (a - t_0)/n$ of each other, Euler's formula translates into

$$\begin{cases} x_{n+1} = x_n + hy_n \\ y_{n+1} = y_n + hf(t_n, x_n, y_n). \end{cases} \qquad (3)$$

The steps of the method are the same as in Section 2.6, except that we have to work with two equations simultaneously. The second equation has a supporting role. Computing y_n is necessary for obtaining x_{n+1}.

EXAMPLE 1 Let us use Euler's method to solve the initial value problem

$$x'' = -\frac{2}{t}x' + \frac{2}{t^2}x, \qquad x(1) = 1, \ x'(1) = 2, \qquad (4)$$

numerically in the interval $[1, 2.5]$ and with step size $h = 0.1$.

From Example 5 of Section 3.2, the above equation has the general solution $x(t) = c_1 t + c_2/t^2$, where c_1 and c_2 are constants. We will use the solution $\varphi(t) = (4t^3 - 1)/(3t^2)$ of (4) for comparison. Let us now apply Euler's method to the equivalent initial value problem

$$\begin{cases} x' = y \\ y' = -\dfrac{2}{t}y + \dfrac{2}{t^2}x, \end{cases} \qquad x(1) = 1, \ y(1) = 2. \qquad (5)$$

Step 1. From the above we see that

$$f(t, x, y) = -\frac{2}{t}y + \frac{2}{t^2}x$$

$t_0 = 1$, $a = 2.5$, $n = 15$, $t_{n+1} = t_n + h$, $h = 0.1$, $x_0 = 1$, and $y_0 = 2$.

Step 2. $x_1 = x_0 + hy_0 = 1 + 0.1 \cdot 2 = 1.2$
$y_1 = y_0 + hf(t_0, x_0, y_0) = 2 + 0.1 \cdot f(1, 1, 2) = 2 + 0.1(-2) = 1.8$

Step 3. $x_2 = x_1 + hy_1 = 1.2 + 0.1 \cdot 1.8 = 1.38$
$y_2 = y_1 + hf(t_1, x_1, y_1) = 1.8 + 0.1 \cdot f(1.1, 1.2, 1.8) = 1.8 + 0.1$
$(-1.2892562) = 1.6710744$, computed to seven decimal places.

Step 4. Analogously we obtain $x_3, \dots, x_{15}$. The results are summarized in Table 3.6.1 together with an approximation of the exact values to seven decimal places.

We can now compare the values obtained for x_n and the corresponding values of $\varphi(t_n)$.

TABLE 3.6.1

n	t_n	x_n	y_n	$\varphi(t_n)$
0	1.0	1.0000000	2.0000000	1.0000000
1	1.1	1.2000000	1.8000000	1.1911846
2	1.2	1.3800000	1.6710744	1.3685185
3	1.3	1.5471074	1.5842287	1.5360947
4	1.4	1.7055303	1.5235908	1.6965986
5	1.5	1.8578894	1.4799687	1.8518519
6	1.6	2.0058863	1.4477852	2.0031250
7	1.7	2.1506648	1.4235220	2.1513264
8	1.8	2.2930170	1.4048837	2.2971193
9	1.9	2.4335053	1.3903298	2.4409972
10	2.0	2.5725383	1.3787995	2.5833333
11	2.1	2.7104183	1.3695465	2.7244142
12	2.2	2.8473729	1.3620349	2.8644628
13	2.3	2.9835764	1.3558736	3.0036547
14	2.4	3.1191638	1.3507722	3.1421296
15	2.5	3.2542410	1.3465121	3.2800000

The Second-Order Runge-Kutta Method

The corresponding Heun formula that allows us to solve problem (2) using the second-order Runge-Kutta method is

$$\begin{cases} x_{n+1} = x_n + (h/2)(y_n + y_{n+1}^E) \\ y_{n+1} = y_n + (h/2)[f(t_n, x_n, y_n) + f(t_{n+1}, x_{n+1}^E, y_{n+1}^E)]. \end{cases} \tag{6}$$

As in Section 2.5, we first compute x_{n+1}^E, y_{n+1}^E using Euler's formula (3) and then find improved values x_{n+1}, y_{n+1} using (6).

EXAMPLE 2   To solve the initial value problem (5), proceed at Step 1 and Step 2 as in Example 1, where Euler's formula yields $x_1^E = 1.2$ and $y_1^E = 1.8$.

Step 3. With Heun's formula (6), the improved values of x_1 and y_1 are:

$$x_1 = x_0 + (h/2)(y_0 + y_1^E) = 1 + 0.05(2 + 1.8) = 1.19$$

$$\begin{aligned} y_1 = y_0 + (h/2)[f(t_0, x_0, y_0) + f(t_1, x_1^E, y_1^E)] &= 2 + 0.05[f(1, 1, 2) \\ + f(1.1, 1.2, 1.8)] &= 2 + 0.05(-2 - 1.2892562) = 1.8355372 \end{aligned}$$

Step 4. Compute x_2^E and y_2^E using Euler's formula (3):

$$x_2^E = x_1 + hy_1 = 1.19 + 0.1 \cdot 1.8355372 = 1.3735537$$

$$\begin{aligned} y_2^E = y_1 + hf(t_1, x_1, y_1) &= 1.8355372 + 0.1 \cdot f(1.1, 1.19, 1.8355372) \\ &= 1.8355372 + 0.1 \cdot (-1.3703983) = 1.6984974 \end{aligned}$$

Step 5. Compute x_2 and y_2 using Heun's formula (6):

$$\begin{aligned} x_2 = x_1 + (h/2)(y_1 + y_2^E) \\ = 1.19 + 0.05(1.8355372 + 1.6984974) = 1.3667017 \end{aligned}$$

$$\begin{aligned} y_2 = y_1 + (h/2)[f(t_1, x_1, y_1) + f(t_2, x_2^E, y_2^E)] &= 1.8355372 \\ + 0.05 \cdot [f(1.1, 1.19, 1.8355372) + f(1.2, 1.3735537, 1.6984974)] \\ = 1.8355372 + 0.05 \cdot (-1.3703983 - 0.9231155) = 1.7208615 \end{aligned}$$

TABLE 3.6.2

n	t_n	x_n	y_n	$\varphi(t_n)$
0	1.0	1.0000000	2.0000000	1.0000000
1	1.1	1.1900000	1.8355372	1.1911846
2	1.2	1.3667017	1.7208616	1.3685185
3	1.3	1.5339384	1.6385056	1.5360947
4	1.4	1.6942616	1.5778571	1.6965986
5	1.5	1.8494211	1.5322193	1.8518519
6	1.6	2.0006479	1.4972263	2.0031250
7	1.7	2.1488279	1.4699500	2.1513264
8	1.8	2.2946115	1.4483771	2.2971193
9	1.9	2.4384848	1.4310926	2.4409972
10	2.0	2.5808168	1.4170825	2.5833333
11	2.1	2.7218917	1.4056071	2.7244142
12	2.2	2.8619311	1.3961186	2.8644628
13	2.3	3.0011100	1.3882050	3.0036547
14	2.4	3.1395680	1.3815529	3.1421296
15	2.5	3.2774175	1.3759207	3.2800000

Step 6. Analogously we obtain $x_3, \ldots, x_{15}$. The results are summarized in Table 3.6.2 together with an approximation of the exact values to seven decimal places.

We can now compare the values obtained for x_n and the corresponding values of $\varphi(t_n)$. Also, we can see that these values of x_n are closer to the exact solution than the ones obtained in Table 3.6.1 with the help of Euler's method.

Applications

Bungee jumping[1]

A bungee jumper.

Jill, a young engineer with entrepreneurial skills, opens a bungee-jumping business. One of the problems she faces is the choice of cords for people of different weights. She wants to attempt the jump herself and thinks of a way for finding the right cord. Jill knows that an elastic cord acts like a spring, so she uses as a model the damped harmonic oscillator

$$x'' = -\frac{\nu}{m}x' - \frac{k}{m}x.$$

However, the resistance force of the air is not reflected in the equation, and this is an aspect she cannot neglect. She remembers from physics that for large objects, like the human body, this force opposes the direction of motion and is roughly $\rho S v^2/2$, where v is the velocity, $\rho = 1 \text{ kg/m}^3$ is the density of air, and S the area of the transversal section, measured in square meters. Taking into account the internal-friction coefficient of the cord, she finds that a good approximation of the resistance force is $(\nu/m)(x')^2$, where $\nu = 0.1$ and m is the mass in kilograms. But the resistance force changes sign with the direction of motion, so to add this feature to the model she writes the equation as

$$x'' = -\frac{\nu}{m}x'|x'| - \frac{k}{m}x.$$

Jill's hope to find a neat solution formula vanishes quickly. The equation is not linear, so the exact theory of our previous sections is of little use. Her only chance is to try some numerical procedure. She decides to apply the second-order Runge-Kutta method and uses a computer program similar to the ones in Section 3.7.

Jill is now at the stage of computer experimentation using precise data. She weighs 50 kg, $\nu = 0.1$, she can use cords with damping coefficients k from 20 to 50 N/m (newtons/meter), and she will jump from 20 m above the equilibrium position (i.e., the point she will reach when the oscillations come to an end, which she can determine by finding the elongation L from the formula $k = mg/L$). This means that the initial conditions are $x(0) = -20$ (the coordinate system is oriented downward, since we consider downward velocities as positive) and $x'(0) = 0$, since Jill will dive with zero initial velocity.

---

[1]This is only a "toy" model that has not been tested in practice. **Do not** use it for bungee-jumping purposes without further investigations.

TABLE 3.6.3

n	t_n	x_n	y_n
1	.1	−19.95000000	0.9999000000
2	.2	−19.80014500	1.994401389
3	.3	−19.55124428	2.978134678
22	2.2	−0.531569666	13.78063298
23	2.3	0.845923498	13.73481013
44	4.4	18.97909544	0.3829498805
45	4.5	18.96994122	−0.5669448920
66	6.6	1.021130869	−13.09796153
67	6.7	−0.289502545	−13.08191796
88	8.8	−18.04216191	−0.7277214650
89	8.9	−18.06982336	0.1762558475
111	11.1	−0.215104956	12.47961625
112	11.2	1.031837023	12.42807515
133	13.3	17.23869267	0.1778537455
134	13.4	17.21338100	−0.6844818880
155	15.5	0.674509833	−11.92121862
156	15.6	−0.517877149	−11.89670533
177	17.7	−16.46751779	−0.5000480025
178	17.8	−16.47635130	0.3245925550
199	19.9	−1.094145763	11.40117288
200	20.0	0.047407022	11.40131434

For different lengths of the chord, Jill obtains different numerical results. In the end she understands that the best choice is the 25-N/m cord. The initial value problem describing this choice is

$$x'' = -0.002x'|x'| - 0.5x, \qquad x(0) = -20, \ x'(0) = 0,$$

or, in the form in which the numerical method is applicable,

$$\begin{cases} x_{n+1} = x_n + 0.05(y_n + y_{n+1}^E) \\ y_{n+1} = y_n - 0.0001y_n|y_n| - 0.025x_n \\ \qquad - 0.0001y_{n+1}^E|y_{n+1}^E| - 0.025x_{n+1}^E, \end{cases} \qquad x_0 = -20, \ y_0 = 0.$$

Jill does 200 iterations with step size 0.1, which correspond to the first 20 seconds of motion. A selection of her results is shown in Table 3.6.3.

We can see that after 2.2 seconds Jill would be close to the origin of the coordinate system (her future equilibrium position) and have a downward speed of almost 14 m/s; after 4.4 seconds from start she would almost reach the lowest point, 19 m below the origin, near the surface of the river, with almost zero downward speed; then the cord pulls her up, she passes through the origin 6.6 seconds after start with an upward speed of 13 m/s, reaches a high point about 18 m above the origin, then falls again, and so on. These results fit her security measurements at the jumping bridge, so she decides to make her first practical experiment with the 25-N/m cord.

The graph and the phase-plane representation of the solution, obtained with one of the programs in Section 3.7, are given in Figure 3.6.1.

Figure 3.6.1. (*a*) The graph and (*b*) the phase-plane representation of Jill's bungee jump.

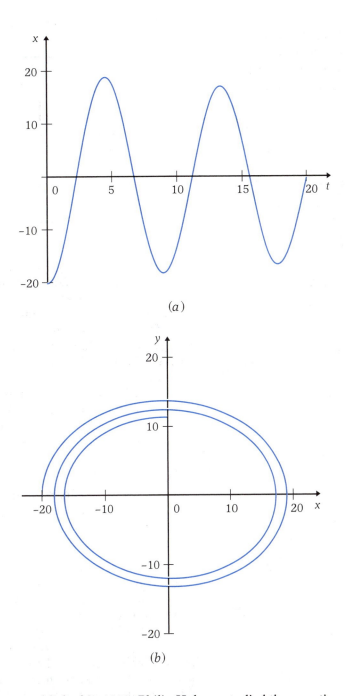

(*a*)

(*b*)

The cantilever beam   In a paper published in 1979, Philip Holmes studied the equation

$$x'' = -\delta x' + x - x^3 + \gamma \cos \omega t,$$

where δ, γ, and ω are positive constants. In two subsequent joint papers with Francis Moon, Holmes showed that this was the simplest possible model for the forced vibrations of a cantilever beam in the nonuniform field of two permanent magnets, when the whole system is shaken periodically by an electromagnetic vibrator, as shown in Figure 3.6.2. The linear velocity-dependent term models the dissipation due to friction, viscous damping from the surrounding air, and magnetic damping. The only difference from the harmonic oscillator is the cubic term,

Figure 3.6.2. The magneto-elastic cantilever beam, built and studied experimentally and theoretically by Holmes and Moon.

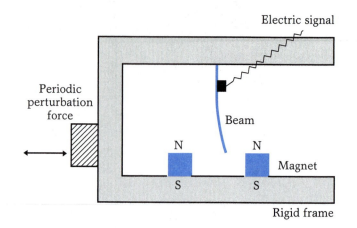

introduced (with positive instead of negative sign) in 1918 by the German engineer Georg Düffing to account for the hardening spring effect observed in many mechanical systems. This term, however, is enough to make any exact method fail.

Düffing's initial equation modeled oscillatory systems with springs that get stiffer as the displacement increases, such as the leaf springs on trucks, which are designed so that more leaves come into contact as the load increases.

Let us investigate numerically Holmes's version of Düffing's equation above in the interval $[0, 120]$ with step size 0.1, for constants $\delta = 0.25$, $\gamma = 0.3$, $\omega = 1$ and initial conditions $x(0) = 0.5, x'(0) = 0$. Using the second-order Runge-Kutta method, the numerical problem becomes

$$\begin{cases} x_{n+1} = x_n + 0.05(y_n + y_{n+1}^E) \\ y_{n+1} = y_n - 0.0125y_n + 0.05x_n - 0.05x_n^3 \\ \quad + 0.015\cos t_n - 0.0125y_{n+1}^E + 0.05x_{n+1}^E \\ \quad - 0.05(x_{n+1}^E)^3 + 0.015\cos t_{n+1}, \end{cases} \qquad x_0 = 0.5, \; y_0 = 0.$$

Since this means 1200 iterations, instead of printing all data we use one of the programs in Section 3.7 to draw the graph and the phase-plane representation of the numerical solution in the interval $[0, 120]$. They appear in Figure 3.6.3.

It is hard to see any pattern in the behavior of the solution, which does not appear to be periodic (unless the period is larger than 120). It is impossible to predict the beam's next move if the computations stop here. We do not know how long it will vibrate near one magnet and when it will move close to the other. If for this choice of the constants we try other initial conditions close to the above ones (and we suggest you do), we will obtain other irregular graphs, but which differ quite a bit from the one in Figure 3.6.3(a). This is surprising, because we would expect that close initial conditions lead to somewhat different but similar behavior. This phenomenon, called *chaos*, will be the topic of Section 5.5. We will see there that the equations whose solutions exhibit this kind of behavior are more common than we expect.

The errors of the methods are the same as the ones we discussed in Section 2.6. It is in fact interesting to remark that the field of numerical methods in differential equations has developed many numerical solvers,

Figure 3.6.3. (*a*) The graph and (*b*) the phase-plane representation of the numerical solution associated to the initial value problem with data $x(0) = 0.5$, $y(0) = 0$ for Duffing's equation with $\delta = 0.25$, $\gamma = 0.3$, $\omega = 1$.

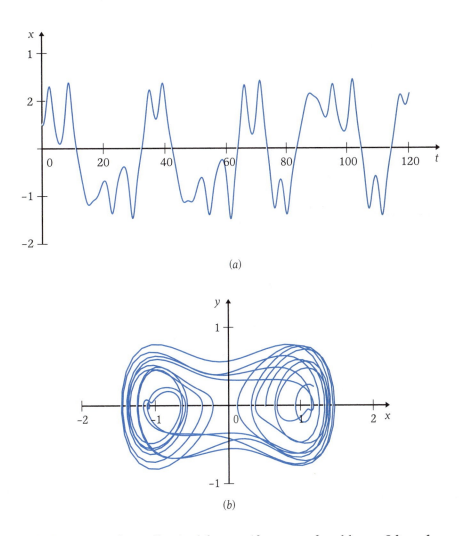

(*a*)

(*b*)

which prove to be well suited for specific types of problems. Often, for the sake of solving one particular equation, a new method is designed. In case of chaotic behavior, however, no numerical method remains accurate on long intervals of time. This is a difficulty that modern research in the field is trying to overcome.

PROBLEMS

Solve Problems 1 through 8 numerically using

   (a) *Euler's method with step size $h = 0.1$*
   (b) *Euler's method with step size $h = 0.05$*
   (c) *the second-order Runge-Kutta method with step size $h = 0.1$*
   (d) *the second-order Runge-Kutta method with step size $h = 0.05$*

to find approximate values of the solution for the first five iterations starting with $t = 0.1$. For each equation, compare the results at (a), (b), (c), and (d). If you wish, use one of the programs of Section 3.7.

1. $x'' = 7x' - 10x$, $\quad x(0) = 1$, $x'(0) = 2$

2. $u'' = 10u' - 25u$, $\quad u(0) = u'(0) = e$

3. $v'' = -7v' - 7v - 1$, $\quad v(0) = v'(0) = -2$

4. $x'' = -x' - x + 1$, $\quad x(0) = x'(0) = 1$

5. $r'' = \dfrac{\theta + 2}{\theta - 1}r' - \dfrac{\theta + 2}{\theta^2 - 1}r$,

$\quad r(0) = 1$, $r'(0) = 2$

6. $w'' = \dfrac{x}{x - 1}w' + \dfrac{w}{x - 1}$,

$\quad w(0) = 0$, $w'(0) = 1$

7. $y'' = -\dfrac{1}{t-2}y' - \dfrac{4t^2-1}{t^2-4}y + 2(1-t)$,

$y(0) = y'(0) = 1$

8. $y'' = \dfrac{x}{x-1}y' + \dfrac{y}{1-x}$, $y(0) = 0,\ y'(0) = 1$

9. In Example 2 at Step 5 we calculated $y_2 = 1.7208615$. For this we used a calculator that can accommodate numbers with a maximum of 8 figures. On the other hand, we obtained in Table 3.4.2 $y_2 = 1.7208616$. For this we used a calculator with a maximum of 10 figures and then rounded off the answer to an 8-digit number. Why did we obtain different results? Will there be any difference in the value of x_{15} obtained with each calculator? If so, what would be the maximum possible difference between the two results?

10. For the bungee-jumping problem, use one of the programs in Section 3.7 to see what cord makes an 80-kg man not exceed 18 m below the equilibrium in his jump, assuming he jumps from 20 m above the equilibrium, with 4 m/s initial velocity.

Find the numerical solutions of Problems 11 through 18 using one of the programs in Section 3.7. In each case use the corresponding program to draw the graph and the phase-space representation of the numerical solution.

11. $x'' = 11x' - 24x$, $x(0) = -1,\ x'(0) = 1$ on $[0, 10]$ with $h = 0.05$

12. $x'' = 16x' - 64x$, $x(0) = x'(0) = 0$ on $[0, 25]$ with $h = 0.1$

13. $x'' = x' - \frac{1}{4}$, $x(0) = 1,\ x'(0) = 2$ on $[0, 20]$ with $h = 0.1$

14. $x'' = \pi x' - \pi x$, $x(0) = x'(0) = \pi$ on $[0, 10]$ with $h = 0.05$

15. $x'' = tx' - x$, $x(1) = 0,\ x'(1) = 1$ on $[1, 10]$ with $h = 0.025$

16. $x'' = \dfrac{1}{t}x' + \dfrac{x}{t}$, $x(1) = 0,\ x'(1) = 1$ on $[1, 50]$ with $h = 0.05$

17. $x'' = -\dfrac{t}{t-2}x' + \dfrac{4t^2}{t^2-4}x + 2(1-t)$,

$x(3) = x'(3) = 1$ on $[3, 20]$ with $h = 0.04$

18. $x'' = \dfrac{t}{t-1}x' - \dfrac{x}{1-t}$, $x(2) = 0,\ x'(2) = 1$ on $[2, 100]$ with $h = 0.1$

19. Find the numerical solution for Holmes's version of Düffing's equation in the interval $[0, 200]$ with step size 0.1 for the following values of the constants:

  (a) $\delta = 0.3,\ \gamma = 0.33,\ \omega = 1$
  (b) $\delta = 0.4,\ \gamma = 0.5,\ \omega = 1.2$
  (c) $\delta = 0.5,\ \gamma = 0.1,\ \omega = 0.9$
  (d) $\delta = 0.1,\ \gamma = 0.2,\ \omega = 0.8$

20. Find the numerical solution of the *Van der Pol equation*,

$$x'' = -\alpha(x^2 - 1)x' - x + \beta \cos \omega t,$$
$$x(0) = 0.5,\ x'(0) = 0$$

in the interval $[0, 150]$ with step size 0.1 for the following values of the constants:

  (a) $\alpha = 1,\ \beta = 1,\ \omega = \pi$
  (b) $\alpha = 1.2,\ \beta = 0.5,\ \omega = 2\pi$
  (c) $\alpha = 0.9,\ \beta = 1.8,\ \omega = \pi/2$
  (d) $\alpha = 0.5,\ \beta = 2,\ \omega = \pi$

3.7 Computer Applications

In this appendix section we will show through examples how to use Maple, *Mathematica*, and MATLAB to obtain exact solutions of second-order equations, numerical solutions, graphs, and phase-plane representations with the help of the methods discussed in Section 3.6 and to draw vector and direction fields. All these are of help in understanding different aspects of the types of second-order equations treated in the previous sections of this chapter.

Maple

Exact Solutions   The command `dsolve`, encountered in Section 2.7, can also be used to obtain the solutions of some second-order equations. Maple solves all homogeneous equations with constant coefficients and some homogeneous equations with variable coefficients; also, it can apply the method of variation of parameters for nonhomogeneous equations once it can compute the solutions of the corresponding homogeneous equations.

EXAMPLE 1 To solve the homogeneous equation with constant coefficients

$$x'' = 3x' - 2x,$$

type at the prompt:

```
> dsolve(diff(x(t),t$2)=3*diff(x(t),t)-2*x(t),x(t));
```

and Maple will display the solution

```
x(t) = _C1 exp(2 t) + _C2 exp(t)
```

i.e., $x(t) = C_1 e^{2t} + C_2 e^t$.

EXAMPLE 2 Maple also solves initial value problems. If we attach to the equation in Example 1 the initial conditions $x(0) = 2$, $x'(0) = 1$, we can solve the corresponding initial value problem as follows:

```
> dsolve({diff(x(t),t$2)=3*diff(x(t),t)-2*x(t),
    x(0)=2, D(x)(0)=1},x(t));
```

and Maple displays the solution

```
x(t) = - exp(2 t) + 3 exp(t)
```

EXAMPLE 3 Since Maple knows the solution of the above equation, it can solve the nonhomogeneous equation

$$x'' = 3x' - 2x + t.$$

```
> dsolve(diff(x(t),t$2)=3*diff(x(t),t)-2*x(t)+t,x(t));
```

Maple displays the solution

```
x(t) = 1/2 t + 3/4 + _C1 exp(2 t) + _C2 exp(t)
```

EXAMPLE 4 The homogeneous equation with variable coefficients

$$x'' = -(1/t)x' + (4/t^2)x$$

can be solved by typing

```
> dsolve(diff(x(t),t$2)=-(1/t)*diff(x(t),t)+
    (4/t^2)*x(t),x(t));
```

Maple displays the solution

$$x(t) = \frac{\_C1 t^4 + \_C2}{t^2}$$

> **EXAMPLE 5**   The nonhomogeneous equation
>
> $$x'' = -(1/t)x' + (4/t^2)x + t^2$$
>
> can now be solved by typing
>
> ```
> > dsolve(diff(x(t),t$2)=-(1/t)*diff(x(t),t)+
>   (4/t^2)*x(t)+t^2,x(t));
> ```
>
> after which Maple displays the solution
>
> $$x(t) = 1/12 \frac{t^6 + 12\_C1t^4 + 12\_C2}{t^2}$$

Numerical solutions   With Maple's help we will first apply Euler's method to solve the initial value problem used in the examples of Section 3.6,

$$x'' = -(2/t)x' + (2/t^2)x, \qquad x(1) = 1, x'(1) = 2, \tag{1}$$

in the interval $[1, 2.5]$, with step size $h = 0.1$. Then we will solve the same initial value problem using the second-order Runge-Kutta method. In each case we write a program, explain how it works, and outline the logic behind it.

Euler's method

```
> f := (t, x, y) → -(2/t) * y + (2/t^2) * x;
```

$$f := (t, x, y) \rightarrow \quad -2\ y/t + 2\frac{x}{t^2}$$

```
> t0:=1: x0:=1: y0:=2:
> h:=0.1:
> n:=15:
> t:=t0: x:=x0: y:=y0:
> for i from 1 to n do
  u:=y:   v:=f(t,x,y):
  x:=x+h*u:   y:=y+h*v:
  t:=t+h:
  print(t,x,y);
  od:
```

and Maple displays the numerical result

```
        1.1, 1.2, 1.80
       1.2, 1.380, 1.671074380
     1.3, 1.547107438, 1.584228650
     1.4, 1.705530303, 1.523590803
     1.5, 1.857889383, 1.479968678
     1.6, 2.005886251, 1.447785244
     1.7, 2.150664775, 1.423521952
     1.8, 2.293016970, 1.404883714
     1.9, 2.433505341, 1.390329781
     2.0, 2.572538319, 1.378799518
     2.1, 2.710418271, 1.369546482
```

```
2.2, 2.847372919, 1.362034947
2.3, 2.983576414, 1.355873626
2.4, 3.119163777, 1.350772192
2.5, 3.254240996, 1.346512141
```

We first wrote the vector field, which Maple displays on the screen. Then we wrote the initial conditions $t_0 = 1, x_0 = 1, y_0 = 2$, the step size $h = 0.1$, the number of steps $n = 15$, and the starting point of the numerical procedure: $t = t_0, x = x_0, y = y_0$. With do, we asked Maple to apply Euler's formula (equations (3), Section 3.6) from $i = 1$ to $i = n$. For this we assigned the variables u to y and v to $f(t, x, y)$, because by directly applying Euler's formula, x:=x+h*y and y:=y+h*f(t,x,y), Maple computes a new value for x and uses it in the second equation. Introducing u and v, Maple assigns the same value of x in the right-hand side of each equation, as Euler's method requires.

Second-order Runge-Kutta method The only changes from the program used for Euler's method appear for do. Instead of using only equations (3), we now also need equations (6) of Section 3.6. We first apply Euler's formula to compute first approximations, denoted by u and v, and then use formula (6) to compute x and y.

```
> for i from 1 to n do
  a:=y:   b:=f(t,x,y):
  u:=x+h*a:   v:=y+h*b:
  x:=x+h*(a+v)/2:
  y:=y+h*(b+f(t+h,u,v))/2:
  t:=t+h:
  print(t,x,y);
  od:
```

Maple displays the numerical result

```
1.1, 1.190000000, 1.835537190
1.2, 1.366701728, 1.720861508
1.3, 1.533938351, 1.638505613
1.4, 1.694261581, 1.577857115
1.5, 1.849421076, 1.532219313
1.6, 2.000647861, 1.497226319
1.7, 2.148827859, 1.469949992
1.8, 2.294611484, 1.448377064
1.9, 2.438484785, 1.431092575
2.0, 2.580816782, 1.417082451
2.1, 2.721891657, 1.405607103
2.2, 2.861931090, 1.396118565
2.3, 3.001110034, 1.388205019
2.4, 3.139568038, 1.381552927
2.5, 3.277417499, 1.375920733
```

Graphs and phase-plane representations Instead of (or additionally to) writing the numerical solution, we can draw its graph as well as the phase-plane representation of $(x(t), y(t))$. We can obtain the graph of x with the following Runge-Kutta program, which varies slightly from the above one.

```
> f := (t, x, y) → - (2/t) * y + (2/t²) * x;
```

$$f := (t, x, y) \to -2\ y/t + 2\frac{x}{t^2}$$

```
> t0:=1: x0:=1: y0:=2:
> M0:=[t0,x0]:
> h:=0.1:
> n:=15:
> t:=t0: x:=x0: y:=y0: M:=M0:
> for i from 1 to n do
  a:=y:   b:=f(t,x,y):
  u:=x+h*a:   v:=y+h*b:
  x:=x+h*(a+v)/2:
  y:=y+h*(b+f(t+h,u,v))/2:
  M:=(M,[t,x]):
  t:=t+h:
  od:
  plot([M]);
```

and Maple displays the picture in Figure 3.7.1(*a*). If we want the graph
of *y*(*t*), then we need to change M0:=[t0,x0]: to M0:=[t0,y0]: and
M:=(M,[t,x]): to M:=(M,[t,y]):. If we would like to have the phase-

Figure 3.7.1. The graphs (*a*) of *x*, (*b*) of *y*, and (*c*) the phase-plane representation of the solution (*x*, *y*) for the initial value problem (1).

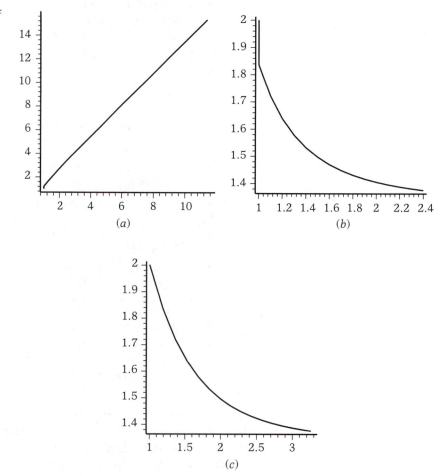

Figure 3.7.2. The direction field of system (3).

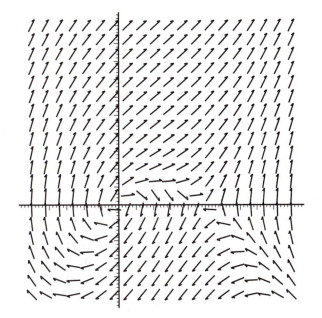

plane representation of the solution, then we must replace `M0:=[t0,x0]:` with `M0:=[x0,y0]:` and `M:=(M,[t,x]):` with `M:=(M,[x,y]):`.

Maple also has a built-in program that draws the graph of the solution. It is called `DEplot` and can be used within the package `DEtools`. For more details see Section 4.6 and the on-line help of Maple.

Direction fields To understand the phase-plane portrait of a second-order equation, it is often useful to draw the direction field. The package `DEtools` has for this purpose a special command called `dfieldplot`. For equation (4) of Section 3.5,

$$x'' = x' + x^2 - x, \tag{2}$$

the direction field can be drawn as follows. We write (2) as a system,

$$\begin{cases} x' = y \\ y' = y + x^2 - x, \end{cases} \tag{3}$$

and proceed as below. Then Maple produces the picture in Figure 3.7.2.

```
> with(DEtools):
> dfieldplot([diff(x(t),t)=y(t), diff(y(t),t)=y(t)+
  (x(t))^2-x(t)],[x(t),y(t)], t=-2..2, x=-1..2, y=-1..2);
```

Mathematica

Exact solutions The command `DSolve` can also be used to obtain the solutions of some second-order equations. *Mathematica* can solve all homogeneous equations with constant coefficients and some homogeneous equations with variable coefficients; also, in some cases it can apply the method of variation of parameters for nonhomogeneous equations once it can compute the solutions of the corresponding homogeneous equations.

EXAMPLE 1   To solve the homogeneous equation with constant coefficients

$$x'' = 3x' - 2x,$$

type at the prompt

    DSolve[x''[t]==3x'[t]-2x[t], x[t], t]

and *Mathematica* will display the solution

$$\{\{x[t] \rightarrow E^t 9C[1] + E^t C[2])\}\}$$

EXAMPLE 2   *Mathematica* also solves initial value problems. If we attach to the equation in Example 1 the initial conditions $x(0) = 2$, $x'(0) = 1$, we can solve the corresponding initial value problem as follows:

    DSolve[{x''[t]==3x'[t]-2x[t], x[0]==2, x'[0]==1},
    x[t], t]

Mathematica displays the solution

$$\{\{x[t] \rightarrow E^t(3 - E^t)\}\}$$

EXAMPLE 3   Since *Mathematica* knows the solution of the above equation, it can solve the nonhomogeneous equation

$$x'' = 3x' - 2x + t.$$

    DSolve[x''[t]==3x'[t]-2x[t]+t, x[t], t]

Mathematica displays the solution

$$\left\{\left\{x[t] \rightarrow \frac{3}{4} + \frac{t}{2} + E^t C[1] + E^{2t} C[2]\right\}\right\}$$

EXAMPLE 4   The homogeneous equation with variable coefficients

$$x'' = -(1/t)x' + (4/t^2)x$$

can be solved by typing

    DSolve[x''[t]==-(1/t)x'[t]+(4/t^2)x[t], x[t], t]

Mathematica displays the solution

$$\left\{\left\{x[t] \rightarrow \frac{4C[1] + t^4 C[2] - C[1] \, 4C[2]}{4t^2}\right\}\right\}$$

> **EXAMPLE 5** The nonhomogeneous equation
>
> $$x'' = -(1/t)x' + (4/t^2)x + t^2$$
>
> can be solved as follows:
>
> ```
> DSolve[x''[t]==-(1/t)x'[t]+(4/t^2)x[t]+t^2, x[t], t]
> ```
>
> *Mathematica* displays the solution
>
> $$\{\{x[t] \to \frac{t^4}{12} + \frac{C[1]}{t^2} + t^2 C[2]\}\}$$

Numerical solutions

With *Mathematica*'s help we will first apply Euler's method to solve the initial value problem used in the examples of Section 3.6,

$$x'' = -(2/t)x' + (2/t^2)x, \qquad x(1) = 1, x'(1) = 2, \qquad (4)$$

in the interval $[1, 2.5]$, with step size $h = 0.1$. Then we will solve the same initial value problem using the second-order Runge-Kutta method. In each case we write a program, explain how it works, and outline the logic behind it.

Euler's method

```
f[t_,x_,y_]:=-(2/t)*y+(2/t^2)*x
t0=1; x0=1; y0=2;
h=0.1;
n=15;
t=t0; x=x0; y=y0;
Do[u=y; v=f[t,x,y];
  x=x+h*u; y=y+h*v;
  t=t+h;
  Print[t," ", x," ", y],
  {i,1,n}]
```

Mathematica displays the numerical result:

```
1.1 1.2 1.8
1.2 1.38 1.67107
1.3 1.54711 1.58423
1.4 1.70553 1.52359
1.5 1.85789 1.47997
1.6 2.00589 1.44779
1.7 2.15066 1.42352
1.8 2.29302 1.40488
1.9 2.43351 1.39033
2.  2.57254 1.3788
2.1 2.71042 1.36955
2.2 2.84737 1.36203
2.3 2.98358 1.35587
2.4 3.11916 1.35077
2.5 3.25424 1.34651
```

We first wrote the vector field; the initial conditions $t_0 = 1$, $x_0 = 1$, $y_0 = 2$; the step size $h = 0.1$; the number of steps $n = 15$; and the starting point of the numerical procedure: $t = t_0$, $x = x_0$, $y = y_0$. With Do, we asked *Mathematica* to apply Euler's formula (equations (3), Section 3.6) from $i = 1$ to $i = n$. For this we assigned the variables u to y and v to $f(t, x, y)$, because by directly applying Euler's formula, x=x+h*y and y=y+h*f[t,x,y], *Mathematica* computes a new value for x and uses it in the second equation. Introducing u and v, *Mathematica* assigns the same value of x in the right-hand side of each equation, as Euler's method requires.

Second-order Runge-Kutta method The only changes from the program used for Euler's method appear for Do. Instead of using only equations (3), we now also need equations (6) of Section 3.6. We first apply Euler's formula to compute first approximations, denoted by u and v, and then use equation (6) to compute x and y.

```
Do[a=y; b=f[t,x,y];
   u=x+h*a; v=y+h*b;
   x=x+h*(a+v)/2;
   y=y+h*(b+f[t+h,u,v])/2;
   t=t+h;
   Print[t," ", x," ", y],
   {i,1,n}]
```

Mathematica displays the numerical result:

```
1.1 1.19 1.83554
1.2 1.3667 1.72086
1.3 1.53394 1.63851
1.4 1.69426 1.57786
1.5 1.84942 1.53222
1.6 2.00065 1.49723
1.7 2.14883 1.46995
1.8 2.29461 1.44838
1.9 2.43848 1.43109
2.  2.58082 1.41708
2.1 2.72189 1.40561
2.2 2.86193 1.39612
2.3 3.00111 1.38821
2.4 3.13957 1.38155
2.5 3.27742 1.37592
```

Vector fields   To understand the phase-plane portrait of a second-order equation, it may be useful to draw the vector field. The package <<Graphics`PlotField' has for this purpose a special command called PlotVectorField. For equation (4) of Section 3.5,

$$x'' = x' + x^2 - x, \qquad (5)$$

the vector field can be drawn as follows:

```
<<Graphics`PlotField'
PlotVectorField[{y,y+x^2-x}, {x,-1,2}, {y,-1,2},
Axes -> True];
```

Figure 3.7.3. The vector field of equation (5).

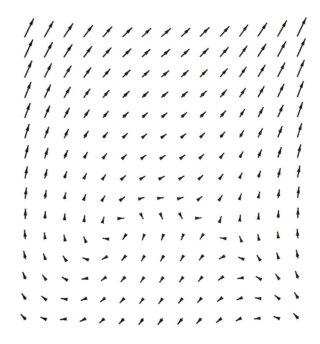

Mathematica produces the picture in Figure 3.7.3. Notice that the vectors are not normalized (i.e., they do not all have the same length). This has the advantage of better illustrating the equilibrium solutions, since the vectors have length zero at such points.

MATLAB

Exact solutions

The command `dsolve` can also be used to solve some second-order equations. MATLAB solves all homogeneous equations with constant coefficients and some homogeneous equations with variable coefficients; also, it applies the method of variation of parameters for nonhomogeneous equations once it can compute the solutions of the corresponding homogeneous equations.

EXAMPLE 1   To solve the homogeneous equation with constant coefficients

$$x'' = 3x' - 2x,$$

type at the prompt

```
>> dsolve('D2x=3*Dx-2*x')
```

and MATLAB will display the answer

```
ans =
C1*exp(2*t)+C2*exp(t)
```

EXAMPLE 2   MATLAB also solves initial value problems. If we attach the initial conditions $x(0) = 2, x'(0) = 1$ to the equation in Example 1, we can solve the corresponding initial value problem as follows:

```
>> dsolve('D2x=3*Dx-2*x' , 'x(0)=2, Dx(0)=1')
```

MATLAB will display the answer

```
ans =
-exp(2*t)+3*exp(t)
```

EXAMPLE 3   Since MATLAB knows the solution of the above equation, it can solve the nonhomogeneous equation,

$$x'' = 3x' - 2x + t.$$

```
>> dsolve('D2x=3*Dx-2*x+t')
```

and MATLAB displays the answer

```
ans =
1/2*t+3/4+C1*exp(2*t)+C2*exp(t)
```

EXAMPLE 4   The homogeneous equation with variable coefficients

$$x'' = -(1/t)x' + (4/t^2)x$$

can be solved by typing

```
>> dsolve('D2x=-(1/t)*Dx+(4/t^2)*x')
```

MATLAB displays the solution

```
ans =
t^2*C1+1/t^2*C
```

EXAMPLE 5   The nonhomogeneous equation

$$x'' = -(1/t)x' + (4/t^2)x + t^2$$

can now be solved by typing

```
>> dsolve('D2x=-(1/t)*Dx+(4/t^2)*x+t^2')
```

MATLAB displays the solution

```
ans =
1/12*t^4+t^2*C1+1/t^2*C2
```

Numerical solutions   With MATLAB's help we will first apply Euler's method to solve the initial value problem used in the examples of Section 3.6,

$$x'' = -(2/t)x' + (2/t^2)x, \qquad x(1) = 1, x'(1) = 2, \tag{6}$$

in the interval $[1, 2.5]$, with step size $h = 0.1$. Then we will solve the same initial value problem using the second-order Runge-Kutta method. In each case we write a program, explain how it works, and outline the logic behind it.

Euler's method   As in Section 2.7, we first create a file f.m, in which we write the vector field

```
function F=f(t,x,y)
F=-(2/t).*y+(2/t.^2)*x;
```

Then we write the program

```
>> t0=1; x0=1; y0=2;
>> h=0.1;
>> n=15;
>> t=t0; x=x0; y=y0; T=t; X=x; Y=y;
>> for i=1:n
    u=y; v=f(t,x,y);
    x=x+h*u;
    y=y+h*v;
    T=[T;t];
    X=[X;x];
    Y=[Y;y];
    t=t+h;
    end
>> [T,X,Y]
```

and MATLAB displays the numerical result

```
ans =
    1.0000    1.0000    2.0000
    1.1000    1.2000    1.8000
    1.2000    1.3800    1.6711
    1.3000    1.5471    1.5842
    1.4000    1.7055    1.5236
    1.5000    1.8579    1.4800
    1.6000    2.0059    1.4478
    1.7000    2.1507    1.4235
    1.8000    2.2930    1.4049
    1.9000    2.4335    1.3903
    2.0000    2.5725    1.3788
    2.1000    2.7104    1.3695
    2.2000    2.8474    1.3620
    2.3000    2.9836    1.3559
    2.4000    3.1192    1.3508
    2.5000    3.2542    1.3465
```

We first wrote the vector field, the initial conditions $t_0 = 1$, $x_0 = 1$, $y_0 = 2$, the step size $h = 0.1$, the number of steps $n = 15$, and the starting point of the numerical procedure: $t = t_0$, $x = x_0$, $y = y_0$. We then asked MATLAB to apply Euler's formula (equations (3), Section 3.6) from $i = 1$ to $i = n$. For this we assigned the variables u to y and v to $f(t, x, y)$, because by directly applying Euler's formula, x=x+h*y and y=y+h*f(t,x,y), MATLAB computes a new value for x and uses it in the second equation. Introducing u and v, MATLAB assigns the same value of x in the right-hand side of each equation, as Euler's method requires.

Second-order Runge-Kutta method The only changes from the program used for Euler's method appear in iterating from $i = 1$ to $i = n$. Instead of using just equations (3), we now also need equations (6) of Section 3.6. We first apply Euler's formula to compute first approximations, denoted by u and v, and then use equation (6) to compute x and y.

```
>> t=t0;  x=x0;  y=y0;  z=x0;  T=t;  X=x;  Y=y;
>> for i=1:n
     a=y;  b=f(t,x,y);
     u=x+h*a;  v=y+h*b;
     x=x+h*(a+v)/2;
     y=y+h*(b+f(t+h,u,v))/2;
     t=t+h;
     T=[T;t];  X=[X;x];  Y=[Y;y];
   end
>> [T,X,Y]
```

MATLAB displays the numerical result

```
ans =
      1.0000      1.0000      2.0000
      1.1000      1.1900      1.8355
      1.2000      1.3667      1.7209
      1.3000      1.5339      1.6385
      1.4000      1.6943      1.5779
      1.5000      1.8494      1.5322
      1.6000      2.0006      1.4972
      1.7000      2.1488      1.4699
      1.8000      2.2946      1.4484
      1.9000      2.4385      1.4311
      2.0000      2.5808      1.4171
      2.1000      2.7219      1.4056
      2.2000      2.8619      1.3961
      2.3000      3.0011      1.3882
      2.4000      3.1396      1.3816
      2.5000      3.2774      1.3759
```

If we want to plot the graphs of $x(t)$, $y(t)$, or the phase-plane representation of the solution, we just have to use, instead of [T,X,Y] the command plot(T,X), plot(T,Y), or plot(X,Y), respectively, at the end of the program.

Vector fields   To understand the phase-plane portrait of a second-order equation, it is often useful to draw the vector field. For equation (4) of Section 3.5,

$$x'' = x' + x^2 - x, \tag{7}$$

the vector field can be drawn as follows. Write (1) as two first-order equations,

$$\begin{cases} x' = y \\ y' = y + x^2 - x \end{cases} \tag{8}$$

and proceed as below. MATLAB then produces the picture in Figure 3.7.4.

```
>> x1=-1;  x2=2;  y1=-1;  y2=2;
>> axis([x1 x2 y1 y2]);
>> [x,y]=meshgrid(x1:0.2:x2, y1:0.2:y2);
>> dx=y;
>> dy=y+x.^2-x;
>> quiver(x,y,dx,dy)
```

Figure 3.7.4. The direction field of system (8).

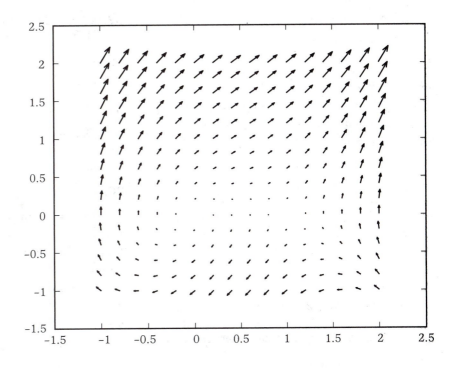

This program is very similar to that for drawing slope fields (see Section 2.7), the only difference being that the vectors are not normalized (i.e., of the same length). This has the advantage of making the identification of equilibria easier, since the vectors at such points have zero length.

PROBLEMS

Use Maple, Mathematica, or MATLAB to find the general solutions of the second-order equations in Problems 1 through 6, which have constant coefficients.

1. $x'' = x' + x$

2. $x'' = -5x' + 3x$

3. $x'' = \frac{1}{3}x' + 4x$

4. $x'' = 15x' - \frac{2}{3}x$

5. $x'' = 7x' + 12x$

6. $x'' = -4x' + x$

Use Maple, Mathematica, or MATLAB to find the general solutions of the nonhomogeneous second-order equations in Problems 7 through 12, whose homogeneous parts have constant coefficients.

7. $x'' = x' + x + 5t$

8. $x'' = -5x' + 3x - e^t$

9. $x'' = \frac{1}{3}x' + 4x + \sin t$

10. $x'' = 15x' - \frac{2}{3}x - 2\cos t$

11. $x'' = 7x' + 12x + t^2$

12. $x'' = -4x' + x - \frac{2}{5}t^3$

Use Maple, Mathematica, or MATLAB to find the general solutions of the homogeneous second-order equations in Problems 13 through 18.

13. $x'' = \frac{1}{t}x' + t^2x$

14. $x'' = e^t x' + x$

15. $x'' = (\sin t)x' + (\cos t)x$

16. $x'' = -\frac{1}{3}x' - x$

17. $x'' = \frac{1}{16}x' + 2x$

18. $x'' = (\cosh t)x' + (\sinh t)x$

Use Maple, Mathematica, or MATLAB to find the general solutions of the nonhomogeneous second-order equations in Problems 19 through 24.

19. $x'' = \frac{1}{t}x' + t^2x + t^2$

20. $x'' = e^t x' + x + 5$

21. $x'' = \dfrac{t}{4}x' + x + t$

22. $x'' = \frac{1}{3}x' - x - t$

23. $x'' = \frac{1}{16}x' + 2x + 1$

24. $x'' = (\cosh t)x' + (\sinh t)x + 1$

Use Maple, Mathematica, or MATLAB to solve the initial value problems in Problems 25 through 30.

25. $x'' = 6x' + 9x, \quad x(0) = 1, \ x'(0) = 4$

26. $x'' = -3x' + 2x, \quad x(1) = 2, \ x'(1) = 3$

27. $x'' = \dfrac{t}{2}x' + 2x, \quad x(1) = x'(1) = 2$

28. $x'' = 4x' - 5x, \quad x(0) = 2, \ x'(0) = 1$

29. $x'' = x' + 2x + t, \quad x(2) = x'(2) = 5$

30. $x'' = tx' + 4x + t, \quad x(0) = 0,$
$\qquad x'(0) = 1$

Use Maple, Mathematica, or MATLAB to solve the equations in Problems 21 through 36 numerically in the specified intervals, with step sizes 0.1 and 0.05, using all numerical methods described in this section.

31. $x'' = 2x' + 4x, \quad x(0) = 1, \ x'(0) = -1$ in
$\qquad [0, 2]$

32. $x'' = -2x' + 3tx, \quad x(1) = 1, \ x'(1) = -2$ in
$\qquad [1, 3]$

33. $x'' = \dfrac{2t}{3}x' - x, \quad x(1) = 0, \ x'(1) = 1$ in
$\qquad [-1, 1]$

34. $x'' = -t^2x' + \dfrac{t}{6}x, \quad x(0) = 2, \ x'(0) = 1$ in
$\qquad [0, 1.5]$

35. $x'' = 3(\sin t)x' + 2t^3x + t, \quad x(1) = x'(1) = 1$
$\qquad$ in $[1, 4]$

36. $x'' = 2tx' + \dfrac{3}{2t}x + \sin t, \quad x(1) = \pi,$
$\qquad x'(1) = 2\pi$ in $[1, 2]$

Use Maple, Mathematica, or MATLAB to draw the direction fields of the equations in Problems 37 through 42. In each case try to sketch the flow.

37. $x'' = 2x' + 3x^2 + x$

38. $x'' = -5x' + 4x + x^3$

39. $x'' = \frac{1}{3}(x')^2 + 4x^3 - 1$

40. $x'' = 15x' - \sin\left(\frac{2}{3}x\right)$

41. $x'' = 7x' + 12x - 2$

42. $x'' = -4xx' + x^4$

3.8 Modeling Experiments

Before attempting to work on any of the modeling experiments below, read the introductory paragraph of Section 2.8. This will give you an idea about what is expected from you, how to proceed, and what to emphasize.

Galileo's Pendulum

Take a piece of string at least 2 meters long and a small but heavy bob, find a suitable support device, and build a pendulum. To avoid frictions, make sure that you strongly fasten the string to the support. Then do several experiments, first with small oscillations (smaller than 15 arc degrees) and then with larger ones. Find a good way to measure the initial angle. To measure the time of an oscillation more accurately, let the pendulum swing several times and then divide the measured time by the number of oscillations. Also measure the time that passes between the start and the moment the pendulum passes the 10th time, then the 20th time, then the 30th time, etc., through the vertical position. Repeat the same experiment several times until you are sure that your measurements are accurate. Compare your measurements with the theoretical results obtained in Section 3.2. In other words, check whether the angle given by the formula

$$\phi(t) = c_1 \cos \omega t + c_2 \sin \omega t, \tag{1}$$

agrees with the vertical position at the measured times. Then, using one of the computer programs in Section 3.7, numerically solve the equation

$$\phi'' = -\frac{g}{L} \sin \phi. \tag{2}$$

For every type of experiment, compare the measured results with the theoretical ones obtained with the help of formula (1) and with the numerical solution of equation (2). Use different lengths for the pendulum and go through the whole process again. Make sure that you are consistent with the physical units: mass, length, time. What conclusions can you draw? Can you use graphical representations? Can you state some theorem? What are the arguments that support it? How did Galileo perform his experiments? Can you learn something from his experience?

Bungee Jumping[1]

Consider the second-order equation introduced in Section 3.6,

$$x'' = -\frac{\nu}{m}x'|x'| - \frac{k}{m}x,$$

which is suggested as a model in bungee jumping. This model assumes that the elastic cord acts like a spring. We did not discuss the fact that the cord does not always resemble a vertical line, but is often bent and curved in the air. At such times the jumper seems to be detached from the cord. For example, during the first part of her way down, she is free-falling. Does this observation change anything? Is it still all right to use the vertical spring as a model? You might want to consult some physics textbooks in this sense. Bring arguments in favor of your conclusion. If your answer is negative, find a better model or explain why you still prefer the one above.

Use the new model, the old model, or both to perform further numerical experiments with the help of the computer programs in Section 3.7. What results do you obtain if varying the coefficients? In case you work with several models, compare the results. What conclusions can you draw? Is the motion always described by an oscillating curve with decreasing amplitude? Can you state some theorem in this sense? Can you prove it? How do people in the bungee-jumping business determine the length and strength of the cords? Is their choice based on scientific research done by somebody else? Find out more about the way this is done and decide what approach you would take. Bring arguments in favor of your choice.

Suspension Bridges

Recall from Section 3.4 the model of the Tacoma Narrows Bridge, which is given by the equation

$$x'' = -bx' - cx + w(t), \tag{3}$$

where $b, \delta > 0$ are constants,

$$c = \begin{cases} \delta + g, & \text{if } x < 0 \\ g, & \text{if } x \geq 0, \end{cases}$$

and $g \approx 9.8$ m/s^2 is the gravitational acceleration. First study the homogeneous equation

$$x'' + bx' + cx = 0. \tag{4}$$

For this, assign some values to $b, c,$ and δ and take some initial data. Study the cases $x > 0$ and $x < 0$ separately and see what solutions you obtain.

---

[1]This "experiment" is purely theoretical. **Do not** attempt to apply it in practice.

Then decide what would be a good way to approximate this solution through a differentiable function. Is this approximation unique? Study equation (4) numerically. Find a way to apply the computer programs of Section 3.7 to this problem.

Assume now that the external force is periodic (due to marching, for example), i.e., of the form $w(t) = a \cos \omega t$, for a and ω constants. Choose values of the constants that lead to resonance, and then some values close to those. In both cases, study equation (3) numerically. Apply to this more complicated case the experience you gained while studying equation (4). Graph your results. What conclusions can you draw? Can you state a theorem? What arguments can you bring in its favor?

What other models have been used to study the Tacoma Narrows Bridge disaster? What were the conclusions of those models? How do they compare with your results? If you were an engineer, what kind of suspension bridge would you build? What are the lessons learned from pursuing this study?

An Electric Circuit

Read carefully the subsection of Section 3.5 on electric circuits and then consult a physics textbook to learn more about the laws of Ohm, Faraday, and Kirchhoff. Consider an electric circuit formed with an inductor, a capacitor, a resistor, and a generator, all connected in series (see Figure 3.8.1). Assuming that the variation of the generator in time is given by a function $g(t)$, write down the equation describing the dynamics of the charge in the plates.

Study the case when $g(t) = a \cos \omega t$, where a and ω are positive constants, i.e., when the voltage source varies periodically in time. Solve the equation and discuss all possible outcomes relative to the choice of the constants. Draw graphs. Give the physical interpretation in each case. Do you encounter beats and resonance? If so, what do they mean in terms of the circuit? Make an analogy with the spring-mass system. Write a dictionary that translates the properties of the circuit into those of the spring.

Consider other differentiable functions g of importance in engineering. Why are they important? Where do they apply? Consult some engineering manuals or talk to an electrical engineer to learn more. Can you integrate the new equations? If so, give the physical interpretation of the possible solutions. If not, take some (physically realistic) values of the constants and use one of the programs in Section 3.7 to obtain numerical solutions for various initial value problems. Vary the initial conditions only a little bit and see whether the solutions are similar over longer time intervals. Draw graphs. Do patterns occur? Are the solutions totally irregular? Can you draw any conclusions? Can you state any theorem? If so, what other mathematical and physical arguments can you bring in its favor?

Figure 3.8.1. A series *RLC* electric circuit with a generator.

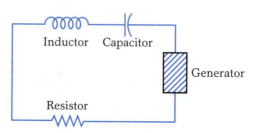

4 Linear Systems

The convenience of mathematical symbolism for handling certain deductive inferences is, I think, indisputable.

PAUL SAMUELSON
Nobel Prize for Economics, 1970

4.1 Linear Algebra

In this section we will introduce some elements of linear algebra, which will later help us solve certain systems of linear differential equations. We will start with linear algebraic systems, define the concepts of *matrix* and *determinant*, extend the notion of linear independence to vectors, and finally describe a method for computing the *eigenvalues* and the *eigenvectors* of a matrix. To facilitate understanding, we will restrict our presentation to two- and three-dimensional systems, the extension to n following without difficulty.

Linear Algebraic Systems

High-school algebra deals with simple two-dimensional linear systems, which we used in Chapter 3 to solve two-dimensional linear systems of differential equations derived from linear second-order differential equations. In this chapter we will generalize this idea and show that certain linear systems of differential equations can be reduced to algebraic linear systems. For this, however, we first need to present some elements of linear algebra.

The general form of a linear algebraic system of three equations and three unknowns is

$$\begin{cases} a_{11}x_1 + a_{12}x_2 + a_{13}x_3 = b_1 \\ a_{21}x_1 + a_{22}x_2 + a_{23}x_3 = b_2 \\ a_{31}x_1 + a_{32}x_2 + a_{33}x_3 = b_3, \end{cases} \tag{1}$$

where a_{ij} and b_i, $i, j = 1, 2, 3$, are given numbers, called coefficients, and x_1, x_2, x_3 are the unknowns. If $b_1 = b_2 = b_3 = 0$, the system is called *homogeneous*; otherwise it is called *nonhomogeneous*.

Symbolism plays an important role in mathematics. A good notation can often help us understand things better. For this reason let us give other forms to system (1). We can write it in *vector form*:[1] If

$$\mathbf{a}_i = \begin{pmatrix} a_{1i} \\ a_{2i} \\ a_{3i} \end{pmatrix}, \quad i = 1, 2, 3, \quad \text{and} \quad \mathbf{b} = \begin{pmatrix} b_1 \\ b_2 \\ b_3 \end{pmatrix},$$

---

[1] It is more convenient in this case to write the vectors as *columns* (i.e., vertically) instead of representing them as *rows* (i.e., horizontally).

system (1) becomes

$$\mathbf{a}_1 x_1 + \mathbf{a}_2 x_2 + \mathbf{a}_3 x_3 = \mathbf{b}. \tag{2}$$

An even more compact notation is the *matrix form*

$$\mathbf{A}\mathbf{x} = \mathbf{b}, \tag{3}$$

where

$$\mathbf{x} = \begin{pmatrix} x_1 \\ x_2 \\ x_3 \end{pmatrix} \quad \text{and} \quad \mathbf{A} = \begin{pmatrix} a_{11} & a_{12} & a_{13} \\ a_{21} & a_{22} & a_{23} \\ a_{31} & a_{32} & a_{33} \end{pmatrix}$$

represent the vector of the unknowns and the *matrix* of the system, which is the table of the coefficients a_{ij}. Notice that $\mathbf{A}$ has three *rows* and three *columns*, so we say that it is three-dimensional. The only meaning we assign here to equation (3) is that of symbolically representing system (1) or linear systems in general. The advantages of this notation are its brevity and independence of dimension: It can stand for two-, three-, or n-dimensional systems; it also separates the unknowns and the coefficients.

Determinants

An important notion in linear algebra is the *determinant* of a matrix $\mathbf{A}$ (denoted by $|\mathbf{A}|$ or by $\det \mathbf{A}$), obtained from a formula involving the coefficients a_{ij}. For a two-dimensional matrix

$$\mathbf{A} = \begin{pmatrix} a_{11} & a_{12} \\ a_{21} & a_{22} \end{pmatrix}$$

we define the determinant as $|\mathbf{A}| = a_{11}a_{22} - a_{12}a_{21}$.

EXAMPLE 1   The determinant of the matrix

$$\mathbf{A} = \begin{pmatrix} 3 & -5 \\ 1 & 2 \end{pmatrix}$$

is $|\mathbf{A}| = 3 \cdot 2 - (-5) \cdot 1 = 11$.

In preparation for defining the determinant of a three-dimensional system, let us interpret the above definition as follows: Take the coefficient a_{11}, eliminate all the coefficients of the row and column containing a_{11} (i.e., eliminate a_{12} and a_{21}), and multiply a_{11} by what is left, i.e., by a_{22}. Now take a_{12} with opposite sign, eliminate all the coefficients of the row and column containing a_{12} (i.e., a_{11} and a_{22}), and multiply by what is left, i.e., by a_{21}. Then add the two products.

We now define the determinant of the three-dimensional matrix of system (3) as

$$|\mathbf{A}| = a_{11}|\mathbf{A}_{11}| - a_{12}|\mathbf{A}_{12}| + a_{13}|\mathbf{A}_{13}|, \tag{4}$$

where $|\mathbf{A}_{11}|$ is the determinant of the matrix obtained by eliminating all the elements of the row and column containing a_{11}. This means that

$$\mathbf{A}_{11} = \begin{pmatrix} a_{22} & a_{23} \\ a_{32} & a_{33} \end{pmatrix}.$$

We define $|\mathbf{A}_{12}|$ and $|\mathbf{A}_{13}|$ similarly. Notice that the signs alternate: plus for a_{11}, minus for a_{12}, plus for a_{13}. So relation (4) can be rewritten as

$$|\mathbf{A}| = a_{11}(a_{12}a_{23} - a_{13}a_{22}) - a_{12}(a_{11}a_{33} - a_{13}a_{31}) + a_{13}(a_{11}a_{22} - a_{12}a_{21}). \tag{5}$$

EXAMPLE 2 According to formula (5), the determinant of the matrix

$$\mathbf{A} = \begin{pmatrix} 2 & 5 & -2 \\ 4 & -3 & 2 \\ 6 & 2 & -2 \end{pmatrix}$$

is $|\mathbf{A}| = 2[(-3)(-2) - 2 \cdot 2] - 5[4(-2) - 2 \cdot 6] + (-2)[4 \cdot 2 - (-3) \cdot 6] = 52$.

Following the above rule, the determinant of an n-dimensional matrix is defined as

$$|\mathbf{A}| = \sum_{i=1}^{n} (-1)^{1+i} a_{1i} |\mathbf{A}_{1i}|,$$

where $|\mathbf{A}_{1i}|$ is the determinant of the matrix obtained by eliminating the row and the column involving a_{1i}. The factor $(-1)^{1+i}$ makes signs alternate.

The Solution of a Linear System

Recall from high school that a system can be solved by reduction: Multiply the equations pairwise by suitable numbers and then add them in order to eliminate an unknown. The system reduces to a lower-dimensional one. And so on.

EXAMPLE 3 To solve the system

$$\begin{cases} 2x + y = 1 \\ 3x - 2y = 5, \end{cases}$$

multiply the first equation by 2, obtain $4x + 2y = 2$, add this new equation to the second one above and obtain $7x = 7$, which yields $x = 1$. Substitute this value of x into either of the initial equations and find $y = -1$. So $(x, y) = (1, -1)$ is the solution of the system.

We have thus proved that the above algebraic system has a unique solution. But this is not the case with all systems. Some linear systems have no solutions at all, whereas others have infinitely many. This has the following geometric interpretation. Each linear equation of a two-dimensional system represents a line in a plane. The solution is unique when the two lines intersect and does not exist when the lines are parallel. Infinitely many solutions appear when the lines coincide (see Figure 4.1.1). What is the corresponding geometric interpretation for linear three-dimensional systems of algebraic equations?

Figure 4.1.1. A linear two-dimensional algebraic system can have (*a*) a unique solution, (*b*) no solution, or (*c*) infinitely many solutions.

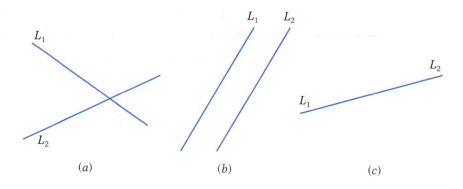

(a) (b) (c)

EXAMPLE 4 The system

$$\begin{cases} \dfrac{1}{2}x - 3y = 8 \\[2mm] \dfrac{3}{2}x - 9y = -2 \end{cases}$$

has no solutions. Indeed, multiply the first equation by -3 and add it to the second equation. This leads to the relation $0 = -26$, which is absurd.

EXAMPLE 5 The homogeneous system

$$\begin{cases} 2x + 3y = 0 \\ 6x + 9y = 0 \end{cases}$$

has infinitely many solutions, since the two equations are identical; indeed, by multiplying the first equation by 3, we obtain the second equation. For $x = \alpha$ we obtain $y = -\frac{2}{3}\alpha$, so $(x, y) = (\alpha, -\frac{2}{3}\alpha)$ is a solution for every real value of α. Can you find a nonhomogeneous system with infinitely many solutions?

It seems quite difficult to decide in advance whether a given linear algebraic system has a unique solution or not. But linear algebra theory gives the answer in terms of determinants. The proof of the following result is beyond our scope.

Theorem 4.1.1

The system $\mathbf{Ax} = \mathbf{b}$ has a unique solution if and only if $|\mathbf{A}| \neq 0$.

Remark 1. In case of a homogeneous system $\mathbf{Ax} = \mathbf{0}$, the condition $|\mathbf{A}| \neq 0$ implies that $\mathbf{x} = \mathbf{0}$ is the only solution. (Can you prove this using Theorem 4.1.1?)

Remark 2. The necessary and sufficient condition that the homogeneous system $\mathbf{Ax} = \mathbf{0}$ has solutions other than $\mathbf{x} = \mathbf{0}$ is that $|\mathbf{A}| = 0$.

Compute the determinants of the matrices defining the systems in Examples 3 to 5 and see if the results you obtain agree with the above theorem and remarks.

Linearly Independent Vectors

The notion of linear independence studied in Section 3.1 extends to vectors. For simplicity we will present this extension for three vectors. The generalization to n vectors is obvious.

DEFINITION 4.1.1

The vectors $\mathbf{x}_1, \mathbf{x}_2, \mathbf{x}_3$ are called *linearly dependent* if there exist three constants c_1, c_2, c_3, not all zero, such that

$$c_1\mathbf{x}_1 + c_2\mathbf{x}_2 + c_3\mathbf{x}_3 = \mathbf{0}. \tag{6}$$

They are called *linearly independent* if (6) is satisfied only for $c_1 = c_2 = c_3 = 0$. The left-hand side of relation (6) is called a *linear combination* of vectors.

Notice that two linearly dependent vectors have the same direction; just their lengths and orientations may differ (see Figure 4.1.2(a)). The geometric interpretation of linear dependence for three vectors is that the sum of two vectors (each multiplied by a suitable constant) is a vector that has the same direction as the third vector (see Figure 4.1.2(b)).

Relation (6) can be also viewed as a linear homogeneous system of unknowns c_1, c_2, c_3 (compare to relation (2)); we can write (6) as

$$\mathbf{Xc} = \mathbf{0},$$

where $\mathbf{X}$ represents the matrix of the nine coefficients (three for each column vector $\mathbf{x}_i$, $i = 1, 2, 3$) and $\mathbf{c} = (c_1, c_2, c_3)$. From Remark 1 we can draw the following conclusion, helpful for deciding whether some given vectors are linearly independent.

Figure 4.1.2. (*a*) Two linearly dependent vectors: $\mathbf{x} = 2\mathbf{y}$. (*b*) Three linearly dependent vectors: $\mathbf{x} + \mathbf{y} = 2\mathbf{z}$.

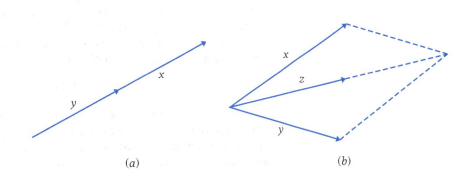

(*a*)　　　　　　(*b*)

Theorem 4.1.2

If $\mathbf{X}$ denotes the matrix of the three-dimensional vectors $\mathbf{x}_1, \mathbf{x}_2, \mathbf{x}_3,$ then these vectors are linearly independent only if $|\mathbf{X}| \neq 0$. (Obviously, in case of n-dimensional vectors, this is true for n given vectors.)

EXAMPLE 6   The vectors

$$\mathbf{x}_1 = \begin{pmatrix} 1 \\ 2 \\ -1 \end{pmatrix}, \qquad \mathbf{x}_2 = \begin{pmatrix} 5 \\ 3 \\ 1 \end{pmatrix}, \qquad \text{and} \qquad \mathbf{x}_3 = \begin{pmatrix} 2 \\ 7 \\ 0 \end{pmatrix}$$

are linearly independent. Indeed,

$$\mathbf{X} = \begin{pmatrix} 1 & 5 & 2 \\ 2 & 3 & 7 \\ -1 & 1 & 0 \end{pmatrix}$$

is their corresponding matrix and $|\mathbf{X}| = -32 \neq 0$. Does the value of the determinant change if we take row vectors instead of column vectors, i.e., if $\mathbf{x}_1 = (1, 2, -1)$, etc.? (See also Problems 9 and 10.)

Eigenvalues and Eigenvectors

The reason for introducing the *eigenvalues* and *eigenvectors* of a matrix will become clear in Section 4.3, when we learn how to solve linear systems of differential equations with constant coefficients. The terminology is borrowed from German, in which *eigen* means "own." As we will now see, we can attach to every matrix "its own" values and vectors.

To define the *eigenvalues* of a matrix, we consider along with the homogeneous system $\mathbf{Ax} = \mathbf{0}$ another homogeneous system,

$$\mathbf{A}_\lambda \mathbf{x} = \mathbf{0}, \tag{7}$$

where λ is a parameter and (for $n = 3$)

$$\mathbf{A}_\lambda = \begin{pmatrix} a_{11} - \lambda & a_{12} & a_{13} \\ a_{21} & a_{22} - \lambda & a_{23} \\ a_{31} & a_{32} & a_{33} - \lambda \end{pmatrix}. \tag{8}$$

Obviously, system (7) has the solution $\mathbf{x} = \mathbf{0}$. We want to answer the following question: What values of λ provide additional solutions for system (7)? According to Remark 2 above, other solutions appear only if $|\mathbf{A}_\lambda| = 0$. The values λ that provide additional solutions can therefore be obtained by solving the equation in λ,

$$|\mathbf{A}_\lambda| = 0,$$

called the *characteristic equation* of $\mathbf{A}$. It is important to note that $|\mathbf{A}_\lambda|$ is a polynomial in λ, as can be seen in Examples 7, 8, and 9. This is good news because polynomial equations are easier to handle than other kinds of equations encountered in algebra.

DEFINITION 4.1.2

The solutions λ of the equation $|\mathbf{A}_\lambda| = 0$ are called *eigenvalues* of $\mathbf{A}$. For an eigenvalue λ, every nonzero solution $\mathbf{x}$ of the algebraic system $\mathbf{A}_\lambda \mathbf{x} = \mathbf{0}$ is called an *eigenvector*.

EXAMPLE 7   To determine the eigenvalues of the matrix

$$\mathbf{A} = \begin{pmatrix} 1 & 3 \\ 1 & -1 \end{pmatrix},$$

solve the characteristic equation $|\mathbf{A}_\lambda| = \mathbf{0}$, i.e.,

$$\det \begin{pmatrix} 1 - \lambda & 3 \\ 1 & -1 - \lambda \end{pmatrix} = 0,$$

which is equivalent to $\lambda^2 - 4 = 0$. Thus the eigenvalues are $\lambda_1 = 2$ and $\lambda_2 = -2$.

To find the eigenvectors corresponding to $\lambda_1 = 2$, solve the system $\mathbf{A}_\lambda \mathbf{x} = \mathbf{0}$ for $\lambda = 2$, which in this case takes the form

$$\begin{cases} -x_1 + 3x_2 = 0 \\ x_1 - 3x_2 = 0. \end{cases}$$

Since for $x_2 = \alpha$, where α is any real constant, it follows that $x_1 = 3\alpha$, any vector of the form

$$\mathbf{u}_\alpha = \begin{pmatrix} 3\alpha \\ \alpha \end{pmatrix}$$

is an eigenvector corresponding to the eigenvalue $\lambda_1 = 2$.

To find the eigenvectors corresponding to $\lambda_2 = -2$, solve the system $\mathbf{A}_\lambda \mathbf{x} = \mathbf{0}$ for $\lambda = -2$, which in this case takes the form

$$\begin{cases} 3x_1 + 3x_2 = 0 \\ x_1 + x_2 = 0. \end{cases}$$

By taking $x_2 = \beta$, where β is any real constant, it follows that $x_1 = 3 - \beta$. Therefore any vector of the form

$$\mathbf{v}_\beta = \begin{pmatrix} -\beta \\ \beta \end{pmatrix}$$

is an eigenvector corresponding to the eigenvalue $\lambda_1 = -2$.

It can be shown that every eigenvalue leads to infinitely many eigenvectors, all linearly dependent. In Section 4.3, however, we will be interested in finding only linearly independent eigenvectors and, if possible, as many as the dimensionality of the system. In Example 7 this goal can be achieved, because any nonzero eigenvector $\mathbf{u}_\alpha$ is linearly independent from any eigenvector $\mathbf{v}_\beta$. In fact, if the eigenvalues are real and distinct, the corresponding eigenvectors are linearly independent. But for multiple eigenvalues, linear independence may be impossible to achieve (see Examples 8 and 9).

EXAMPLE 8   To determine the eigenvalues of the matrix

$$\mathbf{A} = \begin{pmatrix} 1 & -1 \\ 1 & 3 \end{pmatrix},$$

solve the characteristic equation $|\mathbf{A}_\lambda| = \mathbf{0}$, i.e.,

$$\det \begin{pmatrix} 1-\lambda & -1 \\ 1 & 3-\lambda \end{pmatrix} = 0,$$

which is equivalent to $\lambda^2 - 4\lambda + 4 = 0$. The eigenvalues of $\mathbf{A}$ are thus the roots $\lambda_1 = \lambda_2 = 2$ of this equation. To determine the eigenvectors, return to the system $|\mathbf{A}_\lambda \mathbf{x}| = \mathbf{0}$, which for $\lambda = 2$ becomes

$$\begin{cases} -x_1 - x_2 = 0 \\ x_1 + x_2 = 0. \end{cases}$$

Take, say, $x_1 = \alpha$, where α is real, and obtain $x_2 = -\alpha$, so the system has infinitely many solutions of the form $x_1 = \alpha, x_2 = -\alpha$. Therefore every eigenvector has the form

$$\mathbf{u} = \begin{pmatrix} \alpha \\ -\alpha \end{pmatrix}.$$

But whatever distinct values $\alpha_1, \alpha_2 \neq 0$ we choose, the corresponding eigenvectors

$$\mathbf{u}_1 = \begin{pmatrix} \alpha_1 \\ -\alpha_1 \end{pmatrix} \quad \text{and} \quad \mathbf{u}_2 = \begin{pmatrix} \alpha_2 \\ -\alpha_2 \end{pmatrix}$$

are linearly dependent. (Can you show this?) Therefore, in this case, there are no two linearly independent eigenvectors corresponding to the double eigenvalue. There are, however, situations in which a double eigenvalue leads to linearly independent eigenvectors (see Example 9 and Problem 28).

EXAMPLE 9   To determine the eigenvalues of the matrix

$$\mathbf{A} = \begin{pmatrix} 3 & 2 & 4 \\ 2 & 0 & 2 \\ 4 & 2 & 3 \end{pmatrix},$$

solve the characteristic equation $|\mathbf{A}_\lambda| = \mathbf{0}$, i.e.,

$$\det \begin{pmatrix} 3-\lambda & 2 & 4 \\ 2 & -\lambda & 2 \\ 4 & 2 & 3-\lambda \end{pmatrix} = 0,$$

which is equivalent to the polynomial equation $-\lambda^3 + 6\lambda^2 + 15\lambda + 8 = 0$, whose solutions[2] are the eigenvalues of $\mathbf{A}$: $\lambda_1 = 8, \lambda_2 = -1$, and

---

[2]Recall that if a polynomial equation of the form $\lambda^3 + a\lambda^2 + b\lambda + c = 0$ has integer roots, then they are among the divisors of c.

$\lambda_3 = -1$. To find the corresponding eigenvectors, first substitute $\lambda_1 = 8$ into $\mathbf{A}_\lambda \mathbf{x} = \mathbf{0}$ and obtain the system

$$\begin{cases} -5x_1 + 2x_2 + 4x_3 = 0 \\ 2x_1 - 8x_2 + 2x_3 = 0 \\ 4x_1 + 2x_2 - 5x_3 = 0. \end{cases}$$

Solving this system, we find that it has infinitely many solutions, $x_1 = \alpha$, $x_2 = \alpha/2$, $x_3 = \alpha$, one for each real value of α. So the eigenvalue $\lambda_1 = 8$ corresponds to infinitely many eigenvectors, all linearly dependent. Choose one, say

$$\mathbf{u}_1 = \begin{pmatrix} 2 \\ 1 \\ 2 \end{pmatrix},$$

which corresponds to $\alpha = 2$. To find other eigenvectors, linearly independent from the chosen one, substitute the double eigenvalue $\lambda_2 = \lambda_3 = -1$ into the system $\mathbf{A}_\lambda \mathbf{x} = \mathbf{0}$ and obtain

$$\begin{cases} 4x_1 + 2x_2 + 4x_3 = 0 \\ 2x_1 + x_2 + 2x_3 = 0 \\ 4x_1 + 2x_2 + 4x_3 = 0, \end{cases}$$

which is equivalent to the single equation $2x_1 + x_2 + 2x_3 = 0$. Again, this equation has infinitely many solutions, but they depend on two parameters. Indeed, if we take, say, $x_3 = \beta$ and $x_2 = \gamma$, where β and γ are constants, the system has the solutions $x_1 = -(\gamma + 2\beta)/2$, $x_2 = \gamma$, $x_3 = \beta$. For suitable values of γ and β, say $(\gamma, \beta) = (2, 0)$ for λ_2 and $(\gamma, \beta) = (0, 1)$ for λ_3, obtain the linearly independent eigenvectors

$$\mathbf{u}_2 = \begin{pmatrix} -1 \\ 2 \\ 0 \end{pmatrix} \quad \text{and} \quad \mathbf{u}_3 = \begin{pmatrix} -1 \\ 0 \\ 1 \end{pmatrix}.$$

Can you check that $\mathbf{u}_1$, $\mathbf{u}_2$, and $\mathbf{u}_3$ are linearly independent?

PROBLEMS

Write the linear systems in Problems 1 to 6 in vector form and then in matrix form.

1. $\begin{cases} 5x - 8y = 4 \\ 3x + \pi y = 2\pi \end{cases}$

2. $\begin{cases} -2u + v = 0 \\ -4u - v = 0 \end{cases}$

3. $\begin{cases} 3x - 2y + 4z = -1 \\ 5x + \frac{1}{2}y + 2z = -2 \\ 2x - 4y + 2z = -3 \end{cases}$

4. $\begin{cases} u + v - w = 0 \\ w - u + v = 0 \\ u - v + w = 0 \end{cases}$

5. $\begin{cases} 5x - 2y + 7z - 2w = 2 \\ 4x - 3y - 2z + 3w = 1 \\ 2x + 3y - 4z + 5w = 4 \\ 3x - 5y + 2z - 4w = 8 \end{cases}$

6. $\begin{cases} a + b + c - d = 1 \\ d - c - b - a = 2 \\ b - c - d + a = 3 \\ c + a + d + b = 0 \end{cases}$

7. Compute the determinants of the following matrices:

$$A = \begin{pmatrix} 1 & 2 \\ 2 & -4 \end{pmatrix},$$

$$B = \begin{pmatrix} 0 & 1 \\ 1 & 0 \end{pmatrix},$$

$$C = \begin{pmatrix} 2i & 4 - 3i \\ 6 & -8 + i \end{pmatrix},$$

$$D = \begin{pmatrix} 3 & 1 & 0 \\ 1 & -2 & -1 \\ 1 & 4 & 0 \end{pmatrix},$$

$$E = \begin{pmatrix} 1 & -1 & 2 \\ 0 & 2 & 0 \\ 3 & -2 & 1 \end{pmatrix},$$

$$F = \begin{pmatrix} 0 & 0 & 0 \\ 2 & 2 & 2 \\ -i & 3 + i & 2i - 1 \end{pmatrix}.$$

8. From the definition of a determinant it is obvious that for a given matrix there is a unique determinant. But is the converse true? Can two different matrices have the same determinant?

9. We have defined the determinant of a matrix using the elements of the first line. In fact, we obtain the same result if we use any other line, so if A is the n-dimensional matrix of elements a_{ij}, $i, j = 1, \ldots, n$, then for any i fixed, $1 \leq i \leq n$,

$$|A| = \sum_{j=1}^{n} (-1)^{i+j} a_{ij} |A_{ij}|.$$

Prove this relation for $n = 3$.

10. We have defined the determinant of a matrix using the elements of the first line. In fact we obtain the same result if we use any other column, so if A is the n-dimensional matrix of elements a_{ij}, $i, j = 1, \ldots, n$, then for any j fixed, $1 \leq j \leq n$,

$$|A| = \sum_{i=1}^{n} (-1)^{i+j} a_{ij} |A_{ij}|.$$

Prove this relation for $n = 3$.

Determine whether the systems in Problems 11 through 16 have solutions and whether they are unique. What is the geometrical interpretation of a linear equation and of the solutions in each case?

11. $\begin{cases} x - y = 1 \\ 2x + 3y = -1 \end{cases}$

12. $\begin{cases} x_1 + 3x_2 = 0 \\ 3x_1 - 6x_2 = 0 \end{cases}$

13. $\begin{cases} 2x_1 - x_2 - x_3 = -3 \\ x_1 - 4x_2 + \frac{1}{2}x_3 = 1 \\ 6x_1 - x_2 + 2x_3 = 2 \end{cases}$

14. $\begin{cases} x - y + z = 1 \\ x + y - z = 1 \\ -x + y + z = 1 \end{cases}$

15. $\begin{cases} 6x_1 - 3x_2 + 9x_3 = 18 \\ -2x_1 + x_2 - 3x_3 = -6 \\ x_1 + x_2 + x_3 = 3 \end{cases}$

16. $\begin{cases} 3x_1 - x_2 - x_3 + x_4 = 16 \\ x_1 - 2x_2 + 2x_3 - 2x_4 = 8 \\ 4x_1 + x_2 + 5x_3 - 6x_4 = -1 \\ -x_1 + 3x_2 + x_3 - 2x_4 = 4 \end{cases}$

In each of Problems 17 through 21, find out whether the given vectors are linearly independent.

17. $x_1 = \begin{pmatrix} 2 \\ 1 \\ 3 \end{pmatrix},$ $x_2 = \begin{pmatrix} 3 \\ 1 \\ -1 \end{pmatrix},$ $x_3 = \begin{pmatrix} 0 \\ 1 \\ -1 \end{pmatrix}$

18. $x_1 = \begin{pmatrix} 0 \\ 0 \\ 2 \end{pmatrix},$ $x_2 = \begin{pmatrix} -1 \\ 0 \\ 0 \end{pmatrix},$ $x_3 = \begin{pmatrix} 0 \\ 1 \\ 0 \end{pmatrix}$

19. $x_1 = \begin{pmatrix} 3 \\ 3 \\ 3 \end{pmatrix},$ $x_2 = \begin{pmatrix} -1 \\ -1 \\ -1 \end{pmatrix},$ $x_3 = \begin{pmatrix} 0 \\ 0 \\ 1 \end{pmatrix}$

20. $x_1 = \begin{pmatrix} 3 \\ 5 \\ 7 \end{pmatrix},$ $x_2 = \begin{pmatrix} 2 \\ 4 \\ -6 \end{pmatrix},$ $x_3 = \begin{pmatrix} 1 \\ 2 \\ 3 \end{pmatrix}$

21. $x_1 = \begin{pmatrix} 0 \\ 1 \\ 5 \end{pmatrix},$ $x_2 = \begin{pmatrix} 3 \\ 1 \\ 2 \end{pmatrix},$ $x_3 = \begin{pmatrix} 6 \\ -1 \\ 3 \end{pmatrix}$

Find the eigenvalues of the matrices in Problems 22 through 27. Then find as many linearly independent eigenvectors as possible.

22. $\begin{pmatrix} 3 & 2 \\ 1 & 5 \end{pmatrix}$

23. $\begin{pmatrix} 1 & 0 \\ 0 & -2 \end{pmatrix}$

24. $\begin{pmatrix} 3 & 0 & 1 \\ 4 & 0 & 1 \\ 0 & 1 & 2 \end{pmatrix}$

25. $\begin{pmatrix} 1 & 0 & 1 \\ 0 & 1 & 0 \\ 1 & -1 & 1 \end{pmatrix}$

26. $\begin{pmatrix} 1 & 1 & 1 \\ 0 & 1 & 0 \\ 0 & 1 & 0 \end{pmatrix}$

27. $\begin{pmatrix} -1 & 1 & 0 \\ 1 & -1 & 0 \\ 0 & -1 & 1 \end{pmatrix}$

28. Compute the eigenvalues and three linearly independent eigenvectors of the matrix
$$\begin{pmatrix} 0 & 1 & 1 \\ 1 & 0 & 1 \\ 1 & 1 & 0 \end{pmatrix}$$

29. Show that if $\mathbf{A}$ is a *symmetric* two- or three-dimensional matrix (i.e., such that $a_{ij} = a_{ji}$ for all i and j), then the matrix has only real eigenvalues.

30. For the matrix in Example 8, show that any eigenvector $\mathbf{x}$ is a *linear combination* of $\mathbf{x}_1, \mathbf{x}_2,$ and $\mathbf{x}_3$, i.e., that there are constants $c_1, c_2, c_3,$ not all zero, such that
$$\mathbf{x} = c_1\mathbf{x}_1 + c_2\mathbf{x}_2 + c_3\mathbf{x}_3.$$

4.2 Fundamental Results

This section introduces some fundamental results regarding linear systems with variable coefficients. We will start with an existence and uniqueness theorem, and then discuss linear independence for vector functions, describe the structure of the general solution for homogeneous and nonhomogeneous linear systems, and finally present the method of undetermined coefficients. These generalize the results obtained in Sections 3.1 and 3.3 for linear second-order equations. As in Section 4.1, we will restrict our presentation to two- and three-dimensional systems, the extension to n dimensions following without difficulty.

Existence and Uniqueness

The general form of a three-dimensional first-order *linear system* of differential equations is
$$\begin{cases} x_1' = a_{11}(t)x_1 + a_{12}(t)x_2 + a_{13}(t)x_3 + b_1(t) \\ x_2' = a_{21}(t)x_1 + a_{22}(t)x_2 + a_{23}(t)x_3 + b_2(t) \\ x_3' = a_{31}(t)x_1 + a_{32}(t)x_2 + a_{33}(t)x_3 + b_3(t), \end{cases} \tag{1}$$

where $a_{ij}, b_i, \ i, j = 1, 2, 3,$ are functions of t. We can write this system in *matrix form*,
$$\mathbf{x}' = \mathbf{A}(t)\mathbf{x} + \mathbf{b}(t), \tag{2}$$

where t indicates the dependence of the coefficients on the independent variable. System (2) is called *linear homogeneous* if $\mathbf{b}(t) = \mathbf{0}$ and *linear nonhomogeneous* if $\mathbf{b}(t) \neq \mathbf{0}$. An initial value problem is defined by imposing the initial conditions
$$\mathbf{x}(t_0) = \mathbf{x}_0, \tag{3}$$

at t_0, where $\mathbf{x}_0$ is a given initial vector.

EXAMPLE 1   The linear homogeneous system

$$\begin{cases} x_1' = tx_2 + tx_3 \\ x_2' = tx_1 + tx_3 \\ x_3' = tx_1 + tx_2 \end{cases} \tag{4}$$

has the coefficient matrix

$$\mathbf{A}(t) = \begin{pmatrix} 0 & t & t \\ t & 0 & t \\ t & t & 0 \end{pmatrix}.$$

Initial conditions are imposed by picking a value for t_0 and for the vector $\mathbf{x}(t_0)$, for example, $t_0 = 0$ and

$$\mathbf{x}(0) = \begin{pmatrix} 1 \\ -1 \\ 0 \end{pmatrix}. \tag{5}$$

A *solution* of the initial value problem (2)–(3) is a *vector function* φ (i.e., a function formed by a vector that has functions as components) that is differentiable, that satisfies the property

$$\varphi'(t) = \mathbf{A}(t)\varphi(t) + \mathbf{b}(t)$$

for all t for which it is defined, and for which

$$\varphi(t_0) = \mathbf{x}_0.$$

Can you explain why these definitions generalize the ones of Section 3.1?

EXAMPLE 2   A solution of the initial value problem (4)–(5) of Example 1 is given by the vector function

$$\varphi(t) = \begin{pmatrix} e^{-t^2/2} \\ -e^{-t^2/2} \\ 0 \end{pmatrix}.$$

This means that if substituted for x_1, x_2, and x_3 in (4), the components $\varphi_1(t) = e^{-t^2/2}$, $\varphi_2(t) = -e^{-t^2/2}$, and $\varphi_3(t) = 0$ of the vector function φ satisfy the system (can you show this?) as well as the initial conditions, i.e., $\varphi_1(0) = 1$, $\varphi_2(0) = -1$, and $\varphi_3(0) = 0$.

The following is an extension of Cauchy's theorem. As for linear second-order equations, this is a global result; i.e., the solution exists and is unique in the whole interval in which the coefficient functions are continuous.

Existence and Uniqueness Theorem

If the functions a_{ij} and b_i, $i, j = 1, 2, 3$, are continuous in some interval I that contains t_0, then the initial value problem (2)–(3) has a unique solution φ defined on I.

Linearly Independent Vector Functions

In Sections 3.1 and 4.1 we defined linear independence for functions and for vectors, respectively. We will now give a more general definition that includes both previous notions.

DEFINITION 4.2.1

The vector functions $\mathbf{v}_1, \mathbf{v}_2, \mathbf{v}_3$ are called *linearly dependent* on some interval I if there exist three constants c_1, c_2, c_3, not all zero, such that

$$c_1\mathbf{v}_1(t) + c_2\mathbf{v}_2(t) + c_3\mathbf{v}_3(t) = \mathbf{0} \tag{6}$$

for all t in I. The vector functions $\mathbf{v}_1, \mathbf{v}_2, \mathbf{v}_3$ are called *linearly independent* on I if relation (6) is satisfied only for $c_1 = c_2 = c_3 = 0$. The left-hand side of relation (6) is called a *linear combination* of vector functions.

EXAMPLE 3   The vector functions

$$\mathbf{u}(t) = \begin{pmatrix} t \\ 2t^2 \end{pmatrix}, \qquad \mathbf{v}(t) = \begin{pmatrix} -2t \\ -4t^2 \end{pmatrix}, \qquad \mathbf{w}(t) = \begin{pmatrix} 1 \\ t^3 \end{pmatrix}$$

are linearly dependent. Indeed, $2 \cdot \mathbf{u} + 1 \cdot \mathbf{v} + 0 \cdot \mathbf{w} = \mathbf{0}$.

EXAMPLE 4   The vector functions

$$\mathbf{u}(t) = \begin{pmatrix} t \\ t^2 \\ 1 \end{pmatrix}, \qquad \mathbf{v}(t) = \begin{pmatrix} 0 \\ t^3 \\ 2 \end{pmatrix}$$

are linearly independent. Can you prove it? (See Example 7 of Section 3.1.)

There exist vector functions that are linearly dependent on some intervals and linearly independent on others. To construct such vector functions, see Example 8 of Section 3.1.

If the number of vector functions equals the dimension of the vectors, we can introduce a new function, called *Wronskian*, after the Polish mathematician Josef Maria Hoëné-Wronski (1776–1853). If

$$\mathbf{u}(t) = \begin{pmatrix} u_1(t) \\ u_2(t) \\ u_3(t) \end{pmatrix}, \qquad \mathbf{v}(t) = \begin{pmatrix} v_1(t) \\ v_2(t) \\ v_3(t) \end{pmatrix}, \qquad \mathbf{w}(t) = \begin{pmatrix} w_1(t) \\ w_2(t) \\ w_3(t) \end{pmatrix}$$

are vector functions, their *Wronskian* is the function $W(t) = |\mathbf{X}(t)|$, where

$$\mathbf{X}(t) = \begin{pmatrix} u_1(t) & v_1(t) & w_1(t) \\ u_2(t) & v_2(t) & w_2(t) \\ u_3(t) & v_3(t) & w_3(t) \end{pmatrix}.$$

As the following result shows, the Wronskian is useful for checking the linear independence of differentiable vector functions. Its proof is similar to that of Theorem 3.1.1 (see Problem 18).

Theorem 4.2.2

If $\mathbf{u}, \mathbf{v}, \mathbf{w}$ are vector functions defined on some open interval I and if there is a t_0 in I such that $W(t_0) \neq 0$, then $\mathbf{u}, \mathbf{v}, \mathbf{w}$ are linearly independent on I.

Notice that in Theorem 3.1.1 the condition $W(t_0) \neq 0$ was represented in a different way. Try to express that condition by writing the second-order equation as a system and by using the Wronskian of a two-dimensional matrix.

The General Solution of a Homogeneous System

We now restrict our study to the homogeneous case $\mathbf{b} = \mathbf{0}$ and state a result analogous to Theorem 3.1.2, in which we connect the solutions of (2) with their Wronskian. The proof is left as an exercise (see Problem 19).

Theorem 4.2.3

If the vector functions $\mathbf{u}, \mathbf{v}, \mathbf{w}$, defined on an open interval I, are solutions of the system $\mathbf{x}' = \mathbf{A}(t)\mathbf{x}$, in which the coefficients a_{ij} of $\mathbf{A}$ are continuous on I, then $\mathbf{u}, \mathbf{v}, \mathbf{w}$ are linearly independent on I if and only if $W(t) \neq 0$ for all t in I.

We can summarize Theorems 4.2.2 and 4.2.3 as follows. If $\mathbf{u}, \mathbf{v}, \mathbf{w}$ are solutions of the homogeneous system $\mathbf{x}' = \mathbf{A}(t)\mathbf{x}$, then the following statements are equivalent:

1. $\mathbf{u}, \mathbf{v}, \mathbf{w}$ are linearly independent solutions on I.
2. There is a t_0 in I such that $W(t_0) \neq 0$.
3. $W(t) \neq 0$ for all t in I.

The structure of the general solution of a homogeneous system is given by the following result, which also emphasizes that any linear combination of solutions is a solution itself. This important property, characteristic only of linear systems, is known in the literature as the *superposition principle*. The following theorem also shows that the number of linearly independent solutions that completely describe the entire set of solutions equals the dimension of the system. This implies that in order to determine all the solutions of a linear homogeneous system, it is enough to find as many linearly independent solutions as the dimension of the system.

Fundamental Theorem for Homogeneous Systems

Any solution of the system $\mathbf{x}' = \mathbf{A}(t)\mathbf{x}$, in which the components a_{ij} of $\mathbf{A}$ are continuous functions on some open interval I, is of the form

$$\varphi(t) = c_1\mathbf{u}(t) + c_2\mathbf{v}(t) + c_3\mathbf{w}(t),$$

where $\mathbf{u}, \mathbf{v}, \mathbf{w}$ are solutions of the system, linearly independent on I, and c_1, c_2, c_3 are real constants.

PROOF

Let t_0 be some point in I, denote $\mathbf{x}_0 = \varphi(t_0)$, where φ is some solution of $\mathbf{x}' = \mathbf{A}(t)\mathbf{x}$, and consider the initial value problem

$$\mathbf{x}' = \mathbf{A}(t)\mathbf{x}, \qquad \mathbf{x}(t_0) = \mathbf{x}_0. \qquad (7)$$

Obviously φ is a solution of the initial value problem (7). But on the other hand there exist constants c_1, c_2, c_3, not all zero, such that the function $\varphi(t) = c_1\mathbf{u} + c_2\mathbf{v} + c_3\mathbf{w}$ is also a solution of (5). Indeed, we can find appropriate constants c_1, c_2, c_3 by solving the linear algebraic system of unknowns c_1, c_2, c_3,

$$\begin{cases} u_1(t_0)c_1 + v_1(t_0)c_2 + w_1(t_0)c_3 = x_{01} \\ u_2(t_0)c_1 + v_2(t_0)c_2 + w_2(t_0)c_3 = x_{02} \\ u_3(t_0)c_1 + v_3(t_0)c_2 + w_3(t_0)c_3 = x_{03}, \end{cases}$$

where x_{01}, x_{02}, x_{03} are the components of $\mathbf{x}_0$. This system has a nonzero solution because $\mathbf{u}, \mathbf{v}, \mathbf{w}$ are linearly independent, so $W(t) \neq 0$ for all t in I, in particular for $t = t_0$. From the uniqueness of the solution of (5), it follows that $\varphi = \phi$, so $\varphi(t) = c_1\mathbf{u}(t) + c_2\mathbf{v}(t) + c_3\mathbf{w}(t)$ for all t in I. This completes the proof.•

Any three linearly independent vector functions that are solutions of a three-dimensional linear homogeneous system of differential equations are said to form a *fundamental set of solutions*. Such a set is like a "skeleton" on which all the solutions of the system are based. Let us emphasize that *any* three linearly independent solutions can form such a set. This means that the general solution may take different forms if expressed with respect to different fundamental sets. Of course, this does not change the general solution, only its representation. This is like looking at the same object from different perspectives.

EXAMPLE 5 The general solution of system (4) in Example 1 can be written as

$$\varphi(t) = c_1\mathbf{u}(t) + c_2\mathbf{v}(t) + c_3\mathbf{w}(t),$$

where

$$\mathbf{u}(t) = \begin{pmatrix} e^{t^2} \\ e^{t^2} \\ e^{t^2} \end{pmatrix}, \qquad \mathbf{v}(t) = \begin{pmatrix} e^{-t^2/2} \\ -e^{-t^2/2} \\ 0 \end{pmatrix}, \qquad \mathbf{w}(t) = \begin{pmatrix} e^{-t^2/2} \\ 0 \\ -e^{-t^2/2} \end{pmatrix}$$

is a fundamental system of solutions. The particular solution of Example 2 corresponds to choosing $c_1 = c_3 = 0$ and $c_2 = 1$. We can also write the general solution as

$$\varphi(t) = k_1\overline{\mathbf{u}}(t) + k_2\overline{\mathbf{v}}(t) + k_3\overline{\mathbf{w}}(t),$$

where

$$\overline{\mathbf{u}}(t) = \begin{pmatrix} -e^{-t^2/2} \\ e^{-t^2/2} \\ 0 \end{pmatrix}, \qquad \overline{\mathbf{v}}(t) = \begin{pmatrix} e^{t^2} \\ e^{t^2} \\ e^{t^2} \end{pmatrix}, \qquad \overline{\mathbf{w}}(t) = \begin{pmatrix} 2e^{-t^2/2} \\ -e^{-t^2/2} \\ -e^{-t^2/2} \end{pmatrix}$$

is another fundamental system of solutions. The particular solution of Example 2 now corresponds to the choice $k_1 = -1, k_2 = k_3 = 0$.(Can you show that $\mathbf{u}, \mathbf{v}, \mathbf{w}$ and $\overline{\mathbf{u}}, \overline{\mathbf{v}}, \overline{\mathbf{w}}$ are indeed solutions of the system and that their corresponding Wronskians are nonzero?)

The General Solution of a Nonhomogeneous System

The following result shows that finding the general solution of a linear nonhomogeneous system is equivalent to finding one particular solution of it and a fundamental system of solutions of the corresponding homogeneous system. Its proof is similar to that of the fundamental theorem for second-order equations in Section 3.3 (see Problem 30).

Fundamental Theorem for Nonhomogeneous Systems

Any solution of the nonhomogeneous system $\mathbf{x}' = \mathbf{A}(t)\mathbf{x} + \mathbf{b}(t)$, in which the components a_{ij} of the matrix $\mathbf{A}$ and the components b_i of the vector $\mathbf{b}$ are continuous functions, is of the form

$$\varphi(t) = c_1\mathbf{u}(t) + c_2\mathbf{v}(t) + c_3\mathbf{w}(t) + \varphi_0(t),$$

where $\mathbf{u}, \mathbf{v}, \mathbf{w}$ are linearly independent solutions of the corresponding homogeneous system, φ_0 is a particular solution of the nonhomogeneous system, and c_1, c_2, c_3 are real constants.

Unfortunately there is no general method for finding a fundamental system of solutions for the homogeneous system, unless the coefficients a_{ij} are constant, a case we will address in our next section. However, after obtaining in some way enough linearly independent solutions of the homogeneous system, we can use the method of variation of parameters to find a particular solution φ_0 of the nonhomogeneous system. But for systems, the variation of parameters leads to tedious computations. Therefore we will use here the more practical, but also more restrictive, method of undetermined coefficients.

Method for Nonhomogeneous Systems: Undetermined Coefficients

As in Section 3.3, the *method of undetermined coefficients* applies when the coefficients of the homogeneous equation are constant and the components $b_i(t)$ of the vector $\mathbf{b}(t)$ are constants, exponential functions, sines, cosines, polynomials, or any sum or product of these. The table (reproduced below) as well as Rule 1 and Rule 2 of Section 3.3 apply in this case too. For systems, however, there is a supplementary rule (see Problem 29 for an amendment to it).

RULE 3 If one of the coefficients b_i is $\alpha e^{\lambda t}$, where λ is a simple (i.e., not multiple) eigenvalue of the matrix defining the homogeneous system, then each component of the solution to check should contain a term of the form $ate^{\lambda t} + be^{\lambda t}$ instead of only $ate^{\lambda t}$.

TABLE 4.2.1

$b_i(t)$	Attempted solution
$\alpha e^{\beta t}$	$a e^{\beta t}$
$\alpha \cos \omega t + \beta \sin \omega t$	$a \cos \omega t + b \sin \omega t$
$\alpha \ (\neq 0)$	a
$\alpha + \beta t$	$a + bt$
$\alpha + \beta t + \gamma t^2$	$a + bt + ct^2$
$\alpha + \beta t + \gamma t^2 + \delta t^3$	$a + bt + ct^2 + dt^3$
$\vdots$	$\vdots$

EXAMPLE 6 Let us find a particular solution of the linear nonhomogeneous system

$$\begin{cases} x' = y + 2t \\ y' = -x + 15e^{-t}\cos t, \end{cases}$$

knowing that the general solution of the corresponding homogeneous system is

$$x_h(t) = c_1 \cos t + c_2 \sin t, \qquad y_h(t) = c_2 \cos t - c_1 \sin t,$$

where c_1, c_2 are constants. The components of the vector $\mathbf{b}(t)$ are in this case $2t$ and $15e^{-t}\cos t$. According to Table 4.2.1 we will seek a particular solution $\varphi_0(t) = (x_0(t), y_0(t))$ of the nonhomogeneous system, which involves on one hand a linear function in t and on the other hand the product of an exponential with trigonometric functions. So we will check a solution of the form

$$x_0(t) = a_0 + a_1 t + (a \cos t + b \sin t)e^{-t},$$
$$y_0(t) = \bar{a}_0 + \bar{a}_1 t + (\bar{a} \cos t + \bar{b} \sin t)e^{-t}.$$

Since $a_0 + a_1 t, ae^{-t}\cos t, be^{-t}\sin t$ (and their analogues with coefficients $\bar{a}_0, \bar{a}_1, \bar{a}, \bar{b}$) are not among the solutions of the homogeneous equation, the above choice of $x_0(t)$ and $y_0(t)$ is suitable. Substituting x_0 and y_0 for x and y into the nonhomogeneous system and grouping the terms, we obtain

$$(a_1 - \bar{a}_0) - (\bar{a}_1 + 2)t - (a + b + \bar{b})e^{-t}\sin t + (b - a - \bar{a})e^{-t}\cos t = 0,$$
$$(\bar{a}_1 + a_0) + a_1 t - (\bar{a} + \bar{b} - b)e^{-t}\sin t + (\bar{b} - \bar{a} + a - 15)e^{-t}\cos t = 0.$$

These relations can be satisfied for all t if the coefficients vanish, i.e., if $a_1 - \bar{a}_0 = 0, \bar{a}_1 + 2 = 0, \bar{a}_1 + a_0 = 0, a_1 = 0, a + b + \bar{b} = 0, b - a - \bar{a} = 0, \bar{a} + \bar{b} - b = 0$, and $\bar{b} - \bar{a} + a - 15 = 0$. These lead to a linear algebraic system whose solution is $a_0 = 2, a_1 = 0, a = 3, b = -6, \bar{a}_0 = 0, \bar{a}_1 = -2, \bar{a} = -9$, and $\bar{b} = 3$. Thus a particular solution of the nonhomogeneous system is given by

$$\begin{cases} x_0(t) = 2 + 3(-3\cos t + \sin t)e^{-t} \\ y_0(t) = -2t + 3(\cos t - 2\sin t)e^{-t}. \end{cases}$$

EXAMPLE 7   Let us find the solution of the initial value problem

$$\begin{cases} x' = y \\ y' = -x - 2y + 12(t + t^2)e^{-t}, \end{cases} \qquad x(0) = 1, \; y(0) = -1,$$

knowing that the general solution of the corresponding homogeneous system is

$$x_h(t) = c_1 e^{-t} + c_2 t e^{-t}, \qquad y_h(t) = (c_2 - c_1)e^{-t} - c_2 t e^{-t},$$

where c_1, c_2 are constants. The components of the vector $\mathbf{b}(t)$ are in this case 0 and $12(t + t^2)e^{-t}$. According to Table 4.2.1 we should seek a particular solution $\varphi_0(t) = (x_0(t), y_0(t))$ of the nonhomogeneous system involving an expression of the form $(a_0 + a_1 t + a_2 t^2)e^{-t}$. But $a_0 e^{-t}$ and $a_1 t e^{-t}$ resemble the solutions of the homogeneous system, so the next try is a solution involving $t(a_0 + a_1 t + a_2 t^2)e^{-t}$. However, this is still not a good choice since $a_0 t e^{-t}$ resembles a solution of the homogeneous system. The next choice involves the term $t^2(a_0 + a_1 t + a_2 t^2)e^{-t}$. Since no further resemblances occur, we can now check a solution of the form

$$x_0(t) = (a_0 t^2 + a_1 t^3 + a_2 t^4)e^{-t},$$
$$y_0(t) = (\bar{a}_0 t^2 + \bar{a}_1 t^3 + \bar{a}_2 t^4)e^{-t}.$$

Substituting x_0 and y_0 into the system and grouping the terms, we obtain

$$[2a_0 t + (3a_1 - a_0)t^2 + (4a_2 - a_1)t^3 - a_2 t^4]e^{-t} = (\bar{a}_0 t^2 + \bar{a}_1 t^3 + \bar{a}_2 t^4)e^{-t},$$

$$[2\bar{a}_0 t + (3\bar{a}_1 - \bar{a}_0)t^2 + (4\bar{a}_2 - \bar{a}_1)t^3 - \bar{a}_2 t^4]e^{-t}$$
$$= [12t + (12 - a_0 - 2\bar{a}_0)t^2 - (a_1 + 2\bar{a}_1)t^3 - (a_2 + 2\bar{a}_2)t^4]e^{-t}.$$

Since these relations are satisfied for all t only if the corresponding coefficients are equal, we have $a_0 = 0$, $3a_1 - a_0 = \bar{a}_0$, $4a_2 - a_1 = \bar{a}_1$, $-a_2 = \bar{a}_2$, $2\bar{a}_0 = 12$, $3\bar{a}_1 - \bar{a}_0 = -a_0 - 2\bar{a}_1$, $-\bar{a}_2 = -a_2 - 2\bar{a}_2$. This linear algebraic system has the solution $a_0 = 0$, $a_1 = 2$, $a_2 = 1$, $\bar{a}_0 = 6$, $\bar{a}_1 = 2$, and $\bar{a}_2 = -1$, so the particular solution $\varphi_0(t) = (x_0(t), y_0(t))$ has the form

$$\begin{cases} x_0(t) = (2t^3 + t^4)e^{-t} \\ y_0(t) = (6t^2 + 2t^3 - t^4)e^{-t}. \end{cases}$$

The general solution is then

$$\begin{cases} x(t) = c_1 e^{-t} + c_2 t e^{-t} + (2t^3 + t^4)e^{-t} \\ y(t) = (c_2 - c_1)e^{-t} - c_2 t e^{-t} + (6t^2 + 2t^3 - t^4)e^{-t}. \end{cases}$$

Taking $t = 0$ and using the initial conditions $x(0) = 1$, $y(0) = -1$, we obtain $c_1 = 1$ and $c_2 = 0$, so the solution of the initial value problem is

$$\begin{cases} x(t) = (1 + 2t^3 + t^4)e^{-t} \\ y(t) = (-1 + 6t^2 + 2t^3 - t^4)e^{-t}. \end{cases}$$

PROBLEMS

For each of the linear systems in Problems 1 through 8 find an interval in which the general solution is defined.

1. $\begin{cases} x' = x + 2ty - t \\ y' = (\sin t)x + t^2 y + 3\cos t \end{cases}$

2. $\begin{cases} u' = \dfrac{1}{t}u + \sqrt{3}v + t \\ v' = \sqrt{2t}u + 5v + 1 \end{cases}$

3. $\begin{cases} z' = (\ln t)z + tw \\ w' = \dfrac{1}{\cos t}z + t^3 w \end{cases}$

4. $\begin{cases} r' = (\tan t)r + (\cot t)\theta \\ \theta' = (\sin t)r + (\cos t)\theta \end{cases}$

5. $\begin{cases} x' = x + 2ty - 6t^2 z + 1 \\ y' = 2x + 7ty + 9t^2 z + 2t \\ z' = 6x - 3ty - 4t^2 z + 3t^2 \end{cases}$

6. $\begin{cases} u' = -tu + (\ln t)v + (\sin^2 t)w \\ v' = (\cos^2 t)u + \dfrac{1}{t}v + t^2 w + 1 \\ w' = 5u - t^2 w + t \end{cases}$

7. $\begin{cases} r' = \dfrac{8}{\tan t}r + \dfrac{1}{(1-t)^2}\theta + \dfrac{7}{t^2}\varphi \\ \theta' = (\cot t)r + t\theta \\ \varphi' = \dfrac{\tan t}{\cot t}r + t^{-(1/2)}\theta + t^{3/2}\varphi \end{cases}$

8. $\begin{cases} \alpha' = \dfrac{2}{(1 - \ln t)^2}\beta \\ \beta' = \alpha + \dfrac{2-t}{2+t}\beta + \gamma \\ \gamma' = 7t^{-(5/2)}\alpha + 6t^2\gamma \end{cases}$

Determine whether in Problems 9 through 16 the given vector functions are linearly independent on the specified intervals.

9. $\mathbf{v}_1(t) = \begin{pmatrix} t \\ 1 \end{pmatrix}$,

 $\mathbf{v}_2(t) = \begin{pmatrix} t^2 \\ 1 \end{pmatrix}$ on $(-\infty, \infty)$

10. $\mathbf{v}_1(t) = \begin{pmatrix} t^2 \\ 0 \end{pmatrix}$, $\mathbf{v}_2(t) = \begin{pmatrix} t \\ 0 \end{pmatrix}$ on $(-\infty, \infty)$

11. $\mathbf{v}_1(t) = \begin{pmatrix} \sin t \\ \cos t \end{pmatrix}$,

 $\mathbf{v}_2(t) = \begin{pmatrix} \tan t \\ \cot t \end{pmatrix}$ on $(0, \pi/2)$

12. $\mathbf{v}_1(t) = \begin{pmatrix} 2 \\ t \\ t \end{pmatrix}$, $\mathbf{v}_2(t) = \begin{pmatrix} -\dfrac{1}{t^2} \\ 6 \end{pmatrix}$ on $(0, \infty)$

13. $\mathbf{v}_1(t) = \begin{pmatrix} t \\ 1 \\ 0 \end{pmatrix}$, $\mathbf{v}_2(t) = \begin{pmatrix} -2t^2 \\ -2t \\ 0 \end{pmatrix}$,

 $\mathbf{v}_3(t) = \begin{pmatrix} 1 \\ 1 \\ t \\ 0 \end{pmatrix}$ on $(-\infty, 0)$

14. $\mathbf{v}_1(t) = \begin{pmatrix} 1 \\ \ln t \\ \sin t \end{pmatrix}$, $\mathbf{v}_2(t) = \begin{pmatrix} \dfrac{1}{\cos t} \\ \dfrac{\ln t}{\cos t}t \\ \tan t \end{pmatrix}$,

 $\mathbf{v}_3(t) = \begin{pmatrix} t \\ t\ln t \\ t\sin t \end{pmatrix}$ on $(0, \pi/2)$

15. $\mathbf{v}_1(t) = \begin{pmatrix} t^2 \\ 1 \\ 0 \end{pmatrix}$, $\mathbf{v}_2(t) = \begin{pmatrix} 2t^2 \\ 2 \\ 0 \end{pmatrix}$,

 $\mathbf{v}_3(t) = \begin{pmatrix} -t^2 \\ -1 \\ 0 \end{pmatrix}$ on $(-\infty, \infty)$

16. $\mathbf{v}_1(t) = \begin{pmatrix} 0 \\ 1 \\ t \end{pmatrix}$, $\mathbf{v}_2(t) = \begin{pmatrix} t \\ 0 \\ 1 \end{pmatrix}$,

 $\mathbf{v}_3(t) = \begin{pmatrix} 1 \\ t \\ 0 \end{pmatrix}$ on $(-\infty, \infty)$

17. Are the definitions of the Wronskian given in Section 3.1 (can you recover this definition from Theorem 3.1.1?) and Section 4.2 consistent with each other? In other words, if we consider second-order equations as two-dimensional systems, do the two definitions agree?

18. Using the linear algebra results of Section 4.1 and the idea of proof in Theorem 3.1.1, prove Theorem 4.2.2.

19. Using the linear algebra results of Section 4.1 and the idea of proof in Theorem 3.1.2, prove Theorem 4.2.3.

For Problems 20 through 23 show that the specified vector functions form a fundamental system of solutions for the given linear homogeneous system.

20. $\begin{cases} x' = -3x + \sqrt{2}y \\ y' = \sqrt{2}x - 2y \end{cases}$

$$\mathbf{v} = \begin{pmatrix} e^{-t} \\ \sqrt{2}e^{-t} \end{pmatrix}, \quad \mathbf{w} = \begin{pmatrix} -\sqrt{2}e^{-4t} \\ e^{-4t} \end{pmatrix}$$

21. $\begin{cases} u' = u + v \\ v' = 4u + v \end{cases}$

$$\mathbf{x}(t) = \begin{pmatrix} e^{-t} \\ -2e^{-t} \end{pmatrix}, \quad \mathbf{y}(t) = \begin{pmatrix} e^{3t} \\ 2e^{3t} \end{pmatrix}$$

22. $\begin{cases} r' = 2r + (i + 2)\rho \\ \rho' = -r - (i + 1)\rho \end{cases}$

$$\mathbf{z} = \begin{pmatrix} e^{-it} \\ -e^{-it} \end{pmatrix}, \quad \mathbf{w}(t) = \begin{pmatrix} (i + 2)e^t \\ -e^t \end{pmatrix}$$

23. $\begin{cases} x' = x - y + 4z \\ y' = 3x + 2y - z \\ z' = 2x + y - z \end{cases}$

$$\mathbf{u}(t) = \begin{pmatrix} e^{-2t} \\ -e^{-2t} \\ -e^{-2t} \end{pmatrix}, \quad \mathbf{v}(t) = \begin{pmatrix} e^t \\ -4e^t \\ -e^t \end{pmatrix},$$

$$\mathbf{w}(t) = \begin{pmatrix} e^{3t} \\ 2e^{3t} \\ e^{3t} \end{pmatrix}$$

Using the method of undetermined coefficients, find a particular solution of the nonhomogeneous systems in Problems 24 through 29, knowing that in each case the given vector functions form a fundamental system of solutions of the corresponding homogeneous system.

24. $\begin{cases} x' = 2x - y + 2t \\ y' = 3x - 2y - t^2 \end{cases}$

$$\mathbf{u}(t) = \begin{pmatrix} e^t \\ e^t \end{pmatrix}, \quad \mathbf{v}(t) = \begin{pmatrix} e^{-t} \\ 3e^{-t} \end{pmatrix}$$

25. $\begin{cases} r' = 4r - 3s + \sin t \\ s' = 8r - 6s - \cos t \end{cases}$

$$\mathbf{x}(t) = \begin{pmatrix} e^{-2t} \\ 2e^{-2t} \end{pmatrix}, \quad \mathbf{y}(t) = \begin{pmatrix} 3 \\ 4 \end{pmatrix}$$

26. $\begin{cases} \rho' = \rho - 2\theta + e^{-t} - 1 \\ \theta' = 3\rho - 4\theta - 1 \end{cases}$

$$\mathbf{v}(t) = \begin{pmatrix} e^{-t} \\ e^{-t} \end{pmatrix}, \quad \mathbf{w}(t) = \begin{pmatrix} 2e^{-2t} \\ 3e^{-2t} \end{pmatrix}$$

27. $\begin{cases} x' = x + y + 2z + t \\ y' = x + 2y + z - 2t \\ z' = 2x + y + z + t^2 \end{cases}$

$$\mathbf{u}(t) = \begin{pmatrix} e^{-t} \\ 0 \\ -e^{-t} \end{pmatrix}, \quad \mathbf{v}(t) = \begin{pmatrix} e^t \\ -2e^t \\ e^t \end{pmatrix},$$

$$\mathbf{w}(t) = \begin{pmatrix} e^{4t} \\ e^{4t} \\ e^{4t} \end{pmatrix}$$

28. $\begin{cases} u' = 3u + 2v + 4w + 2\sin t \\ v' = 2u + 2w + t \\ w' = 4u + 2v + 3w - e^{-t} \end{cases}$

$$\mathbf{x}(t) = \begin{pmatrix} e^{-t} \\ 0 \\ -e^{-t} \end{pmatrix}, \quad \mathbf{y}(t) = \begin{pmatrix} e^{-t} \\ -4e^{-t} \\ e^{-t} \end{pmatrix},$$

$$\mathbf{z}(t) = \begin{pmatrix} 2e^{8t} \\ e^{8t} \\ 2e^{8t} \end{pmatrix}$$

29. $\begin{cases} u' = 3u + 6v + e^t \\ v' = -u - 2v - t \end{cases}$

$$\mathbf{r}(t) = \begin{pmatrix} -2 \\ 1 \end{pmatrix}, \quad \mathbf{s}(t) = \begin{pmatrix} -3e^t \\ e^t \end{pmatrix}$$

(*Hint:* In this case Rule 3 applies, but it does not suffice. Since the functions $3u + 6v$ and $-u - 2v$ are linearly dependent, the natural choice of a first degree polynomial will not work. Try it and you will see what happens. Then use a quadratic polynomial instead.)

30. Using the linear algebra results of Section 4.1 and the idea of proof in the fundamental theorem for second-order nonhomogeneous equations in Section 3.3, prove the fundamental theorem for nonhomogeneous systems.

4.3 **Equations with Constant Coefficients**

As for linear second-order equations, we can find the solution of a linear homogeneous system if the coefficients are constant. In this section we study systems of the type

$$\mathbf{x}' = \mathbf{A}\mathbf{x}, \tag{1}$$

where the components of the matrix $\mathbf{A}$ are real constants, and we will show how to solve them with the help of the linear algebra theory of Section 4.1.

If $\mathbf{A}$ reduces to one element, i.e., $\mathbf{A} = a$, we recover the separable equation $x' = ax$ of Section 2.2, whose solution is $x(t) = ce^{at}$. This suggests that we should seek for system (1) solutions of the form

$$\mathbf{x}(t) = \mathbf{u}e^{\lambda t}, \tag{2}$$

where the constant vector $\mathbf{u}$ and the real constant λ are to be determined. Substituting (2) into (1), we are led to the linear algebraic system

$$\mathbf{A}_\lambda \mathbf{u} = \mathbf{0}, \tag{3}$$

having the components of the vector $\mathbf{u}$ as unknowns. For each eigenvalue λ and corresponding eigenvector $\mathbf{u}$, a solution of (1) is given by $\mathbf{u}e^{\lambda t}$.

For a two-dimensional system, the derivation of (3) from substituting (2) into (1) works as follows. System (1) is

$$\begin{cases} x_1' = a_{11}x_1 + a_{12}x_2 \\ x_2' = a_{21}x_1 + a_{22}x_2, \end{cases}$$

and our trial solution (2) takes the form

$$\begin{cases} x_1(t) = u_1 e^{\lambda t} \\ x_2(t) = u_2 e^{\lambda t}. \end{cases}$$

By substituting this trial solution and its derivative into system (1), we get

$$\begin{cases} \lambda u_1 e^{\lambda t} = a_{11}u_1 e^{\lambda t} + a_{12}u_2 e^{\lambda t} \\ \lambda u_2 e^{\lambda t} = a_{21}u_1 e^{\lambda t} + a_{22}u_2 e^{\lambda t}. \end{cases}$$

Dividing each equation by $e^{\lambda t}$, which is never zero, and collecting all the terms on one side, we obtain

$$\begin{cases} (a_{11} - \lambda)u_1 + a_{12}u_2 = 0 \\ a_{21}u_1 + (a_{22} - \lambda)u_2 = 0, \end{cases}$$

which represents equation (3) in the two-dimensional case. Notice that for three-dimensional systems we obtain system (7), given by matrix (8) in Section 4.1. Can you do the computations for the three-dimensional case?

We have thus reduced the linear system of differential equations (1) to the linear algebraic system (3), which has other solutions than the trivial one only if $|\mathbf{A}_\lambda| = 0$, i.e., for those values λ that are eigenvalues of $\mathbf{A}$. Thus the problem of solving (1) is equivalent to that of finding the eigenvalues and eigenvectors of the matrix $\mathbf{A}$.

Real and Distinct Eigenvalues

When solving a three-dimensional system of the form (1), the simplest case is that of real and distinct roots λ_1, λ_2, λ_3 of the characteristic polynomial $|A_\lambda|$. Whatever corresponding eigenvectors $\mathbf{u}$, $\mathbf{v}$, $\mathbf{w}$ we choose, they are linearly independent (see Problem 16). Thus, according to the fundamental theorem for homogeneous systems stated in Section 4.2, the general solution of (1) is of the form

$$\mathbf{x}(t) = c_1 \mathbf{u} e^{\lambda_1 t} + c_2 \mathbf{v} e^{\lambda_2 t} + c_3 \mathbf{w} e^{\lambda_3 t}, \tag{4}$$

where c_1, c_2, c_3 are real constants.

EXAMPLE 1   To solve the linear system of differential equations

$$\begin{cases} x_1' = x_1 + x_2 + x_3 \\ x_2' = 2x_1 + x_2 - x_3 \\ x_3' = -8x_1 - 5x_2 - 3x_3, \end{cases}$$

consider the characteristic polynomial $|A_\lambda| = -\lambda^3 - \lambda^2 + 4\lambda + 4$, whose roots are $\lambda_1 = -2$, $\lambda_2 = -1$, $\lambda_3 = 2$.

To find an eigenvector for the eigenvalue λ, substitute the value obtained for λ into the algebraic system $A_\lambda \mathbf{u} = 0$ and solve this system for $\mathbf{u}$. For $\lambda_1 = -2$ and $u_3 = c$ (constant), the system

$$\begin{cases} 3u_1 + u_2 + u_3 = 0 \\ 2u_1 + 3u_2 - u_3 = 0 \\ -8u_1 - 5u_2 - u_3 = 0 \end{cases}$$

has the solution $u_1 = -4c/7$, $u_2 = 5c/7$, $u_3 = c$. For $c = 7$,

$$\mathbf{u} = \begin{pmatrix} -4 \\ 5 \\ 7 \end{pmatrix}.$$

Proceeding analogously for $\lambda_2 = -1$ and $\lambda_3 = 2$, we obtain as possible eigenvectors

$$\mathbf{v} = \begin{pmatrix} -3 \\ 4 \\ 2 \end{pmatrix} \quad \text{and} \quad \mathbf{w} = \begin{pmatrix} 0 \\ -1 \\ 1 \end{pmatrix}.$$

So the general solution of the initial system of differential equations can be expressed in vector form as

$$\mathbf{x}(t) = c_1 \begin{pmatrix} -4 \\ 5 \\ 7 \end{pmatrix} e^{-2t} + c_2 \begin{pmatrix} -3 \\ 4 \\ 2 \end{pmatrix} e^{-t} + c_3 \begin{pmatrix} 0 \\ -1 \\ 1 \end{pmatrix} e^{2t},$$

or equivalently,

$$\begin{cases} x_1(t) = -4c_1 e^{-2t} - 3c_2 e^{-t} \\ x_2(t) = 5c_1 e^{-2t} + 4c_2 e^{-t} - c_3 e^{2t} \\ x_3(t) = 7c_1 e^{-2t} + 2c_2 e^{-t} + c_3 e^{2t}. \end{cases}$$

Remark. We stress again the fact that the formal aspect of the general solution depends on the choice of the eigenvectors, but each choice describes the same set of solutions. Using different eigenvectors is like defining the same object in different words.

Double Eigenvalues

If all the roots of the characteristic polynomial are real but one repeats itself, we obtain double or multiple eigenvalues. This means that the number of linearly independent eigenvectors might be smaller than the dimensionality of the system, so we may not obtain a fundamental system of solutions. In this case we need to refine our method.

We will discuss here the case of double eigenvalues. The general situation is beyond the scope of this textbook. So assume that among all the eigenvalues two are real and equal, say $\lambda_1 = \lambda_2$. Then, as in the previous case, we obtain one eigenvector $\mathbf{u}$ by checking a solution of type (2) in the system. The analogy with second-order equations (see Section 3.2) suggests that we check another solution of the form $t\mathbf{v}e^{\lambda t}$. But unfortunately, this choice fails. A luckier attempt is checking a solution of the form

$$\mathbf{x}(t) = \mathbf{v}e^{\lambda_1 t} + t\mathbf{u}e^{\lambda_1 t}, \tag{5}$$

with λ_1 the above eigenvalue, $\mathbf{u}$ the already computed eigenvector, and $\mathbf{v}$ to be determined. The computations (see Problem 17) lead to the linear algebraic system

$$\mathbf{A}_{\lambda_1}\mathbf{v} = \mathbf{u}, \tag{6}$$

which we can solve for $\mathbf{v}$. It can be proved that the vectors $\mathbf{u}$ and $\mathbf{v}$ are linearly independent. For the remaining simple (i.e., not multiple) eigenvalues, we can proceed as in the previous case.

EXAMPLE 2 The system

$$\begin{cases} x_1' = 3x_1 - x_2 \\ x_2' = x_1 + x_2 \end{cases}$$

has the eigenvalues $\lambda_1 = \lambda_2 = 2$. To obtain one eigenvector corresponding to this double eigenvalue, proceed as in the previous example. This leads to the algebraic system

$$\begin{cases} u_1 - u_2 = 0 \\ u_1 - u_2 = 0, \end{cases}$$

which has infinitely many solutions: $u_1 = \alpha, u_2 = \alpha$. The choice $\alpha = 1$ provides the eigenvector

$$\mathbf{u} = \begin{pmatrix} 1 \\ 1 \end{pmatrix},$$

which yields the solution

$$\varphi(t) = \begin{pmatrix} 1 \\ 1 \end{pmatrix} e^{2t}.$$

To obtain a second linearly independent eigenvector $\mathbf{v}$, insert into the original system a solution of form (5), i.e., $\mathbf{v}e^{2t} + \mathbf{u}te^{2t}$, that leads to the algebraic system (6), i.e.,

$$\begin{cases} v_1 - v_2 = 1 \\ v_1 - v_2 = 1. \end{cases}$$

This system has the solution $v_1 = 1 + \beta$, $v_2 = \beta$, so for $\beta = 0$ we obtain the eigenvector

$$\mathbf{v} = \begin{pmatrix} 1 \\ 0 \end{pmatrix},$$

which provides a second linearly independent solution,

$$\psi(t) = \begin{pmatrix} 1 \\ 0 \end{pmatrix} e^{2t} + \begin{pmatrix} 1 \\ 1 \end{pmatrix} te^{2t}.$$

Thus the general solution of the system is

$$\mathbf{x}(t) = c_1 \begin{pmatrix} 1 \\ 1 \end{pmatrix} e^{2t} + c_2 \begin{pmatrix} 1 \\ 0 \end{pmatrix} e^{2t} + c_2 \begin{pmatrix} 1 \\ 1 \end{pmatrix} te^{2t},$$

or equivalently,

$$\begin{cases} x_1(t) = (c_1 + c_2)e^{2t} + c_2 te^{2t} \\ x_2(t) = c_1 e^{2t} + c_2 te^{2t}. \end{cases}$$

Complex Eigenvalues

The characteristic polynomial can have complex roots that, as known from elementary algebra, appear in conjugate pairs. So if $\lambda_1 = \alpha + i\beta$ is an eigenvalue, $\lambda_2 = \alpha - i\beta$ is another one. To these eigenvalues correspond complex conjugate eigenvectors, $\mathbf{u} = \mathbf{a} + i\mathbf{b}$ and $\mathbf{v} = \mathbf{a} - i\mathbf{b}$. We can then show (see Problem 18) that system (1) always has the two linearly independent solutions

$$\mathbf{U}(t) = e^{\alpha t}(\mathbf{a}\cos\beta t - \mathbf{b}\sin\beta t) \qquad \text{and} \qquad \mathbf{V}(t) = e^{\alpha t}(\mathbf{a}\sin\beta t + \mathbf{b}\cos\beta t).$$
$$(7)$$

For other eigenvalues we proceed as in the previous cases.

EXAMPLE 3   The linear system of differential equations

$$\begin{cases} x_1' = x_1 - x_2 \\ x_2' = 5x_1 - 3x_2 \end{cases}$$

leads to the eigenvalues $\lambda_1 = -1 + i$, $\lambda_2 = -1 - i$. To obtain the eigenvector $\mathbf{v}$, substitute λ_2 into the corresponding algebraic system (3). This leads to solving the linear system

$$\begin{cases} (2 + i)v_1 - v_2 = 0 \\ 5v_1 - (2 - i)v_2 = 0, \end{cases}$$

which has infinitely many solutions: $v_1 = (2 - i)\gamma/5$, $v_2 = \gamma$. For $\gamma = 5$ obtain the eigenvector

$$\mathbf{v} = \begin{pmatrix} 2 - i \\ 5 \end{pmatrix}.$$

Writing $\mathbf{v}$ as $\mathbf{a} + i\,\mathbf{b}$, see that

$$\mathbf{a} = \begin{pmatrix} 2 \\ 5 \end{pmatrix} \quad \text{and} \quad \mathbf{b} = \begin{pmatrix} -1 \\ 0 \end{pmatrix}.$$

According to (7), two linearly independent solutions are

$$\mathbf{U}(t) = \begin{pmatrix} 2\cos t + \sin t \\ 5\cos t \end{pmatrix} e^{-t} \quad \text{and} \quad \mathbf{V}(t) = \begin{pmatrix} 2\sin t - \cos t \\ 5\sin t \end{pmatrix} e^{-t},$$

so the general solution is

$$\mathbf{x}(t) = c_1 \begin{pmatrix} 2\cos t + \sin t \\ 5\cos t \end{pmatrix} e^{-t} + c_2 \begin{pmatrix} 2\sin t - \cos t \\ 5\sin t \end{pmatrix} e^{-t},$$

or equivalently,

$$\begin{cases} x_1(t) = [(2c_1 - c_2)\cos t + (c_1 + 2c_2)\sin t]e^{-t} \\ x_2(t) = 5(c_1 \cos t + c_2 \sin t)e^{-t}. \end{cases}$$

Applications

Seeking a vaccine for AIDS   An international research group attempting to find a vaccine against AIDS needs to determine the amount of substance produced in a chemical reaction. The scientists realize that the rates of change of the quantities x and y of two substances depend linearly on x and y. The reaction is described by the initial value problem

$$\begin{cases} x' = x - b^2 y \\ y' = 4x + y, \end{cases} \qquad x(0) = b, \; y(0) = 2000, \tag{8}$$

where b is a parameter that can be slightly varied with each experiment about the value 1. Obtain the formula that calculates the change of x and y over time, and see how the two quantities vary in 1 hour if $b = 1$.

Solution. System (8) can be written as $\mathbf{z}' = \mathbf{A}\mathbf{z}$, where

$$\mathbf{A} = \begin{pmatrix} 1 & -b^2 \\ 4 & 1 \end{pmatrix} \quad \text{and} \quad \mathbf{z} = \begin{pmatrix} x \\ y \end{pmatrix}.$$

The eigenvalues of $\mathbf{A}$ are complex conjugates,

$$\lambda_1 = 1 + 2ib \quad \text{and} \quad \lambda_2 = 1 - 2ib.$$

To obtain an eigenvector, substitute λ_2 into $\mathbf{A}_\lambda \mathbf{u} = \mathbf{0}$ and get the algebraic system

$$\begin{cases} 2ibu_1 - b^2 u_2 = 0 \\ 4u_1 + 2ibu_2 = 0. \end{cases}$$

Take $u_2 = \alpha$ and from the first equation obtain $u_1 = \alpha b/(2i) = -\alpha bi/2$. The choice $\alpha = 2$ leads to the eigenvector

$$\begin{pmatrix} 0 \\ 2 \end{pmatrix} + i \begin{pmatrix} -b \\ 0 \end{pmatrix}.$$

Figure 4.3.1. The variation of x and y during the first hour for $b = 1$.

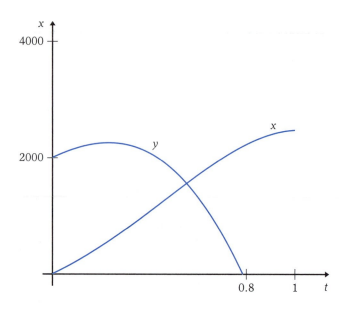

By (7), two linearly independent solutions of (8) are

$$\left[\binom{0}{2}\cos 2bt - \binom{-b}{0}\sin 2bt\right]e^t \quad \text{and} \quad \left[\binom{0}{2}\sin 2bt + \binom{-b}{0}\cos 2bt\right]e^t,$$

so the general solution of (8) is

$$\begin{cases} x(t) = be^t(c_1 \sin 2bt - c_2 \cos 2bt) \\ y(t) = 2e^t(c_1 \cos 2bt + c_2 \sin 2bt). \end{cases}$$

The initial conditions $x(0) = b, y(0) = 2000$ lead to $c_1 = 1000$ and $c_2 = -1$. For $b = 1$ the graphs of x and y show how the quantities vary in 1 hour (see Figure 4.3.1).●

A brewing technique   Since 6000 B.C., when beer was already known in Babylonia and Sumeria, hundreds of brewing techniques have been used. One such technique requires two tanks, A and B, in which water mixes with two kinds of sugar, and a larger tank,C, that collects the combined mixture for fermentation. The tanks A and B in Figure 4.3.2 initially contain 1000 l and 2000 l of water, respectively. From an external source 5 l of liquid sugar flows into tank A every minute, and from another external source 10 l of the second kind of liquid sugar flows into tank B every minute. The mixture from

Figure 4.3.2. The brewing tanks seen from above.

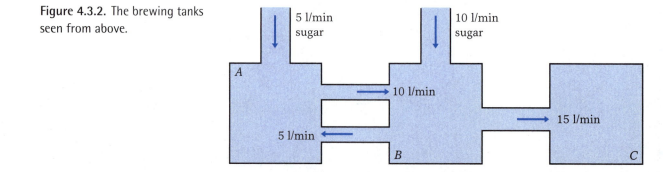

Beer brewing in tanks.

A goes into B at a rate of 10 l per minute, while the mixture from B is pumped into A at a rate of 5 l per minute and into C at a rate of 15 l per minute. The technique requires that the amount of sugar in A and B be known at every moment. Assuming that the liquids mix instantly so that the mixture is homogeneous, derive the formula that gives this amount.

Solution. Denote by u and v the volumes of sugar in tanks A and B, respectively, and notice that $u(0) = v(0) = 0$. The rate of change of u increases with the volume of liquid sugar (5 l), decreases with u in the proportion of $10/1000$ due to the liquid that flows from A to B, and increases with v in the proportion of $5/2000$ due to the liquid pumped from B to A. Similarly, the rate of change of v increases with 10 l, increases with u in the proportion of $10/1000$, and decreases with v in the proportion of $(5 + 15)/2000$ due to the liquid pumped from B into A and C. These remarks lead to the linear nonhomogeneous system

$$\begin{cases} u' = 5 - 0.01u + 0.0025v \\ v' = 10 + 0.01u - 0.01v. \end{cases} \tag{9}$$

To solve it, first consider the corresponding homogeneous system

$$\begin{cases} U' = -0.01U + 0.0025V \\ V' = 0.01U - 0.01V, \end{cases} \tag{10}$$

whose matrix has the eigenvalues $\lambda_1 = -0.005$, $\lambda_2 = -0.015$ and the corresponding linearly independent eigenvectors

$$\begin{pmatrix} 0.5 \\ 1 \end{pmatrix} \quad \text{and} \quad \begin{pmatrix} -0.5 \\ 1 \end{pmatrix}.$$

The general solution of (10) is thus

$$\begin{pmatrix} U(t) \\ V(t) \end{pmatrix} = c_1 e^{-0.005t} \begin{pmatrix} 0.5 \\ 1 \end{pmatrix} + c_2 e^{-0.015t} \begin{pmatrix} -0.5 \\ 1 \end{pmatrix}.$$

Applying the method of undetermined coefficients, we obtain a particular solution of (9),

$$\begin{pmatrix} u_0(t) \\ v_0(t) \end{pmatrix} = \begin{pmatrix} 1000 \\ 2000 \end{pmatrix}.$$

Figure 4.3.3. The graphs of the variation of the amounts of sugar in tanks A and B.

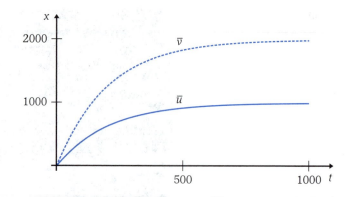

The general solution of (9) is then

$$\begin{pmatrix} u(t) \\ v(t) \end{pmatrix} = c_1 e^{-0.005t} \begin{pmatrix} 0.5 \\ 1 \end{pmatrix} + c_2 e^{-0.015t} \begin{pmatrix} -0.5 \\ 1 \end{pmatrix} + \begin{pmatrix} 1000 \\ 2000 \end{pmatrix}.$$

The initial conditions $u(0) = v(0) = 0$ lead to $c_1 = -2000$ and $c_2 = 0$, so the solution of the corresponding initial value problem is

$$\begin{cases} \bar{u}(t) = -1000e^{-0.005t} + 1000 \\ \bar{v}(t) = -2000e^{-0.005t} + 2000. \end{cases}$$

The graphs of $\bar{u}$ and $\bar{v}$ in Figure 4.3.3 show that the two quantities increase from 0 until they reach a certain saturation threshold.●

PROBLEMS

Solve the systems of differential equations in Problems 1 through 12.

1. $\begin{cases} x' = 2x + y \\ y' = 8x - 5y \end{cases}$

2. $\begin{cases} u' = \frac{5}{3}u + v \\ v' = u + \frac{5}{3}v \end{cases}$

3. $\begin{cases} \alpha' = \alpha + 6\beta \\ \beta' = \alpha \\ \gamma' = 3\beta + \gamma \end{cases}$

4. $\begin{cases} x' = 2x + 3y \\ y' = -4x - y \\ z' = 2x - y + 2z \end{cases}$

5. $\begin{cases} \theta' = 4\theta + r \\ r' = -\theta + 2r \end{cases}$

6. $\begin{cases} v' = 2v + 5w \\ w' = -5v + 2w \end{cases}$

7. $\begin{cases} P' = 7P + 3Q - 9R \\ Q' = 4Q \\ R' = 6Q + 8R \end{cases}$

8. $\begin{cases} u' = v + w \\ v' = w \\ w' = 2v + w \end{cases}$

9. $\begin{cases} x' = 4x + y \\ y' = 3x + 2y \end{cases}$

10. $\begin{cases} p' = 3p - q \\ q' = 2p + 6q \end{cases}$

11. $\begin{cases} z' = -2z - \rho + \psi \\ \rho' = 0 \\ \psi' = -3z + \rho \end{cases}$

12. $\begin{cases} x' = \frac{1}{2}x \\ y' = x + \frac{3}{2}y \\ z' = \frac{1}{2}x + y + z \end{cases}$

13. Rewrite the linear second-order equation

$$x'' = bx' + cx,$$

where b and c are real constants, as a linear system and solve it with the help of the above methods.

14. Consider all possibilities and solve the system

$$\begin{cases} x' = \alpha x + \beta y \\ y' = \gamma x + \delta y, \end{cases}$$

where $\alpha, \beta, \gamma, \delta$ are real constants.

15. Consider all possibilities and solve the system

$$\begin{cases} x' = \alpha x - y - z \\ y' = \beta y - z \\ z' = -y - z, \end{cases}$$

where α and β are real constants.

16. For a three-dimensional linear homogeneous system of differential equations with constant coefficients, show that if the eigenvalues of the coefficient matrix are real and distinct, then the corresponding eigenvectors are linearly independent.

17. Show that in the case of double eigenvalues, the substitution of the vector function $\mathbf{x}(t) = \mathbf{v}e^{\lambda_1 t} + t\mathbf{u}e^{\lambda_1 t}$ into the system $\mathbf{x}' = \mathbf{A}\mathbf{x}$ leads to the algebraic system $\mathbf{A}_{\lambda_1}\mathbf{v} = \mathbf{u}$.

18. Show by directly substituting into the two-dimensional linear homogeneous system $\mathbf{x}' = \mathbf{A}\mathbf{x}$ that if the matrix $\mathbf{A}$ has the complex conjugate eigenvectors $\mathbf{u} = \mathbf{a} + i\mathbf{b}$ and $\mathbf{v} = \mathbf{a} - i\mathbf{b}$ for the eigenvalues $\alpha \pm \beta i$, then the system has as solutions the functions

$$(\mathbf{a}\cos\beta t - \mathbf{b}\sin\beta t)e^{\alpha t} \quad \text{and}$$

$$(\mathbf{a}\cos\beta t + \mathbf{b}\sin\beta t)e^{\alpha t}.$$

Show that these solutions are linearly independent.

19. Tanks A and B (see Figure 4.3.4(a)) initially contain 10,000 l and 5000 l of water, respec-

Figure 4.3.4.
The tanks in (a) Problem 19 and (b) Problem 20.

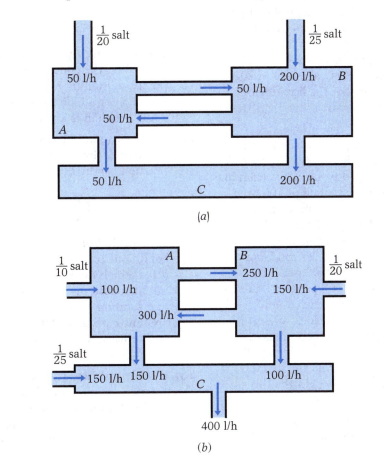

(a)

(b)

Figure 4.3.5. (*a*) The cams of a friend in rock climbing. (*b*) The force F acting in the use of a friend is the sum of the normal component F_n and the vertical component F_v.

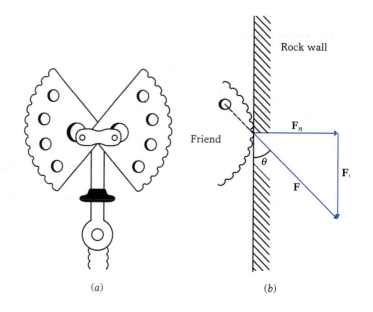

(*a*)   (*b*)

tively. A 1:20 mixture of salt and water flows from an external source into A at a rate of 50 l per hour, and a 1:25 mixture of salt and water flows into B at a rate of 200 l per hour. The mixture from A goes into each B and C at a rate of 50 l per hour, while the mixture from B is pumped into A at a rate of 50 l per hour and into C at a rate of 200 l per hour. Assuming that the liquids mix instantly, derive the formula that gives the amount of salt in A and B.

20. Each of the tanks A, B, and C (see Figure 4.3.4(*b*)) initially contains 1000 l of water. A 1:10 solution of salt and water flows from an external source into A at a rate of 100 l per hour, a 1:20 solution of salt and water flows into B at a rate of 150 l per hour, and a 1:25 solution of salt and water flows into C at a rate of 150 l per hour. The solution from A goes into B at a rate of 250 l per hour and into C at a rate of 150 l per hour, the solution from B is pumped into A at a rate of 300 l per hour and into C at a rate of 100 l per hour, and the solution leaves C at a rate of 400 l per hour. Assuming that the liquids mix instantly, find the formula that gives the amounts of salt in the three tanks.

21. The change in the amounts x and y of two substances that enter a certain chemical

reaction can be described by the initial value problem

$$\begin{cases} x' = -3x + \alpha y \\ y' = \beta x - 2y \end{cases} \quad x(0) = y(0) = 1,$$

where α and β are two parameters that depend on the conditions of reaction (temperature, humidity, etc.). Are there values of α and β for which the solution of the initial value problem is a periodic function of time?

22. The change in the amounts x and y of two substances that enter a certain chemical reaction can be described by the system

$$\begin{cases} x' = ax + y \\ y' = x + ay, \end{cases}$$

where a is a parameter that depends on the conditions of reaction. Are there values of a for which the initial value problem with initial conditions $x(0) = 0, y(0) = 1$ leads to a solution in which after some time the amount x becomes larger than the amount y and remains so if the reaction is continued indefinitely?

23. Rock climbing. The device in Figure 4.3.5(*a*), called a *friend*, was invented in 1973 by Ray Jardine to secure ropes to cracks in rock climbing. A *friend* is made of four cams symmetrically attached to a pivot such that they can rotate by a spring-and-pulley mechanism. In 1998 two undergraduate students at Montana State University in Bozeman, Matthew Bonney and Joshua Coaplen, together with their differential equations course instructor, Erik Doeff, derived a model to determine the best shape of the friend's cams. Taking the pivot point as the origin of a planar coordinate system (see Figure 4.3.5(*b*)), they represented the boundary of the cam that touches the rock as a parametrized curve $(x(t), y(t))$. Using the physical principles of forces (see Figure 4.3.5(*b*)), they were led to the linear system

$$\begin{cases} x' = -ax - y \\ y' = x - ay \end{cases}$$

with $a = \cot \theta$, where θ is the angle of incidence of the force $\mathbf{F}$ against the crack of the wall. What are the possible shapes of the cam? (*Hint:* Represent the solutions of the system in the xy-plane.)

Rock climber with rope secured by device.

4.4 Qualitative Methods

In this section we will present some qualitative methods for the study of autonomous systems of differential equations,

$$\mathbf{x}' = \mathbf{F}(\mathbf{x}), \tag{1}$$

where $\mathbf{F}$ is a vector function of the same dimension as the vector variable $\mathbf{x}$. We will define the *phase space*, classify the equilibria, investigate the notion of *invariant set*, learn how to sketch *flows*, and use them to draw conclusions about the qualitative behavior of the solutions. In the end we will use these results to understand the fish population dynamics of the Tasmanian Sea and the free-market economy described by a linear model.

Phase Space

If the path of a raft floating on the Amazon is a metaphor for a solution in phase plane, the imaginary trace left by a spaceship drifting through the universe is a metaphor for a solution in phase space. But although our imagination is restricted to a three-dimensional world, the phase space can have any number of dimensions. In particular, the phase line and the phase plane are one- and two-dimensional phase spaces, respectively.

The *vector field* consists of all vectors based at points of coordinates $\mathbf{x}$ and oriented in the direction of the vector $\mathbf{F}(\mathbf{x})$ (i.e., of $\mathbf{x}'$). So the vectors representing the velocity of the spaceship, tangent to its trajectory, are part of the vector field. Alternatively, we can draw the *direction field*, when all vectors are adjusted to the same length.

Figure 4.4.1. The vector based at (2, 1, 0) for the vector field defined by system (2).

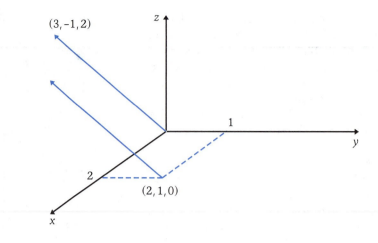

The vector based at $(x, y, z) = (2, 1, 0)$ belonging to the vector field $\mathbf{F}(x, y, z) = (2x - y + z, -x + y - z, x + 3z)$ that defines the system

$$\begin{cases} x' = 2x - y + z \\ y' = -x + y - z \\ z' = x + 3z \end{cases} \qquad (2)$$

is parallel to and has the same length as the vector connecting the origin with the point $\mathbf{F}(2, 1, 0) = (3, -1, 2)$ (see Figure 4.4.1).

To draw a solution curve in phase space, we pick a point that corresponds to some initial conditions and then follow the vector field such that the vectors are tangent to the curve. Since the independent variable t does not show up, we can only guess how fast we move along the curve while t is changing.

For a given system, all possible solution curves in phase space form the *flow*. The goal of the qualitative theory is to understand and describe the flow.

In drawing the flow, a first rule to follow is that, when close together, the curves that represent solutions behave similarly. This property is called *continuity of the solution with respect to initial data*, a notion we will discuss in more detail in Section 5.5. For now we should keep in mind only that most flows behave very much like rivers: While close together, the paths of two rafts are fairly similar (see Figure 4.4.2).

Drawing the flow in phase space for a given system can be a complicated task. Unlike the case in the phase plane, where the restriction

Figure 4.4.2. When close together, solutions behave similarly, as in (*a*), not as in (*b*).

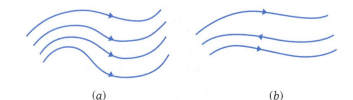

(*a*)   (*b*)

to two dimensions makes things easier, the extra dimension of the phase space can lead to a lot of trouble. Especially for nonlinear systems, whose orbits may be quite erratic (as we will see in Chapter 5), it is often impossible to understand the global flow. But usually we can determine what happens locally. One good starting point is to study the flow near equilibrium solutions. Let us now get some idea about how this can be done in the case of linear systems.

Equilibrium Solutions

In phase space as in the phase plane, equilibrium solutions are represented by fixed, unmoving points. In terms of our metaphor, if we restrict the universe to our rotating galaxy, then its center is an equilibrium. We can now extend Definition 3.4.1 to a more general one.

DEFINITION 4.4.1

An *equilibrium solution* of system (1) is a constant vector function $\mathbf{x}$, such that $\mathbf{F}(\mathbf{x}) = \mathbf{0}$.

EXAMPLE 2   For system (2), the equilibria are the solutions of the algebraic system

$$\begin{cases} 2x - y + z = 0 \\ -x + y - z = 0 \\ x + 3z = 0, \end{cases}$$

which has the unique solution $x = y = z = 0$. So system (2) has only one equilibrium, the zero vector function $(x(t), y(t), z(t)) = (0, 0, 0)$.

The zero vector function is always an equilibrium of a linear system with constant coefficients, $\mathbf{x}' = \mathbf{Ax}$, since $\mathbf{0}$ is a solution of the algebraic system $\mathbf{Ax} = \mathbf{0}$. From Remark 1 of Section 4.1 it follows that if $|\mathbf{A}| \neq 0$, then the zero function is the only equilibrium.

We will deal with nonlinear systems in Chapter 5. For now we will focus on linear systems and study the possible behavior of the flow.

Linear Systems

In Section 3.5 we classified the equilibria of a linear second-order equation

$$x'' + bx' + cx = 0 \tag{3}$$

relative to the roots of the characteristic polynomial $r^2 + br + c$. Following the same pattern, we will now classify the equilibria of a system

$$\mathbf{x}' = \mathbf{Ax} \tag{4}$$

with respect to the roots of the characteristic polynomial $|\mathbf{A}_\lambda|$, i.e., the eigenvalues of $\mathbf{A}$. (Can you show that if $\mathbf{A}$ is the matrix corresponding to the system equivalent to equation (3), then $|\mathbf{A}_\lambda| = \lambda^2 + b\lambda + c$?)

DEFINITION 4.4.2

An equilibrium of the linear system $\mathbf{x}' = \mathbf{Ax}$ is called *hyperbolic* if all eigenvalues of $\mathbf{A}$ are nonzero and have nonzero real part, and it is called *nonhyperbolic* if at least one eigenvalue is zero or has zero real part. A hyperbolic equilibrium is a

(i) **source**, if all eigenvalues are positive or have positive real part,

(ii) **sink**, if all eigenvalues are negative or have negative real part,

(iii) **saddle**, if it is neither a source nor a sink.

The notion of a *nonhyperbolic* equilibrium generalizes that of a *center*. We will continue to use the term *center* for two-dimensional systems when all solutions are periodic around the origin. However, for three- and higher-dimensional systems, the flow near nonhyperbolic equilibria can be very different from the one near sources, sinks, or saddles. Unfortunately, there are no general methods that describe the flow, so every nonhyperbolic equilibrium must be treated as a new problem.

EXAMPLE 3   The nature of the equilibrium $(x, y, z) = (0, 0, 0)$ of the system

$$\begin{cases} x' = ax - y \\ y' = x + by \\ z' = cz \end{cases} \tag{5}$$

depends on the values of the constants $a, b,$ and c. Indeed, the eigenvalues

$$\lambda_1 = \frac{1}{2}[a + b + \sqrt{(a - b)^2 - 4}],$$

$$\lambda_2 = \frac{1}{2}[a + b - \sqrt{(a - b)^2 - 4}], \qquad \lambda_3 = c$$

lead to several situations. Let us describe some of them.

(a) $a = -4, b = 1, c = -2$. The eigenvalues are real, and $\lambda_1, \lambda_2, \lambda_3 < 0$, so the origin is a sink. The flow is sketched in Figure 4.4.3(*a*).

(b) $a = 2, b = 1, c = 3$. The eigenvalues λ_1 and λ_2 are complex, whereas λ_3 is real. Since $\text{Re}\,\lambda_1, \text{Re}\,\lambda_2 > 0$, and $\lambda_3 > 0$, the equilibrium is a source.[1] In the *xy*-plane the solutions spiral away from the origin, so the continuity of the solutions with respect to initial data forces the solutions outside but near the *xy*-plane to spiral too (see Figure 4.4.3(*b*)).

(c) $a = -2, b = -1, c = 3$. The eigenvalues λ_1 and λ_2 are complex, whereas λ_3 is real. Since $\text{Re}\,\lambda_1, \text{Re}\,\lambda_2 < 0$, and $\lambda_3 > 0$, the equilibrium is a saddle. The solutions in the *xy*-plane spiral toward the origin, but the solutions outside this

---

[1] Recall that $\text{Re}\,\lambda$ denotes the *real part* of λ, i.e., if $\lambda = \alpha + \beta i$, then $\text{Re}\,\lambda = \alpha$.

Figure 4.4.3. The flow of system (5) in cases (*a*), (*b*), (*c*), and (*d*).

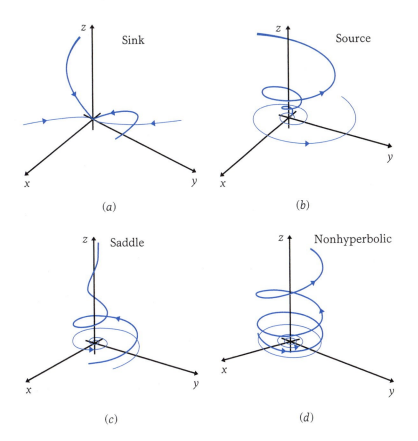

plane spiral away from the origin. Again, the spiraling effect outside and near the *xy*-plane is due to the continuity of the solutions with respect to initial data (see Figure 4.4.3(*c*)).

(d) $a = 2, b = -2, c = 1$. The eigenvalues λ_1 and λ_2 are purely imaginary, whereas $\lambda_3 > 0$ is real. Since Re λ_1 = Re λ_2 = 0, the equilibrium is nonhyperbolic. If the flow is restricted to the *xy*-plane, the equilibrium is a center. Again, the continuity of the solutions with respect to initial data makes the curves outside the *xy*-plane spiral away from the origin (see Figure 4.4.3(*d*)).

Unlike hyperbolic equilibria, nonhyperbolic ones occur even if $|\mathbf{A}| = 0$. In such cases there may exist other equilibria besides the origin, as the following example shows.

EXAMPLE 4 The flow of the system

$$\begin{cases} u' = 0 \\ v' = v \end{cases} \tag{6}$$

is shown in Figure 4.4.4. Since any point with coordinates $(u, 0)$ is an equilibrium, the *u*-axis is filled with equilibria, and since u = constant along any solution, $v' > 0$ for $v > 0$ and $v' < 0$ for $v < 0$, so all the other solutions are represented by lines parallel to the *v*-axis, going upward if $v > 0$ and downward if $v < 0$.

Figure 4.4.4. The flow of system (6).

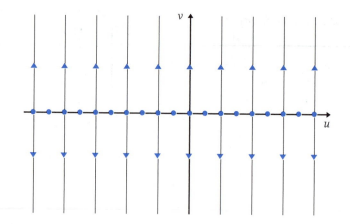

Invariant Sets

In Example 3 we first described the flow in the xy-plane and then outside it. It is natural to ask whether any flow can be decomposed like this. To provide an answer, we first introduce the following definition.

DEFINITION 4.4.3

A set Λ is called *invariant* relative to system (1) if whenever $\mathbf{x}(t_0)$ belongs to Λ for some initial value t_0, then $\mathbf{x}(t)$ belongs to Λ for all t for which the solution $\mathbf{x}$ is defined.

In other words, an invariant set is a union of solutions. In particular, the whole phase space is an invariant set; the same is true about any single solution. But of interest are those nontrivial invariant sets that divide the phase space into components and simplify the study of the flow. Therefore finding invariant sets is an important goal toward understanding the flow.

EXAMPLE 5   The xy-plane is an invariant set for system (5). Indeed, the first two equations involve only the variables x and y, and so are independent of z; therefore they form a separate system having its own phase plane. This further implies that the half-spaces $z > 0$ and $z < 0$ are invariant, so the flow has three components: the xy-plane, the half-space above it, and the half-space below it (see Problem 39). In general, the planes formed by the coordinate axes are not invariant. From this point of view, system (5) is an exception.

EXAMPLE 6   A solution looking like a loop that "connects" an equilibrium with itself (see Figure 4.4.5) is called *homoclinic*. In case of planar flows, homoclinic orbits give rise to natural, nontrivial invariant sets. For the flow in Figure 4.4.5, for example, the union of all periodic solutions inside the loop forms an invariant set. Consequently, the union of all orbits outside the loop forms an invariant set too.

Figure 4.4.5. For planar flows, homoclinic solutions give rise to natural invariant sets.

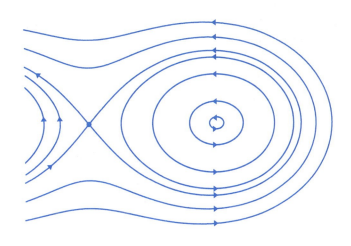

Applications

The fish populations of the Tasmanian Sea

A group of Australian marine ecologists derived a linear model to study the interaction between two fish populations inhabiting the Tasmanian Sea. The two species compete for survival, preying on each other and on plankton. The plankton, however, is almost unaffected by the two species, so it is reasonable to assume that its population remains constant. If x, y, and z denote the three populations, the equations describing the dynamics are found to be

$$\begin{cases} x' = 0.5x - 2.5y + 0.5 \\ y' = 2.6x - 0.5y - 9.9 \\ z' = 0. \end{cases} \tag{7}$$

Since the oceanographers have only a very rough estimate of the sizes of the populations, they would like to see if the conclusion of a qualitative study agrees with their statistical predictions, important for the fish industry. Determine the flow of the above system.

Fishing boat at sea.

Figure 4.4.6. The flow of the fish population model (7).

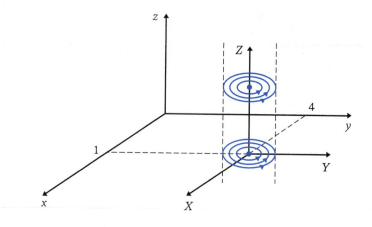

Solution. The above system is not homogeneous, but a suitable transformation of coordinates makes it homogeneous. To find this transformation, notice that the original system has nonhyperbolic equilibria at all points of the form $(x, y, z) = (4, 1, z)$, i.e., the points of the line perpendicular to the xy-plane through the point $(4, 1, 0)$ (see Figure 4.4.6). A suitable transformation must shift the origin of the coordinate system to this point:

$$\begin{cases} X = x - 4 \\ Y = y - 1 \\ Z = z. \end{cases}$$

In the new coordinates, system (7) becomes

$$\begin{cases} X' = 0.5X - 2.5Y \\ Y' = 2.6X - 0.5Y \\ Z' = 0. \end{cases}$$

The first two equations are independent of the third, so the XY-plane is an invariant set for the above system. The eigenvalues of the matrix corresponding to the first two equations are $\lambda_1 = 2.5i$, $\lambda_2 = -2.5i$, so the flow in the XY-plane is formed by periodic orbits. Since $Z' = 0$, every plane $Z = $ constant is invariant, and the flow is the same as in the XY-plane (see Figure 4.4.6). For system (7) this means that any initial conditions $(x(0), y(0), z(0)) = (x_0, y_0, z_0)$ with $x_0, y_0, z_0 > 0$ lead to a periodic change in the population of the two species of fish, a conclusion in agreement with the statistical prediction.•

A linear free-market model

Recall from the introduction that a crude model for a free-market economy is the linear one given by the system

$$\begin{cases} p' = \alpha r \\ r' = -\beta r \end{cases} \tag{8}$$

where p is the *price* of a commodity, r denotes the *excess demand*, i.e., the difference between *demand* and *supply*, and $\alpha, \beta > 0$ are constants. Let us find out whether prices tend to an equilibrium.

Figure 4.4.7. The flow of the linear free-market model (8).

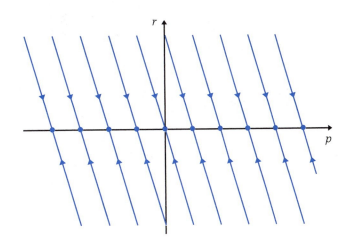

Solution. Notice first that all points $(p, 0)$ are equilibria of system (8); i.e., the p-axis is filled with nonhyperbolic equilibria (see Figure 4.4.7). Of course, since prices are nonnegative, we restrict our phase plane to the half-plane $p > 0$. In (8), dividing the second equation by the first yields $r' = -(\beta/\alpha)p'$, which after integration leads to

$$r = -\frac{\beta}{\alpha}p + c,$$

where c is a constant. This equation represents lines of negative slope $-\beta/\alpha$ in the pr-plane. Since on one hand $p' > 0$ and $r' < 0$ for $r > 0$, and on the other hand $p' < 0$ and $r' > 0$ for $r < 0$, the orientation of the solution lines is as depicted in Figure 4.4.7. This means that every solution leads to an equilibrium. So if demand is larger than supply, the price increases, and if supply is larger than demand, the price decreases, in both cases until the price reaches an equilibrium.●

PROBLEMS

Sketch the vector fields or the direction fields of the systems in Problems 1 through 8. Use, if you wish, one of the programs in Section 4.6.

1. $\begin{cases} x' = 2x + 3y \\ y' = 3x + 2y \end{cases}$

2. $\begin{cases} u' = -u - \frac{1}{5}v \\ v' = \frac{1}{3}u + v \end{cases}$

3. $\begin{cases} z' = z - 4w \\ w' = -\frac{2}{3}z - \frac{1}{6}w \end{cases}$

4. $\begin{cases} r' = 4r - 3\rho \\ \rho' = 3r \end{cases}$

5. $\begin{cases} s' = -s - p + 1 \\ p' = -2s - \frac{1}{3}p \end{cases}$

6. $\begin{cases} X' = -Y \\ Y' = X \end{cases}$

7. $\begin{cases} \theta' = 2 \\ r' = 3\theta - 2 \end{cases}$

8. $\begin{cases} v' = 5v - 1 \\ \mu' = -2v + \frac{1}{3}\mu \end{cases}$

Decide whether the flows in Problems 9 through 16 are continuous with respect to the initial data. For those that are not, explain the reason.

9.

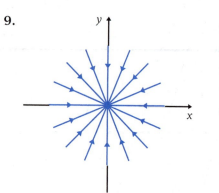

13.

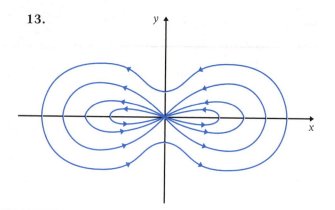

10.

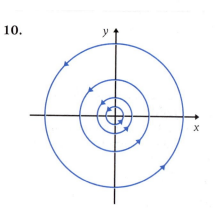

14.

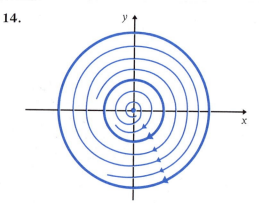

11.

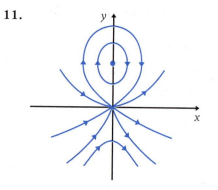

15.

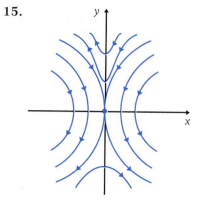

12.

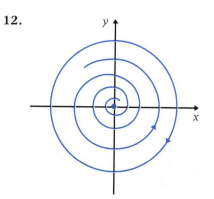

16.

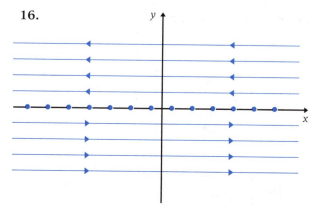

For each of the systems in Problems 17 through 23 determine the equilibria and draw them in phase space.

17. $\begin{cases} x' = x^2 - y^2 \\ y' = 2x - 3y \end{cases}$

18. $\begin{cases} r' = r - 2rp \\ p' = \frac{1}{2}r^2 - p \end{cases}$

19. $\begin{cases} u' = u + s + 1 \\ s' = u^3 + 8s^3 \end{cases}$

20. $\begin{cases} \theta' = \sin\theta - r \\ r' = r - \theta \end{cases}$

21. $\begin{cases} x' = x^2 - y^2 - z^2 \\ y' = 2x - 3y - z \\ z' = 3x + y \end{cases}$

22. $\begin{cases} u' = u - \frac{5}{2}v \\ v' = u^2 + v^2 + w^2 - 9 \\ w' = 0 \end{cases}$

23. $\begin{cases} r' = \cos\theta \\ \rho' = \rho - \theta \\ \theta' = r + \rho + \theta \end{cases}$

Determine the nature of the equilibria of the linear systems in Problems 24 through 30.

24. $\begin{cases} x' = x - y \\ y' = x + y \end{cases}$

25. $\begin{cases} r' = -\frac{7}{3}r + p \\ p' = \frac{3}{4}r - 2p \end{cases}$

26. $\begin{cases} u' = -2u - 7s \\ s' = 8u + 9s \end{cases}$

27. $\begin{cases} \theta' = r - \theta \\ r' = \theta \end{cases}$

28. $\begin{cases} x' = x + y + z \\ y' = x + y - z \\ z' = x - y - z \end{cases}$

29. $\begin{cases} u' = v + 1 \\ v' = w + 1 \\ w' = u + 1 \end{cases}$

30. $\begin{cases} r' = \rho + \theta \\ \rho' = r + \theta \\ \theta' = r + \rho \end{cases}$

In Problems 31 through 36 determine the nature of the equilibria and the invariant sets, and sketch the flows of the linear systems near the equilibria.

31. $\begin{cases} x' = x - 2z \\ y' = 2y \\ z' = -x + \frac{1}{2}z \end{cases}$

32. $\begin{cases} u' = u + v + r \\ v' = 6v - 3r \\ r' = -v + 5r \end{cases}$

33. $\begin{cases} w' = w + \theta \\ \rho' = \rho - \theta \\ \theta' = 6\rho - 2\theta \end{cases}$

34. $\begin{cases} X' = -X \\ Y' = -Y \\ z' = -2Z \end{cases}$

35. $\begin{cases} U' = 3U + V - 2W \\ V' = 2W \\ W' = V \end{cases}$

36. $\begin{cases} r' = 5r + 3\rho \\ \rho' = r - \rho \\ \theta' = 0 \end{cases}$

37. The spreading of cancer cells in the body is called *metastasis*. The Liotta-DeLisi model proposed in 1977 for the metastasis of malignant tumors in mice is given by the linear homogeneous system

$$\begin{cases} x' = -(\alpha + \beta)x \\ y' = \beta x - \gamma y, \end{cases}$$

where x is the number of destroyed cancer cells, y is the number of cells that invade the tissue, and $\alpha, \beta, \gamma > 0$ are constants depending on the type of cancer. Solve the system and give the physical interpretation for all possible choices of $\alpha, \beta,$ and γ.

38. Discuss the behavior of the flow for the linear system

$$\begin{cases} x' = ax + by \\ y' = cx + dy \end{cases}$$

with respect to the parameters $a, b, c,$ and d.

39. In case of system (5) with $a, b, c \neq 0$, do there exist solutions that intersect the xy-plane transversally (i.e., such that the solution curve is not tangent to the plane)?

40. If other forces enter the free-market economy, the linear model describing the dynamics of prices is given by the equations

$$\begin{cases} p' = \alpha r + a_0 \\ r' = -\beta r + b_0, \end{cases}$$

where p represents the price, r the excess demand, and a_0, b_0 are constants. Sketch the flow in this case and see if prices still tend to an equilibrium.

4.5 Numerical Methods

Numerical methods for systems generalize those for second-order equations. In this section we will discuss the Euler and the second-order Runge-Kutta procedures, and then apply them to a model for epidemics with quarantine and to one describing the effect of the forces acting on the buffer springs between the cars of a decelerating train. We consider initial value problems of the type

$$\mathbf{x}' = \mathbf{F}(t, \mathbf{x}), \qquad \mathbf{x}(t_0) = \mathbf{x}_0, \tag{1}$$

in which $\mathbf{F}$ is the vector field and t_0, $\mathbf{x}_0$ are the given initial data. The local truncation errors can be estimated and are the same as for single equations. Numerical methods, however, are more reliable for linear than for nonlinear systems. In this textbook we deal only with systems that raise no difficulties, but even some linear systems can be hard to approach numerically since the global truncation error may be too large. For simplicity we will present the ideas in terms of three-dimensional systems.

Euler's Method

The basic idea in using numerical methods for systems is to apply the first-order-equation procedure to every equation of the system. Thus, for Euler's method, after selection of the points $t_0 < t_1 < \cdots < t_n = a$ in the interval $[t_0, a]$ at a distance $h = (a - t_0)/n$ of each other, Euler's formula becomes

$$\begin{cases} x_{n+1} = x_n + hF_1(t_n, x_n, y_n, z_n) \\ y_{n+1} = y_n + hF_2(t_n, x_n, y_n, z_n) \\ z_{n+1} = z_n + hF_3(t_n, x_n, y_n, z_n). \end{cases} \tag{2}$$

The steps of the method are the same as in Section 2.6, but now we have to work with three equations simultaneously.

EXAMPLE 1   Let us use Euler's method to numerically solve the initial value problem

$$\begin{cases} x' = tx - y + z \\ y' = 3x + y - z \qquad x(0) = 1, \ y(0) = 1, \ z(0) = 0, \\ z' = x + y - tz \end{cases} \tag{3}$$

in the interval $[0, 1.5]$ with step size $h = 0.1$. Formula (2) translates to

$$\begin{cases} x_{n+1} = x_n + h[t_n x_n - y_n + z_n] \\ y_{n+1} = y_n + h[3x_n + y_n - z_n] \\ z_{n+1} = z_n + h[x_n + y_n - t_n z_n]. \end{cases} \tag{4}$$

TABLE 4.5.1

n	t_n	x_n	y_n	z_n
0	0.0	1.000000000	1.000000000	0.000000000
1	0.1	0.900000000	1.400000000	0.200000000
2	0.2	0.789000000	1.790000000	0.428000000
3	0.3	0.668580000	2.162900000	0.677340000
4	0.4	0.540081400	2.512030000	0.940167800
5	0.5	0.404498436	2.831240640	1.207772228
6	0.6	0.262376517	3.114937012	1.470957524
7	0.7	0.113721159	3.358047916	1.720431426
8	0.8	−0.042080009	3.555925913	1.947178134
9	0.9	−0.206321188	3.704176688	2.142788474
10	1.0	−0.381028916	3.798419153	2.299723061
11	1.1	−0.569001417	3.833980087	2.411489779
12	1.2	−0.773840604	3.805528693	2.472723770
13	1.3	−0.999981968	3.706657004	2.479165727
14	1.4	−1.252728752	3.529411541	2.427541686
15	1.5	−1.538297763	3.263779901	2.315354129

With $h = 0.1, n = 15, t_{n+1} = t_n + 0.1$, and initial data $t_0 = 0, x_0 = 1, y_0 = 1, z_0 = 0$, we obtain

$$x_1 = x_0 + h[t_0x_0 - y_0 + z_0] = 0.9,$$

$$y_1 = y_0 + h[3x_0 + y_0 - z_0] = 1.4,$$

$$z_1 = z_0 + h[x_0 + y_0 - t_0z_0] = 0.2,$$

$$x_2 = x_1 + h[t_1x_1 - y_1 + z_1] = 0.789,$$

$$y_2 = y_1 + h[3x_1 + y_1 - z_1] = 1.79,$$

$$z_2 = z_1 + h[x_1 + y_1 - t_1z_1] = 0.428.$$

Continuing like this we end up with the results in Table 4.5.1.

The Second-Order Runge-Kutta Method

The corresponding Heun formula that allows us to solve problem (1) using the second-order Runge-Kutta method is

$$\begin{cases} x_{n+1} = x_n + \dfrac{h}{2}[F_1(t_n, x_n, y_n, z_n) + F_1(t_{n+1}, x_{n+1}^E, y_{n+1}^E, z_{n+1}^E)] \\[2mm] y_{n+1} = y_n + \dfrac{h}{2}[F_2(t_n, x_n, y_n, z_n) + F_2(t_{n+1}, x_{n+1}^E, y_{n+1}^E, z_{n+1}^E)] \\[2mm] z_{n+1} = z_n + \dfrac{h}{2}[F_3(t_n, x_n, y_n, z_n) + F_3(t_{n+1}, x_{n+1}^E, y_{n+1}^E, z_{n+1}^E)], \end{cases} \quad (5)$$

where the upper index E means that the term was computed with Euler's method. As in Section 3.5, after obtaining $x_{n+1}^E, y_{n+1}^E, z_{n+1}^E$ using Euler's method, we get better values $x_{n+1}, y_{n+1}, z_{n+1}$ with Heun's formula above.

EXAMPLE 2   To solve the initial value problem (3) with Heun's formula in the interval $[0, 1.5]$ with step size $h = 0.1$, proceed first as in Example 1 and compute with Euler's formula (4) that $x_1^E = 1$, $y_1^E = 1.4$, $z_1^E = 1$. Formula (5) translates to

$$
\begin{cases}
x_{n+1} = x_0 + \dfrac{h}{2}(t_n x_n - y_n + z_n + t_{n+1} x_{n+1}^E - y_{n+1}^E + z_{n+1}^E) \\[2mm]
y_{n+1} = y_0 + \dfrac{h}{2}(3x_n + y_n - z_n + 3x_{n+1}^E + y_{n+1}^E - z_{n+1}^E) \\[2mm]
z_{n+1} = z_0 + \dfrac{h}{2}(x_n + y_n - z_n + x_{n+1}^E + y_{n+1}^E - t_{n+1} z_{n+1}^E),
\end{cases}
\tag{6}
$$

so with the initial data $t_0 = 0$, $x_0 = 1$, $y_0 = 1$, $z_0 = 0$, we obtain

$$
x_1 = x_0 + \frac{h}{2}[t_0 x_0 - y_0 + z_0 + t_1 x_1^E - y_1^E + z_1^E] = 0.8945,
$$

$$
y_1 = y_0 + \frac{h}{2}[3x_0 + y_0 - z_0 + 3x_1^E + y_1^E - z_1^E] = 1.3795,
$$

$$
z_1 = z_0 + \frac{h}{2}[x_0 + y_0 - t_0 z_0 + x_1^E + y_1^E - t_0 z_1^E] = 0.214.
$$

Then compute x_2^E, y_2^E, z_2^E using the above values of x_1, y_1, z_1 and Euler's formula (2); obtain x_2, y_2, z_2 with Heun's formula (4); and so on. The results are in Table 4.5.2.

If we apply the second-order Runge-Kutta method in the same interval $[0, 1.5]$ with step size 0.01, at $t = 1.5$ we obtain

$$
x_{150} = -1.446517992, \qquad y_{150} = 2.892694838,
$$

$$
z_{150} = 2.071162754,
$$

TABLE 4.5.2

n	t_n	x_n	y_n	z_n
0	0.0	1.000000000	1.000000000	0.000000000
1	0.1	0.894500000	1.395000000	0.214000000
2	0.2	0.780743950	1.773058750	0.451336650
3	0.3	0.660339886	2.127972638	0.704230104
4	0.4	0.534487654	2.454202842	0.964154824
5	0.5	0.403891752	2.746862558	1.222215625
6	0.6	0.268689624	3.001637640	1.469518363
7	0.7	0.128396493	3.214640732	1.697508652
8	0.8	−0.018133699	3.382201772	1.898255436
9	0.9	−0.172733144	3.500599904	2.064661404
10	1.0	−0.337933972	3.565743174	2.190588593
11	1.1	−0.516975145	3.572802772	2.270894318
12	1.2	−0.713807314	3.515808053	2.301379180
13	1.3	−0.933100907	3.387207087	2.278654656
14	1.4	−1.180262830	3.177395277	2.199942240
15	1.5	−1.461467158	2.874211673	2.062818978

and with step size 0.001, at the same point ($t = 1.5$), we obtain

$$x_{1500} = -1.446370922, \qquad y_{1500} = 2.892887650,$$
$$z_{1500} = 2.071246574.$$

For comparison, Euler's method with step size 0.01 leads to

$$x_{150} = -1.455412510, \qquad y_{150} = 2.927176423,$$
$$z_{150} = 2.094350839,$$

and with step size 0.001 yields

$$x_{1500} = -1.447270212, \qquad y_{1500} = 2.896294325,$$
$$z_{1500} = 2.073545784.$$

Applications

Epidemics with quarantine

Red Cross workers.

To act swiftly and efficiently, the International Red Cross is interested in good predictions about the early development of epidemics in underdeveloped countries. Certain epidemics require *quarantine*, i.e., isolation of the infected individuals. Let us assume that such an epidemic breaks out in some remote and isolated village of 1059 people in which 41 infections are reported and 10 individuals are already quarantined. To cope with the quarantine problem and stop the spreading of the disease, the health organization wants to estimate the evolution of the number of infected individuals. For this it uses the following linear model, specific for the disease,

$$\begin{cases} I' = \dfrac{3}{125 - 5t}(3I + 2U - Q) \\[2mm] Q' = \dfrac{1}{5}(I - Q) \\[2mm] U' = -(U + I), \end{cases}$$

where I, Q, and U represent the number of infected, quarantined, and uninfected individuals. The model gives a good prediction for the first 5 days after the outbreak, and it was derived following certain observations and past statistical data:

1. The rate of change in the number of infected individuals increases as aI and bU (in this case $a = 3, b = 2$) but decreases as $-cQ$ (in this case $c = 1$); on the other hand, the factor $3/(125 - 5t)$, which decreases in time, appears since there are individuals who are immune to the disease, others who take cautionary measures, etc.
2. The rate of change in the number of quarantined individuals increases as $d(I - Q)$ (in this case $d = \frac{1}{5}$, which was determined from the conditions of the specific area).
3. The rate of change of uninfected individuals decreases for this disease as $-(U + I)$, i.e., the sum of infected and uninfected individuals.

Figure 4.5.1. The number of infected and quarantined individuals in the first 5 days after the outbreak of the epidemic.

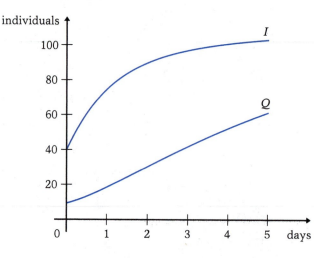

The above system is linear but it has a variable coefficient, so the exact methods are inefficient. Since we need some good estimates of what will happen, the qualitative methods are useless too. But for a short-time prediction, numerical methods are useful. Let us therefore apply the second-order Runge-Kutta method with step size $h = 0.1$ in the interval $[0, 5]$ (days). This translates to

$$\begin{cases} I_{n+1} = I_n + \dfrac{h}{2}\left[\dfrac{3}{125 - 5t_n}(3I_n + 2U_n - Q_n) \right. \\ \qquad\qquad \left. + \dfrac{3}{125 - 5t_{n+1}}(3I_{n+1}^E + 2U_{n+1}^E - Q_{n+1})\right] \\ Q_{n+1} = Q_n + \dfrac{h}{2}(I_n + I_{n+1}^E) \\ U_{n+1} = U_n - \dfrac{h}{2}(U_n + I_n + U_{n+1}^E + I_{n+1}^E), \end{cases}$$

with $I_0 = 41, Q_0 = 10, U_0 = 1018$. Using one of the drawing programs in Section 4.6, we obtain the graphs for I and Q represented in Figure 4.5.1.

Figure 4.5.1 shows that if the health organization manages to secure the quarantine at the pace described by the graph of Q, then the spread of the disease can be stopped after 3 to 4 days, when the number of infected individuals stabilizes at about 100, i.e., approximately 10% of the population. If the quarantine process is less successful (say a rate of change $\frac{1}{10}(I - Q)$), then the infection will continue to spread, but more slowly than during the first couple of days. So the Red Cross knows that the success of the operation depends on how quickly and well it can organize the quarantine.

The decelerating train   A problem faced by a group of researchers who designed the buffer springs that link the cars of a train was the choice of the strength and flexibility of those springs. To prepare the practical tests, they had to understand how the springs behave if the force that acts on them is large,

Figure 4.5.2. An engine with two cars behaves like a system of two coupled oscillators.

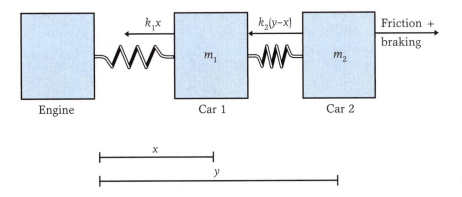

for example, when the train brakes. The researchers proposed a model for an engine with two cars (see Figure 4.5.2), which reduces to a system of two coupled oscillators.

If $x(t)$ and $y(t)$ denote the displacements of the cars, which have masses m_1 and m_2, by Hooke's law (see Section 3.4) we can write $m_1 x'' = -k_1 x + k_2(y - x)$ and $m_2 y'' = -k_2(y - x)$, where k_1 and k_2 are the constants characterizing the two springs. But this is unrealistic in the case of a braking train. We have to take into consideration the friction between wheels and rail, reliably modeled as proportional to the velocity, and the effect of braking, by taking the coefficient of proportionality as $t^3/10$. So the equations that describe the motion of the train during the last seconds before stopping are

$$\begin{cases} m_1 x'' = -k_1 x + k_2(y - x) + \dfrac{t^3}{10} x' + \dfrac{t^3}{10}(y' - x') \\ m_2 y'' = -k_2(y - x) + \dfrac{t^3}{10}(y' - x'). \end{cases}$$

This is a second-order system of two equations, which with the substitutions $u = m_1 x'$ and $v = m_2 y'$ and the choices $k_1 = k_2 = 1$ ton/meter and $m_1 = m_2 = 10$ tons, becomes the first-order four-dimensional linear system

$$\begin{cases} x' = \dfrac{u}{10} \\ y' = \dfrac{v}{10} \\ u' = -2x + y + \dfrac{t^3}{100} v \\ v' = x - y + \dfrac{t^3}{100}(v - u). \end{cases}$$

This system describes the motion until $x(t) = y(t) = 0$, when the buffer disengages and different springs take over the motion. Again, as in the case of the epidemic, a numerical method seems the most suitable for answering the question. Let us therefore solve the above system by using Euler's method with step size 0.01 in the interval $[0, 5]$ (seconds), for initial conditions $x(0) = 1, y(0) = 10$ (meters), $u(0) = v(0) = -1$

Figure 4.5.3. The graphs of
x and y, which describe the
motion of the buffer springs.

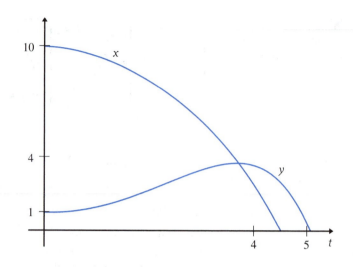

(meters/second), which correspond to the length of the cars and springs
and the deceleration due to braking. Euler's formula leads to

$$
\begin{cases}
x_{n+1} = x_n + \dfrac{hu_n}{10} \\[2mm]
y_{n+1} = y_n + \dfrac{hv_n}{10} \\[2mm]
u_{n+1} = u_n - 2hx_n + hy_n + \dfrac{ht_n^3}{100}v_n \\[2mm]
v_{n+1} = v_n + hx_n - hy_n + \dfrac{ht_n^3}{100}(v_n - u_n),
\end{cases}
$$

with $x_0 = 1$, $y_0 = 10$, $u_0 = v_0 = -1$. Using one of the programs in Section 4.6, the graphs of x and y appear as in Figure 4.5.3. They show that
the first car has one slight oscillation before x reaches 0 after almost 5
seconds, and the second car simply reaches $y = 0$, without swinging, after about 4.4 seconds. This result gives the researchers an idea of what
happens with springs having $k_1 = k_2 = 1$ and a lead in pursuing their
numerical and practical experiments.

PROBLEMS

Solve Problems 1 through 8 employing

   (a) *Euler's method with step size* $h = 0.1$,
   (b) *Euler's method with step size* $h = 0.05$,
   (c) *the second-order Runge-Kutta method with
      step size* $h = 0.1$,
   (d) *the second-order Runge-Kutta method with
      step size* $h = 0.05$

*to find approximate values of the solution in the
given intervals. For each solution compare the
results at (a), (b), (c), and (d). Use, if you wish, one
of the computer programs in Section 4.6.*

1.
$\begin{cases} x' = x + 5y \\ y' = 7x - y, \end{cases}$

    $x(1) = 1$, $y(1) = 2$ in the interval $[1, 2]$

2.
$\begin{cases} u' = -u - 3v \\ v' = \frac{1}{2}u + \frac{1}{3}v, \end{cases}$

    $u(1) = -1$, $v(1) = 1$ in the interval $[1, 2.5]$

3.
$\begin{cases} X' = tX + t^2Y \\ Y' = 2X - 3tY, \end{cases}$

    $X(2) = 5$, $Y(2) = 1$ in the interval $[2, 3]$

4. $\begin{cases} r' = -(\sin t)r + (\cos t)\theta \\ \theta' = -tr - 2t\theta, \end{cases}$

$r(2) = 5, \ \theta(2) = 1$ in the interval $[2, 3.2]$

5. $\begin{cases} x' = 2x - y + \frac{2}{5}z \\ y' = -3x + 4y - \frac{3}{5}z \\ z' = -x - y - z, \end{cases}$

$x(1) = 1, \ y(1) = 2, \ z(1) = 1$ in the interval $[1, 2]$

6. $\begin{cases} u' = 4u - 5v + w \\ v' = \frac{1}{2}u - 3v + \frac{2}{3}w \\ w' = u - v + \frac{5}{7}w, \end{cases}$

$u(3) = 2, \ v(3) = 1, \ w(3) = 2$ in the interval $[3, 4.5]$

7. $\begin{cases} s' = 4ts - 5t^3r + u \\ r' = \dfrac{1}{2t}s - 3tr + \dfrac{t}{3}u \\ u' = (\cos t)s - r + \dfrac{5}{2t}u, \end{cases}$

$s(1) = 1, \ r(1) = 3, \ u(1) = 2$ in the interval $[1, 2]$

8. $\begin{cases} U' = 5U - V + tT \\ V' = tT + 1 \\ T' = V - tT, \end{cases}$

$U(\frac{1}{2}) = 0, \ V(\frac{1}{2}) = 1, \ T(\frac{1}{2}) = 1$ in the interval $[\frac{1}{2}, 2]$

For the quarantine model, solve Problems 9 through 12 by determining the evolution of the disease during the first 5 days for the given initial conditions.

9. $I(0) = 15, \ Q(0) = 3,$
$U(0) = 814$

10. $I(0) = 23, \ Q(0) = 7,$
$U(0) = 1121$

11. $I(0) = 16, \ Q(0) = 11,$
$U(0) = 948$

12. $I(0) = 35, \ Q(0) = 21,$
$U(0) = 1024$

Determine the outcome of the decelerating-train experiments in Problems 13 to 16 under the assumption that the data remain the same except for the given changes.

13. $k_1 = 1, \ k_2 = 1.2,$
$u(0) = v(0) = -1.5$

14. $k_1 = 0.8, \ k_2 = 1.2,$
$u(0) = v(0) = -1.2$

15. $k_1 = 1.2, \ k_2 = 1,$
$u(0) = v(0) = -1.1$

16. $k_1 = k_2 = 0.8,$
$u(0) = v(0) = -2$

Solve the initial value problems in Problems 17 through 24 using exact methods, and then apply successively Euler's and Runge-Kutta's methods with step sizes 0.1, 0.01, and 0.001 in the specified intervals. Compare the results and determine in each case the error at the final step. Use, if you wish, one of the computer programs in Section 4.6.

17. $\begin{cases} x' = 2x - 3y \\ y' = -3x + 7y, \end{cases}$

$x(0) = y(0) = 1$ in $[0, 0.1]$

18. $\begin{cases} u' = -u - 2\pi v \\ v' = \frac{1}{2}u + 2v, \end{cases}$

$u(0) = v(0) = 12$ in $[0, 1.2]$

19. $\begin{cases} r' = \frac{2}{3}r + \rho + 1 \\ \rho' = -r - \rho - 1, \end{cases}$

$r(0) = \rho(0) = -\frac{1}{2}$ in $[0, 2]$

20. $\begin{cases} X' = X - 2Z - 14 \\ Z' = X + 18, \end{cases}$

$X(10) = Z(10) = 11$ in $[10, 11]$

21. $\begin{cases} x' = 2x - 3y + z \\ y' = -3x + 7y - z \\ z' = \frac{5}{7}x - 2y + 4z, \end{cases}$

$x(0) = y(0) = z(0) = 3$ in $[0, 0.8]$

22. $\begin{cases} u' = -u + 3v + 6w \\ v' = u - 2v - 5w \\ w' = \frac{2}{3}u - 5\pi v + w, \end{cases}$

$u(0) = v(0) = w(0) = \pi$ in $[0, 2]$

23. $\begin{cases} \alpha' = \beta + 1 \\ \beta' = \gamma + 1 \\ \gamma' = \alpha + 1, \end{cases}$

$\alpha(0) = \beta(0) = \gamma(0) = 1$ in $[0, 1]$

24. $\begin{cases} X' = Y + Z - 2 \\ Y' = Z + X - 2 \\ Z' = Y + Z - 2, \end{cases}$

$X(0) = Y(0) = Z(0) = 1$ in $[0, 1.5]$

4.6 Computer Applications

In this section we will show through examples how to use Maple, *Mathematica*, and MATLAB to obtain exact and numerical solutions, to represent their graphs, to find eigenvalues and eigenvectors, and to draw flows. Some of the programs generalize the ones used for second-order equations.

Maple

Exact solutions

The command `dsolve` encountered in Section 3.7 can be used for systems too. A convenient way to find the exact solution of a system or of an initial value problem is to first write each equation separately and then apply the command `dsolve`, with or without initial conditions.

> **EXAMPLE 1**   To solve the initial value problem
>
> $$\begin{cases} x' = x + y - z \\ y' = -x + y + z \, , \\ z' = x - y + z \end{cases} \qquad x(0) = 1, \; y(0) = -1, \; z(0) = 0,$$
>
> proceed as shown below. (Here Maple has been set to echo input and show results in typeset format.) Write first the three equations
>
> ```
> > e1:=diff(x(t),t)=x(t)+y(t)-z(t);
> ```
>
> $$e1 := \frac{\partial}{\partial t} x(t) = x(t) + y(t) - z(t)$$
>
> ```
> > e2:=diff(y(t),t)=-x(t)+y(t)+z(t);
> ```
>
> $$e2 := \frac{\partial}{\partial t} y(t) = -x(t) + y(t) + z(t)$$
>
> ```
> > e3:=diff(z(t),t)=x(t)-y(t)+z(t);
> ```
>
> $$e3 := \frac{\partial}{\partial t} z(t) = x(t) - y(t) + z(t)$$
>
> and then use the command `dsolve` as follows:
>
> ```
> > dsolve({e1,e2,e3,x(0)=1,y(0)=-1,z(0)=0},
>   {x(t),y(t),z(t)});
> ```
>
> Maple will display the solution
>
> $$\{z(t) = \frac{2}{3} e^t \sqrt{3} \sin(t\sqrt{3}), \; y(t) = \frac{1}{3} e^t (-3\cos(t\sqrt{3}) - \sqrt{3}\sin(t\sqrt{3})),$$
>
> $$x(t) = -\frac{1}{3} e^t (-3\cos(t\sqrt{3}) + \sqrt{3}\sin(t\sqrt{3}))\}$$

Numerical solutions

With the help of Maple we will numerically solve the initial value problem

$$\begin{cases} x' = tx - y + z \\ y' = 3x + y - z \, , \\ z' = x + y - tz \end{cases} \qquad x(0) = y(0) = 1, \; z(0) = 0, \tag{1}$$

in the interval [0, 1.5] with step size 0.1. We will first use Euler's method and then the second-order Runge-Kutta method.

Euler's method To numerically solve the initial value problem (1) with the help of Euler's method, we first write the three functions defining the vector field and then assign the initial conditions t_0, x_0, y_0, z_0, the step size h, and the number of steps n. The do command then uses Euler's formula to iterate all the steps.

```
> f1:=(t,x,y,z)->t*x-y+z;
```
$$f1 := (t, x, y, z) \rightarrow tx - y + z$$

```
> f2:=(t,x,y,z)->3*x+y-z;
```
$$f2 := (t, x, y, z) \rightarrow 3x + y - z$$

```
> f3:=(t,x,y,z)->x+y-t*z;
```
$$f3 := (t, x, y, z) \rightarrow x + y - tz$$

```
> t0:=0: x0:=1: y0:=1: z0:=0:
> h:=0.1:
> n:=15:
> t:=t0: x:=x0: y:=y0: z:=z0:
> for i from 1 to n do
  u:=f1(t,x,y,z): v:=f2(t,x,y,z):  w:=f3(t,x,y,z):
  x:=x+h*u: y:=y+h*v: z:=z+h*w:
  t:=t+h:
  print(t,x,y,z);
  od:
```

Maple displays the numerical result below:

```
 .1, .9, 1.4, .2
 .2, .789, 1.79, .428
 .3, .66858, 2.1629, .67734
 .4, .5400814, 2.512030, .9401678
 .5, .404498436, 2.83124064, 1.207772228
 .6, .2623765166, 3.114937012, 1.470957524
 .7, .1137211588, 3.358047916, 1.720431426
 .8, -.0420800091, 3.555925913, 1.947178134
 .9, -.2063211877, 3.704176688, 2.142788474
1.0, -.3810289160, 3.798419153, 2.299723061
1.1, -.5690014168, 3.833980087, 2.411489779
1.2, -.7738406035, 3.805528693, 2.472723770
1.3, -.9999819682, 3.706657004, 2.479165727
1.4, -1.252728752, 3.529411541, 2.427541686
1.5, -1.538297763, 3.263779901, 2.315354129
```

The second-order Runge-Kutta method In case of the second-order Runge-Kutta method, the first part of the program is the same as for Euler's method. The difference appears in using the command do. The changes reflect formula (5) of Section 4.5. The program works as follows.

```
> for i from 1 to n do
  a:=f1(t,x,y,z): b:=f2(t,x,y,z): c:=f3(t,x,y,z):
```

```
u:=x+h*a: v:=y+h*b: w:=z+h*c:
x:=x+h*(a+f1(t+h,u,v,w))/2:
y:=y+h*(b+f2(t+h,u,v,w))/2:
z:=z+h*(c+f3(t+h,u,v,w))/2:
t:=t+h:
print(t,x,y,z);
od:
```

Maple displays the numerical result below:

```
 .1, .8945000000, 1.395000000, .2140000000
 .2, .7807439500, 1.773058750, .4513366500
 .3, .6603398859, 2.127972638, .7042301039
 .4, .5344876539, 2.454202842, .9641548238
 .5, .4038917521, 2.746862558, 1.222215625
 .6, .2686896235, 3.001637640, 1.469518363
 .7, .1283964925, 3.214640732, 1.697508652
 .8, -.0181336987, 3.382201772, 1.898255436
 .9, -.1727331441, 3.500599904, 2.064661404
1.0, -.3379339723, 3.565743174, 2.190588593
1.1, -.5169751448, 3.572802772, 2.270894318
1.2, -.7138073142, 3.515808053, 2.301379180
1.3, -.9331009066, 3.387207087, 2.278654656
1.4, -1.180262830, 3.177395277, 2.199942240
1.5, -1.461467158, 2.874211673, 2.062818978
```

To draw graphs of x, y, and z using the Runge-Kutta method, we can proceed as follows:

```
> f1:=(t,x,y,z)->t*x-y+z;
```

$$f1 := (t, x, y, z) \rightarrow tx - y + z$$

```
> f2:=(t,x,y,z)->3*x+y-z;
```

$$f2 := (t, x, y, z) \rightarrow 3x + y - z$$

```
> f3:=(t,x,y,z)->x+y-t*z;
```

$$f3 := (t, x, y, z) \rightarrow x + y - tz$$

```
> t0:=0: x0:=1: y0:=1: z0:=0:
> M0:=[t0,x0]: N0:=[t0,y0]: P0:=[t0,z0]:
> h:=0.1:
> n:=15:
> t:=t0: x:=x0: y:=y0: z:=z0: M:=M0: N:=N0: P:=P0:
> for i from 1 to n do
  a:=f1(t,x,y,z): b:=f2(t,x,y,z): c:=f3(t,x,y,z):
  u:=x+h*a: v:=y+h*b: w:=z+h*c:
  x:=x+h*(a+f1(t+h,u,v,w))/2:
  y:=y+h*(b+f2(t+h,u,v,w))/2:
  z:=z+h*(c+f3(t+h,u,v,w))/2:
  M:=(M,[t,x]): N:=(N,[t,y]): P:=(P,[t,z]):
  t:=t+h:
  od:
PLOT(CURVES([M],[N],[P]));
```

Figure 4.6.1. The graphs of the numerical solutions x, y, z obtained with the Runge-Kutta method for the initial value problem (1).

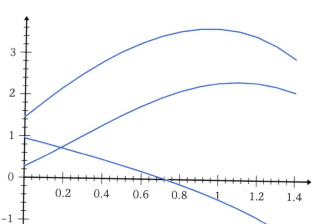

The graphs of the solutions are drawn in Figure 4.6.1. Compare these graphs with the numerical data obtained with the help of the Runge-Kutta method.

Eigenvalues and eigenvectors

Maple is also capable of finding the eigenvalues and eigenvectors of a given matrix. For this it is endowed with a linear algebra package, which we must call with a specific command. Then we need to define the matrix and finally use the commands `eigenvalues` and `eigenvectors`.

EXAMPLE 2   To obtain the eigenvalues and eigenvectors of the matrix

$$\mathbf{A} = \begin{pmatrix} 1 & -3 & 3 \\ 3 & -5 & 3 \\ 6 & -6 & 4 \end{pmatrix},$$

we can proceed as follows. We first call the linear algebra package with the command `with(linalg)`. We then use the command `matrix` to explain to Maple that the matrix we define is three-dimensional (i.e., three rows and three columns), and then write the elements of the matrix in a sequence starting with the first row, then the second row, etc. The command `eigenvalues` will find that the characteristic polynomial has the simple root 4 and the double root -2. The command `eigenvectors` will then obtain for the eigenvalue -2 the two eigenvectors that it writes down, and for the eigenvalue 4 the one eigenvector that it writes down.

```
> with(linalg):
> A := matrix(3,3, [1,-3,3,3,-5,3,6,-6,4]);
```

$$A := \begin{bmatrix} 1 & -3 & 3 \\ 3 & -5 & 3 \\ 6 & -6 & 4 \end{bmatrix}$$

```
> eigenvalues(A);
```

$$4, -2, -2$$

```
> eigenvectors(A);
```

$$[\,-2, 2,\ \{[1, 1, 0], [-1, 0, 1]\}\,],\ \ [\,4, 1,\ \{[1, 1, 2]\}\,]$$

Figure 4.6.2. The direction field and the flow of the system in Example 3.

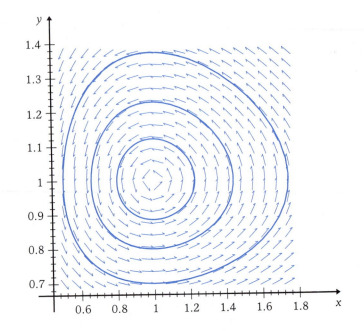

Flows
: With the help of the `DEplot` command we can draw the direction field and the flow of a two-dimensional linear or nonlinear system in the phase plane. To obtain the curves, we need to specify initial conditions and Maple will draw a curve for each specified initial condition.

> **EXAMPLE 3**   To draw the direction field and the solution curves through (1.2, 1.2), (1, 0.7), and (0.8, 0.1) for the two-dimensional system
>
> $$\begin{cases} x' = x(1 - y) \\ y' = 3y(x - 1), \end{cases}$$
>
> proceed as follows. First call the package `DEtools` and then use the command `DEplot` as shown below. Since the curve is obtained numerically, a step size is also specified. Maple then draws the picture in Figure 4.6.2.
>
> ```
> > with(DEtools):
> > DEplot([diff(x(t),t)=x(t)*(1-y(t)),
>   diff(y(t),t)=.3*y(t)*(x(t)-1)],[x(t),y(t)],t=-7..7,
>   [[x(0)=1.2,y(0)=1.2],[x(0)=1,y(0)=.7],
>   [x(0)=0.8,y(0)=1]],stepsize=.2);
> ```

Mathematica

Exact solutions
: The command `Dsolve` encountered in Section 3.7 can be used for systems too. A convenient way to find the exact solution of a system or of an initial value problem is to first write each equation separately and then apply the command `dsolve`, with or without initial conditions.

EXAMPLE 4 To solve the initial value problem

$$\begin{cases} x' = x + y - z \\ y' = -x + y + z, \\ z' = x - y + z \end{cases} \qquad x(0) = 1, \ y(0) = -1, \ z(0) = 0,$$

proceed as shown below. Write first the three equations

```
> e1=D[x[t],t]==x[t]+y[t]-z[t]
```

$$x'[t] == x[t] + y[t] - z[t]$$

```
> e2=D[y[t],t]==-x[t]+y[t]+z[t]
```

$$y'[t] == -x[t] + y[t] + z[t]$$

```
> e3=D[z[t],t]==x[t]-y[t]+z[t]
```

$$z'[t] == x[t] - y[t] + z[t]$$

and then use the command Dsolve as follows:

```
DSolve[{e1,e2,e3,
x[0]==1,y[0]=-1,z[0]=0},{x[t],y[t],z[t]},t]
```

Unfortunately, *Mathematica* does not offer the solution in terms of trigonometric functions, as Maple and MATLAB do, but gives the answer in exponentials to complex powers. Though the result is in essence the same, reaching the trigonometric form requires further work, in particular the use of *Euler's formula*: $e^{\alpha+i\beta} = e^\alpha(\cos\beta + i\sin\beta)$. Therefore *Mathematica* is less suited to this kind of problem if complex eigenvalues show up. For real eigenvalues, however, DSolve of *Mathematica* is as good as dsolve of Maple and MATLAB.

Numerical solutions With the help of *Mathematica* we will now numerically solve the initial value problem

$$\begin{cases} x' = tx - y + z \\ y' = 3x + y - z, \\ z' = x + y - tz \end{cases} \qquad x(0) = y(0) = 1, \ z(0) = 0, \qquad (2)$$

in the interval [0, 1.5] with step size 0.1. First we use Euler's method and then the second-order Runge-Kutta method.

Euler's method To numerically solve the initial value problem (2) with the help of Euler's method, we first write the three functions defining the vector field and then assign the initial conditions t_0, x_0, y_0, z_0, the step size h, and the number of steps n. The Do command then uses Euler's formula to iterate all the steps.

```
> f1[t_,x_,y_,z_]:=t*x-y+z
> f2[t_,x_,y_,z_]:=3*x+y-z
> f3[t_,x_,y_,z_]:=x+y-t*z
> t0=0; x0=1; y0=1; z0=0;
> h=0.1;
```

```
> n=15;
> t=t0; x=x0; y=y0; z=z0;
> Do[u=f1[t,x,y,z]; v=f2[t,x,y,z]; w=f3[t,x,y,z];
  x=x+h*u; y=y+h*v; z=z+h*w;
  t=t+h;
  Print[t," ",x," ",y," ",z],
  {i,1,n}]
```

Mathematica displays the numerical result below:

```
0.1 0.9 1.4 0.2
0.2 0.789 1.79 0.428
0.3 0.66858 2.1629 0.67734
0.4 0.540081 2.51203 0.940168
0.5 0.404498 2.83124 1.20777
0.6 0.262377 3.11494 1.47096
0.7 0.113721 3.35805 1.72043
0.8 -0.04208 3.55593 1.94718
0.9 -0.206321 3.70418 2.14279
1.  -0.381029 3.79842 2.29972
1.1 -0.569001 3.83398 2.41149
1.2 -0.773841 3.80553 2.47272
1.3 -0.999982 3.70666 2.47917
1.4 -1.25273 3.52941 2.42754
1.5 -1.5383 3.26378 2.31535
```

The second-order Runge-Kutta method   In case of the second-order Runge-Kutta method, the first part of the program is the same as for Euler's method. The difference appears in using the command Do. The changes reflect formula (5) of Section 4.5. The program works as follows.

```
> Do[a=f1[t,x,y,z]; b=f2[t,x,y,z]; c=f3[t,x,y,z];
  u=x+h*a; v=y+h*b; w=z+h*c;
  x=x+h*(a+f1[t+h,u,v,w])/2;
  y=y+h*(b+f2[t+h,u,v,w])/2;
  z=z+h*(c+f3[t+h,u,v,w])/2;
  t=t+h;
  Print[t," ",x," ",y," ",z],
  {i,1,n}]
```

Mathematica displays the numerical solution:

```
0.1 0.8945 1.395 0.214
0.2 0.780744 1.77306 0.451337
0.3 0.66034 2.12797 0.70423
0.4 0.534488 2.4542 0.964155
0.5 0.403892 2.74686 1.22222
0.6 0.26869 3.00164 1.46952
0.7 0.128396 3.21464 1.69751
0.8 -0.0181337 3.3822 1.89826
0.9 -0.172733 3.5006 2.06466
1.  -0.337934 3.56574 2.19059
1.1 -0.516975 3.5728 2.27089
1.2 -0.713807 3.51581 2.30138
```

```
1.3 -0.933101 3.38721 2.27865
1.4 -1.18026 3.1774 2.19994
1.5 -1.46147 2.87421 2.06282
```

Eigenvalues and eigenvectors

Mathematica is also capable of finding the eigenvalues and eigenvectors of a given matrix. For this we first have to write the matrix and then use the commands Eigenvalues and Eigenvectors.

EXAMPLE 5 To obtain the eigenvalues and eigenvectors of the matrix

$$\mathbf{A} = \begin{pmatrix} 1 & -3 & 3 \\ 3 & -5 & 3 \\ 6 & -6 & 4 \end{pmatrix},$$

we can proceed as follows. We first define the matrix by rows and then use the commands below. *Mathematica* provides the double eigenvalue -2, the simple eigenvalue 4, and three linearly independent eigenvectors.

```
A={{1,-3,3}, {3,-5,3}, {6,-6,4}}
Eigenvalues[A]
```

$$\{-2, -2, 4\}$$

```
Eigenvectors[A]
```

$$\{\{-1, 0, 1\}, \{1, 1, 0\}, \{1, 1, 2\}\}$$

Flows

Unlike Maple, *Mathematica* has no direct way of drawing flows. However, we can write a program that numerically computes the solutions of a two-dimensional system between 0 and t_1 with the NDSolve (i.e., numerically dsolve) command and then plot them in the phase plane.

EXAMPLE 6 To draw a few solutions in the phase plane for the system

$$\begin{cases} x' = x(1 - y) \\ y' = 3y(x - 1) \end{cases}$$

for initial conditions $x(0), y(0)$ between 0 and 3, proceed as follows. First write the equations

```
e1=x'[t]==x[t](1-y[t]);
e2=y'[t]==3y[t](x[t]-1);
```

and then use the following program:

```
numsol[t0_, a_, b_, t1_]:=({x[t],y[t]} /.
First[NDSolve[{e1,e2, x[0]==a, y[0]==b},
{x[t],y[t]}, {t,0,t1}]]) /. t->t0
flow[t1_]:=ParametricPlot[Evaluate[Flatten[Table[numsol
[t,a,b,t1],{a,0,3}, {b,0,3}], 1]], {t,0,t1}]
flow[6]
```

The last command is the one that draws the flow for a chosen t_1 ($t_1 = 6$ in this case) and creates the picture in Figure 4.6.3.

Figure 4.6.3. The flow of the system in Example 6.

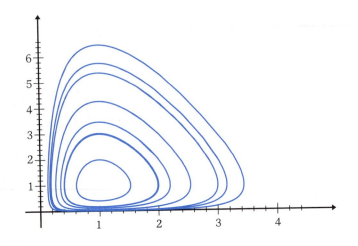

MATLAB

Exact solutions The command `dsolve` encountered in Section 3.7 can be used for systems too. A convenient way to find the exact solution of a system or of an initial value problem is to first write each equation separately and then apply the command `dsolve`, with or without initial conditions.

EXAMPLE 7   To solve the initial value problem

$$\begin{cases} x' = x + y - z \\ y' = -x + y + z, \\ z' = x - y + z \end{cases} \quad x(0) = 1, \ y(0) = -1, \ z(0) = 0,$$

proceed as shown below. Write first the three equations

```
>> e1='Dx=x+y-z' , e2='Dy=-x+y+z' , e3='Dz=x-y+z'
e1=
Dx=x+y-z
e2=
Dy=-x+y+z
e3=
Dz=x-y+z
```

and then use the command `dsolve` as follows:

```
>> [x,y,z]=dsolve(e1,e2,e3,'x(0)=1','y(0)=-1',
    'z(0)=0','t')
```

MATLAB displays the solution

$$x = 1/3 * \exp(t) * (-3 * \cos(\sqrt{3} * t) + \sqrt{3} * \sin(\sqrt{3} * t)$$

$$y = 1/3 * \exp(t) * (-3 * \cos(\sqrt{3} * t) - \sqrt{3} * \sin(\sqrt{3} * t))$$

$$z = 2/3 * \exp(t) * \sqrt{3} * \sin(\sqrt{3} * t)$$

Numerical solutions With MATLAB we will now numerically solve the initial value problem

$$\begin{cases} x' = tx - y + z \\ y' = 3x + y - z, \\ z' = x + y - tz \end{cases} \quad x(0) = y(0) = 1, \ z(0) = 0 \qquad (3)$$

in the interval [0, 1.5] with step size 0.1. First we will use Euler's method and then the second-order Runge-Kutta method.

Euler's method We must first create three files, f1.m, f2.m, and f3.m, in which to define the vector field. In f1.m write the first component:

```
function F1=f1(t,x,y,z)
F1=t.*x-y+z;
```

In f2.m write the second component:

```
function F2=f2(t,x,y,z)
F2=3.*x+y-z;
```

In f3.m write the third component:

```
function F3=f3(t,x,y,z)
F3=x+y-t.*z;
```

To numerically solve the initial value problem (2) with the help of Euler's method, we then assign the initial conditions t_0, x_0, y_0, z_0, the step size h, and the number of steps n. The iteration command that follows uses Euler's formula.

```
>> t0=0; x0=1; y0=1; z0=0;
>> h=0.1;
>> n=15;
>> t=t0; x=x0; y=y0; z=z0;
>> for i=1:n
   u=f1(t,x,y,z); v=f2(t,x,y,z); w=f3(t,x,y,z);
   x=x+h*u; y=y+h*v; z=z+h*w;
   t=t+h;
   T=[T;t]; X=[X;x]; Y=[Y;y]; Z=[Z;z];
   end
>> [T,X,Y,Z]
```

Through the last command, MATLAB displays the numerical solution:

```
ans=
0.1000      0.9000      1.4000      0.2000
0.2000      0.7890      1.7900      0.4280
0.3000      0.6686      2.1629      0.6773
0.4000      0.5401      2.5120      0.9402
0.5000      0.4045      2.8312      1.2078
0.6000      0.2624      3.1149      1.4710
0.7000      0.1137      3.3580      1.7204
0.8000     -0.0421      3.5559      1.9472
0.9000     -0.2063      3.7042      2.1428
1.0000     -0.3810      3.7984      2.2997
1.1000     -0.5690      3.8340      2.4115
1.2000     -0.7738      3.8055      2.4727
1.3000     -1.0000      3.7067      2.4792
1.4000     -1.2527      3.5294      2.4275
1.5000     -1.5383      3.2638      2.3154
```

The second-order Runge-Kutta method In the case of the second-order Runge-Kutta method, the first part of the program is the same as for

Figure 4.6.4. The graphs of the numerical solutions x, y, z obtained with the Runge-Kutta method for the initial value problem (3).

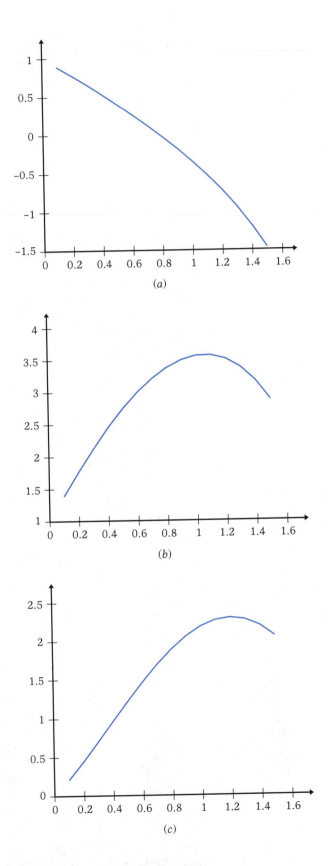

(a)

(b)

(c)

Euler's method. The difference appears in using the iteration command. The changes reflect formula (5) of Section 4.5. The program works as follows.

```
>> t0=0; x0=1; y0=1; z0=0; h=0.1; n=15;
>> t=t0; x=x0; y=y0; z=z0;
>> for i=1:n
     a=f1(t,x,y,z); b=f2(t,x,y,z); c=f3(t,x,y,z);
     u=x+h*a; v=y+h*b; w=z+h*c;
     x=x+h*(a+f1(t+h,u,v,w))/2;
     y=y+h*(b+f2(t+h,u,v,w))/2;
     z=z+h*(c+f3(t+h,u,v,w))/2;
     t=t+h;
     T=[T;t]; X=[X;x]; Y=[Y;y]; Z=[Z;z];
   end
>> [T,X,Y,Z]
ans =
     0.1000     0.8945     1.3950     0.2140
     0.2000     0.7807     1.7731     0.4513
     0.3000     0.6603     2.1280     0.7042
     0.4000     0.5345     2.4542     0.9642
     0.5000     0.4039     2.7469     1.2222
     0.6000     0.2687     3.0016     1.4695
     0.7000     0.1284     3.2146     1.6975
     0.8000    -0.0181     3.3822     1.8983
     0.9000    -0.1727     3.5006     2.0647
     1.0000    -0.3379     3.5657     2.1906
     1.1000    -0.5170     3.5728     2.2709
     1.2000    -0.7138     3.5158     2.3014
     1.3000    -0.9331     3.3872     2.2787
     1.4000    -1.1803     3.1774     2.1999
     1.5000    -1.4615     2.8742     2.0628
```

We can also use the commands >> plot(T,X), >> plot(T,Y), and >> plot(T,Z) for graphing the numerical solutions $x(t)$, $y(t)$, and $z(t)$, respectively (see Figure 4.6.4).

Eigenvalues and eigenvectors

MATLAB can find the eigenvalues and eigenvectors of a given matrix. For this we have to write the matrix, use the eig command to obtain the eigenvalues, and use a bracket command, which displays the eigenvectors.

EXAMPLE 8 To obtain the eigenvalues and eigenvectors of the matrix

$$\mathbf{A} = \begin{pmatrix} 1 & -3 & 3 \\ 3 & -5 & 3 \\ 6 & -6 & 4 \end{pmatrix},$$

we can proceed as follows. We first define the matrix by rows and then use the commands below. MATLAB provides the eigenvalue 4, the double eigenvalue -2, and three linearly independent eigenvectors. Notice that the bracket command [ ] displays the eigenvectors as columns as well as a matrix having the eigenvalues on its main diagonal. Also notice that MATLAB will always come up with as many eigenvectors as the dimension of the matrix, but some may not be linearly independent; this can happen if multiple eigenvalues occur. Therefore it is important to check the linear independence of the vectors.

```
>> A=[[1,-3,3];[3,-5,3];[6,-6,4]]
   A =
          1        -3         3
          3        -5         3
          6        -6         4
>> eig(A)
   ans =
          4.0000
         -2.0000
         -2.0000
>> [V,D]=eig(A)
   V =
         -0.4082        -0.7821        -0.0351
         -0.4082        -0.5941        -0.7240
         -0.8165         0.1881        -0.6889

   D =
          4.0000              0              0
               0        -2.0000              0
               0              0        -2.0000
```

Flows   Unlike Maple, MATLAB has no direct way of drawing flows. However, we can write a program that numerically computes the solutions of a two-dimensional system between 0 and t_1 with the help of some built-in numerical solver, say **ode23**, and then plot the solutions in the phase plane.

EXAMPLE 9   To draw a few solutions in the phase plane for the system

$$\begin{cases} x' = x(1-y) \\ y' = 3y(x-1) \end{cases}$$

for initial conditions $x(0)$, $y(0)$ between 0 and 3, proceed as follows. In a file called, say, f.m, write the above system as follows:

```
function xp=f(t,x)
xp=x;
y=x(2); x=x(1);
xp(1)=x-x.*y;
xp(2)=-3.*y+3.*x.*y;
```

The flow can be produced with the following program, in which we have taken initial conditions of the form $x(0) = y(0) = n$ for n from 1 to 6. We have also taken the variable t from 0 to 20, to make sure that

Figure 4.6.5. The flow of the system in Example 9.

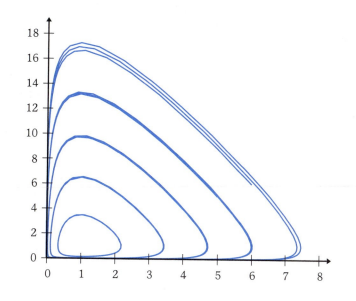

the orbits are continued for a sufficiently long interval. (You can experiment with shorter or longer intervals.) The `'k'` just tells MATLAB to draw the picture in black (different symbols correspond to different colors). The `hold on` command is placed such that it tells MATLAB to stop the drawing until all values of n are taken. Then MATLAB produces the picture in Figure 4.6.5.

```
>> for n=0:6
   [t,x]=ode23('f', [0 20], [n n]);
   plot(x(:,1), x(:,2), 'k'),
   hold on
   end
```

PROBLEMS

Use Maple, Mathematica, or MATLAB to find the general solutions of the linear homogeneous systems in Problems 1 through 8. In case initial conditions are specified, solve the corresponding initial value problem.

1. $\begin{cases} x' = 2.17x + 7.23y \\ y' = \sqrt{5}x - 2\pi y \end{cases}$

2. $\begin{cases} r' = -\frac{3}{45}r + 100.0001\theta \\ \theta' = 55r - 3.289\theta \end{cases}$

3. $\begin{cases} u' = 4.44u + 5.11v \\ v' = 6.66u - 7.12v, \end{cases}$
$u(1.03) = 14.27, \ v(1.03) = -22.56$

4. $\begin{cases} z' = 4.44z + 5.11w \\ w' = 6.66z - 7.12w, \end{cases}$
$z(1.03) = 5, \ w(1.03) = -5$

5. $\begin{cases} x' = 18.023x + 21.987y - 13.135z \\ y' = 43.762x + 84.092y + 55.678z \\ z' = 22.777x + 43.089y - 14.683z \end{cases}$

6. $\begin{cases} u' = 12.25u + 14.52v + 11.22w \\ v' = 17.81u + 25.97v + 76.52w \\ w' = 31.76u + 74.16v + 99.02w \end{cases}$

7. $\begin{cases} \alpha' = 3.1889\alpha - 4.9881\beta + 5.9982\gamma \\ \beta' = 8.3345\alpha - 9.9987\beta + 2.6678\gamma \\ \gamma' = 5.5555\alpha + 6.6677\beta - 4.4443\gamma, \end{cases}$
$\alpha(2.81) = \beta(2.81) = 16.5, \ \gamma(2.81) = 0$

8. $\begin{cases} r' = 2r - \pi\phi + \dfrac{\pi}{3}\theta \\ \phi' = 5r - \dfrac{\pi}{5}\phi + 3\pi\theta \\ \theta' = 8r - \pi\phi - 2\pi\theta, \end{cases}$
$r(\pi/3) = 2, \ \phi(\pi/3) = 1, \ \theta(\pi/3) = 2$

Use Maple, Mathematica, or MATLAB to find the general solutions of the linear nonhomogeneous systems in Problems 9 through 14. In case initial conditions are specified, solve the corresponding initial value problem.

9.
$$\begin{cases} x' = 2.17x + 7.23y + 0.02t \\ y' = \sqrt{5}x - 2\pi y + 5e^{-t} \end{cases}$$

10.
$$\begin{cases} r' = -\frac{3}{45}r + 100.0001 + 2.35\theta \\ \theta' = 55r - 3.289\theta - 2t^2 \end{cases}$$

11.
$$\begin{cases} u' = 4.44u - 5.11v - \sin t \\ v' = 6.66u - 7.12v + \cos t, \end{cases}$$
$$u(1) = 14.27, \ v(1) = -22.56$$

12.
$$\begin{cases} x' = 18.023x + 21.987y - 13.135z - t \\ y' = 43.762x + 84.092y + 55.678z - 2t \\ z' = 22.777x + 43.089y + 14.683z + 3t \end{cases}$$

13.
$$\begin{cases} u' = 12.25u + 14.52v + 11.22w + 1 \\ v' = 17.81u + 25.97v + 76.52w - t \\ w' = 31.76u + 74.16v + 99.02w + t^2 \end{cases}$$

14.
$$\begin{cases} r' = 2r - \pi\phi + \dfrac{\pi}{3}\theta + \sin 2t \\ \phi' = 5r - \dfrac{\pi}{5}\phi + 3\pi\theta - 2t \sin 2t \\ \theta' = 8r - \pi\phi - 2\pi\theta + e^{-t} \cos t, \end{cases}$$
$$r(\pi/3) = 2, \ \phi(\pi/3) = 1, \ \theta(\pi/3) = 2$$

Use Maple, Mathematica, or MATLAB to find the numerical solution of the initial value problems in Problems 15 to 18 in the interval $[1, 100]$ with step size 0.1. Use both Euler's and the second-order Runge-Kutta method. Then draw the graphs of the solutions and the phase plane representation.

15.
$$\begin{cases} x' = 2.17x + 7.23y \\ y' = 5x - 2y \end{cases}, \qquad x(1) = y(1) = 1$$

16.
$$\begin{cases} r' = -\frac{3}{45}r + 100.0001\theta \\ \theta' = 55r - 3.289\theta, \end{cases}$$
$$r(1) = 2, \ \theta(1) = 3.5$$

17.
$$\begin{cases} u' = 4.44u - 5.11v - \sin t \\ v' = 6.66u - 7.12v + \cos t, \end{cases}$$
$$u(1) = 14.27, \ v(1) = -22.56$$

18.
$$\begin{cases} z' = 4.44z - 5.11w - t^2 \\ w' = 6.66z - 7.12w + 2e^{-t}, \end{cases}$$
$$z(1) = 5, \ w(1) = -5$$

4.7 Modeling Experiments

Before attempting to work on any of the modeling experiments below, read the introductory paragraph of Section 2.8. This will give you an idea about what is expected from you, how to proceed, and what to emphasize.

Free-Market Models

Consider the linear free-market model in Section 4.4 and its generalization in Chapter 1. Recall that the generalized model is given by the system

$$\begin{cases} p' = \alpha f(r) \\ r' = -\beta h(r), \end{cases}$$

where p is the price of a commodity, r is the excess demand, α and β are positive constants, and f and h are functions with certain properties that make the model realistic. In Chapter 1 we proposed a set of conditions to be satisfied by f and h. Can you think of other realistic conditions? For your choice of conditions, can f and h be linear functions? In case they are close to being linear (i.e., if their graphs are close to being straight lines), are prices still tending to an equilibrium? What if f and h are not close to being linear? To explore your model, consider particular cases of functions and values of constants and then use any of the computer programs in Section 4.6.

Do some library research and look for economics textbooks in which other models are proposed. How do these models differ from the one proposed here? Do they still reach the conclusion that prices tend to an equilibrium? What happens if other forces enter the market? What general or particular conclusions can you draw about those models?

Malignant Tumors and Metastasis

Recall the Liotta-DeLisi model of Section 4.4 in which the metastasis of malignant tumors in mice is given by the linear homogeneous system

$$\begin{cases} x' = -(\alpha + \beta)x \\ y' = \beta x - \gamma y, \end{cases}$$

where x is the number of destroyed cancer cells, y is the number of cells that invade the tissue, and $\alpha, \beta, \gamma > 0$ are constants depending on the type of cancer. What is the general solution of the above system? How does the solution depend on α, β, and γ? In other words compare the general solutions for different values of the constants. For this you can graph some particular solutions and see what happens. What conclusions can you draw? Can you state and prove some theorem?

Imagine now that the model is given by the linear nonhomogenous system

$$\begin{cases} x' = -(\alpha + \beta)x + \delta \\ y' = \beta x - \gamma y + \eta, \end{cases}$$

where δ and η are constants. How does its general solution differ from the general solution of the previous system? Is this a realistic model? What is the physical interpretation of the constants δ and η? Take some particular values of the coefficients and graph some particular solutions. Then take other coefficients and repeat the procedure. Do you see any patterns? Consult the literature on the Liotta-DeLisi model and learn more about it. Can this model be used for any type of cancer? Does it apply only to mice?

Epidemics with Quarantine

For epidemics with quarantine, consider models of the form

$$\begin{cases} I' = r(t)(aI + bU - cQ) \\ Q' = \alpha I \\ U' = -\beta U - \gamma I, \end{cases}$$

where I, Q, and U represent the number of infected, quarantined, and uninfected individuals; $a, b, c, \alpha, \beta, \gamma > 0$ are constants; and r is a function of time t. Fix some values for the constants and choose different functions r. First try some functions r that are similar to the one in Section 4.5. In each case use one of the programs in Section 4.6 to draw the graphs of the solutions, and see what happens. Do you obtain solutions that make sense from the point of view of an epidemic with quarantine? What happens if you keep r fixed and vary the constants? Vary them just a little first, and then vary them more. Do you see any patterns? Can you state any theorem about the behavior of the solutions? Can you prove it?

The Decelerating Train Recall the system in Section 4.5 that described the motion of a decelerating train,

$$\begin{cases} m_1 x'' = -k_1 x + k_2(y - x) + \dfrac{t^3}{10} y' \\ m_2 y'' = -k_2(y - x) + \dfrac{t^3}{10}(y' - x'), \end{cases}$$

where m_1 and m_2 are the masses of the two cars, and x and y represent the distances between the engine and each of the cars. Write the above system of two second-order equations as a system of four first-order equations. Consider one of the programs in Section 4.6 for obtaining numerical solutions, and adapt it to a system of four equations. Then take different values of the constants m_1, m_2, k_1, and k_2 and run the program in each case. Draw the graphs of the solutions. Keep the constants k_1 and k_2 fixed, and vary only the masses. Then keep the masses fixed and vary the constants. What conclusions can you draw about the behavior of the cars at braking? Is this a realistic model? Can you improve it? Can you state any theorem about the decelerating motion of the cars?

5 Nonlinear Systems

In the past an equation was only considered solved when one had expressed the solution with the aid of a finite number of known functions; but this is hardly possible one time in a hundred. What we should always try to do, is to solve the qualitative problem, that is to find the general form of the curve representing the unknown function.

HENRI POINCARÉ
King Oscar's Prize, 1889

5.1 Linearization

The goal of this chapter is to provide some basic qualitative methods for the study of nonlinear systems of differential equations. In this section we will investigate the behavior of the flow near equilibrium solutions using the *linearization* technique and connect it to the Hartman-Grobman theorem, which relates a nonlinear system to the corresponding linear one near the equilibrium. Using this result we offer a classification of the equilibria for nonlinear systems. Finally we apply the linearization technique to analyze a model of competing species.

Isolated Equilibria

Most nonlinear systems have a remarkable property. In the neighborhood of equilibrium solutions they behave like the linear system that approximates them. The main goal of this section is to analyze this property. For this we first need to see what is meant by the linear system that approximates a nonlinear system near an equilibrium.

Consider the autonomous system of differential equations

$$\mathbf{x}' = \mathbf{F}(\mathbf{x}), \tag{1}$$

where $\mathbf{F}$ is the vector field and $\mathbf{x}$ is the vector variable. Recall from Section 4.4 that an *equilibrium solution* of system (1) is a constant vector function $\mathbf{x}$ such that $\mathbf{F}(\mathbf{x}) = \mathbf{0}$. The phase-space representation of an equilibrium solution is a point. But unlike most linear systems, which have one equilibrium, nonlinear systems can have more than one or none at all.

EXAMPLE 1 The system

$$\begin{cases} x' = y - 2xy \\ y' = x^2 - 1 \end{cases}$$

has two equilibria, $(1, 0)$ and $(-1, 0)$ (see Figure 5.1(a)).

Figure 5.1.1. The equilibria in
(*a*) Example 1 and (*b*) Example 2.

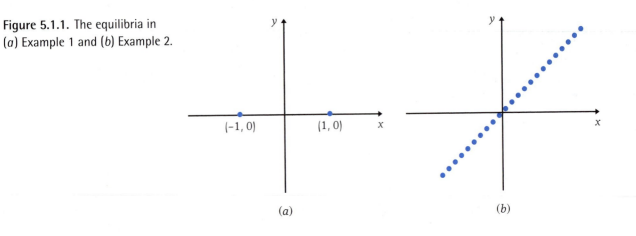

(*a*)   (*b*)

EXAMPLE 2   The system

$$\begin{cases} x' = x^2 - y^2 \\ y' = x - y \end{cases}$$

has infinitely many equilibria, namely, all points for which $x = y$ (see
Figure 5.1(*b*)).

EXAMPLE 3   The system

$$\begin{cases} x' = \dfrac{1}{y} \\ y' = \dfrac{2}{x} \end{cases}$$

has no equilibria at all.

Each equilibrium in Example 2 has other equilibria in all of its neighborhoods. In what follows we would like to avoid such situations, so we
will deal only with *isolated* equilibrium solutions, i.e., those that have a
neighborhood free of other equilibria (as in Example 1).

A Change of Variables

Since we are interested in the behavior of the flow near an isolated equilibrium $\mathbf{x}_0$, let us shift the origin of the frame to $\mathbf{x}_0$. We achieve this by
the change of variables

$$\mathbf{u} = \mathbf{x} - \mathbf{x}_0. \tag{2}$$

In $\mathbf{u}$-coordinates, system (1) takes the form

$$\mathbf{u}' = \mathbf{G}(\mathbf{u}), \tag{3}$$

where $\mathbf{G}$ is the new form of the vector field, and the origin is an equilibrium, i.e., $\mathbf{G}(\mathbf{0}) = \mathbf{0}$.

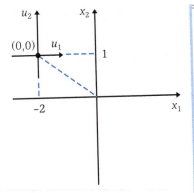

Figure 5.1.2. The shift of the new coordinate system to the equilibrium.

EXAMPLE 4   The nonlinear system

$$\begin{cases} x_1' = (x_1 - 3)(x_2 - 1) \\ x_2' = (x_1 + 2)(x_2 + 5) \end{cases} \tag{4}$$

has the point $(-2, 1)$ as an isolated equilibrium (see Figure 5.1.2). The change of variables that shifts the origin of the frame to the equilibrium is

$$\begin{cases} u_1 = x_1 + 2 \\ u_2 = x_2 - 1. \end{cases} \tag{5}$$

Consequently, $u_1' = x_1'$, $u_2' = x_2'$, $(x_1 - 3)(x_2 - 1) = (u_1 - 5)u_2$, and $(x_1 + 2)(x_2 + 5) = u_1(u_2 + 6)$, so the new system is

$$\begin{cases} u_1' = -5u_2 + u_1u_2 \\ u_2' = 6u_1 + u_1u_2. \end{cases} \tag{6}$$

That is, $G_1(u_1, u_2) = -5u_2 + u_1u_2$, $G_2(u_1, u_2) = 6u_1 + u_1u_2$, and $G_1(0, 0) = G_2(0, 0) = 0$. So in u_1u_2-coordinates, $(0, 0)$ is the equilibrium corresponding to $(-2, 1)$ in x_1x_2-coordinates.

The Linearization Technique

When studying the flow near an isolated equilibrium, we first shift the origin of the frame to the equilibrium and then linearize the new system around **0**. Let us describe this technique in detail.

Step 1. Using the change of variables (2), transform system (1), $\mathbf{x}' = \mathbf{F}(\mathbf{x})$, into the equivalent system (3), $\mathbf{u}' = \mathbf{G}(\mathbf{u})$, which has **0** as an isolated equilibrium.

Step 2. Compute the partial derivatives of **G** with respect to the components of the vector **u** at the equilibrium **0** and form the partial-derivative matrix **A**. For example, if $\mathbf{G} = (G_1, G_2)$ and $\mathbf{u} = (u_1, u_2)$, then

$$\mathbf{A} = \begin{pmatrix} \dfrac{\partial G_1}{\partial u_1}(0, 0) & \dfrac{\partial G_1}{\partial u_2}(0, 0) \\ \dfrac{\partial G_2}{\partial u_1}(0, 0) & \dfrac{\partial G_2}{\partial u_2}(0, 0) \end{pmatrix}.$$

Step 3. Write the linearized system around **0**, given by

$$\mathbf{u}' = \mathbf{A}\mathbf{u}. \tag{7}$$

For the example at Step 2, the linearized system around $(0, 0)$ is

$$\begin{cases} u_1' = \dfrac{\partial G_1}{\partial u_1}(0, 0)u_1 + \dfrac{\partial G_1}{\partial u_2}(0, 0)u_2 \\ u_2' = \dfrac{\partial G_2}{\partial u_1}(0, 0)u_1 + \dfrac{\partial G_2}{\partial u_2}(0, 0)u_2. \end{cases}$$

EXAMPLE 5   Let us linearize the nonlinear system (4) around the equilibrium $(-2, 1)$.

Step 1. The change of variables (5) transforms system (4) into system (6).

Step 2. The corresponding matrix $\mathbf{A}$ takes the form

$$\begin{pmatrix} \dfrac{\partial G_1}{\partial u_1}(0, 0) & \dfrac{\partial G_1}{\partial u_2}(0, 0) \\ \dfrac{\partial G_2}{\partial u_1}(0, 0) & \dfrac{\partial G_2}{\partial u_2}(0, 0) \end{pmatrix} = \begin{pmatrix} 0 & -5 \\ 6 & 0 \end{pmatrix}.$$

Step 3. The linearized system around $(u_1, u_2) = (0, 0)$ is then $\mathbf{u}' = \mathbf{A}\mathbf{u}$; this is

$$\begin{cases} u_1' = -5u_2 \\ u_2' = 6u_1. \end{cases} \tag{8}$$

Notice that system (8) was obtained from (6) by simply neglecting the nonlinear terms. But this can be done only for polynomial vector fields. In most cases the use of partial derivatives is the only way to obtain the linearized system.

EXAMPLE 6   The nonlinear system

$$\begin{cases} x' = 2 \sin y \\ y' = 3 \sin x - \sin y \end{cases} \tag{9}$$

has $(0, 0)$ as an isolated equilibrium, so no change of variables is necessary for linearization. Here $G_1(x, y) = 2 \sin y$ and $G_2(x, y) = 3 \sin x - \sin y$, so

$$\frac{\partial G_1}{\partial x}(0, 0) = 0, \qquad \frac{\partial G_1}{\partial y}(0, 0) = 2,$$

$$\frac{\partial G_2}{\partial x}(0, 0) = 3, \qquad \frac{\partial G_2}{\partial y}(0, 0) = -1,$$

and the linearized system at $(0, 0)$ is

$$\begin{cases} x' = 2y \\ y' = 3x - y. \end{cases} \tag{10}$$

The Hartman–Grobman Theorem

At the end of the 19th century, Poincaré was aware of the connection between linear and nonlinear systems near an equilibrium solution. But a general statement and a full proof of this result had to await the years 1959 and 1960, when the American mathematician Philip Hartman and the Russian mathematician D. M. Grobman independently published the results that we summarize as follows.

Figure 5.1.3. A perturbed center (*a*) can become a spiral sink (*b*) or a spiral source (*c*).

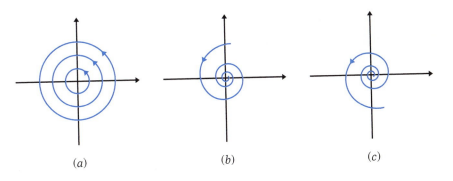

(*a*)   (*b*)   (*c*)

The Hartman–Grobman Theorem

Let $\mathbf{x}_0$ be an isolated equilibrium of the nonlinear system $\mathbf{x}' = \mathbf{F}(\mathbf{x})$, which corresponds to the equilibrium $\mathbf{0}$ of the linearized system $\mathbf{u}' = \mathbf{A}\mathbf{u}$. Then if

(a) $\mathbf{0}$ is a hyperbolic equilibrium of the linear system, the solutions of the nonlinear system near $\mathbf{x}_0$ resemble the behavior of the solutions of the linear system near $\mathbf{0}$.

(b) $\mathbf{0}$ is a nonhyperbolic equilibrium for the linear system, no conclusion can be drawn about the behavior of the flow of the nonlinear system near $\mathbf{x}_0$.

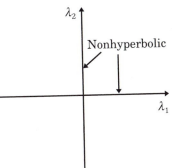

Figure 5.1.4. For two-dimensional systems, in the plane of all possible pairs of eigenvalues, the nonhyperbolic equilibria are given by two lines: $\lambda_1 = 0$ and $\lambda_2 = 0$.

In other words if $\mathbf{0}$ is a source, sink, or saddle for the linearized system, then the solutions of the nonlinear system in some small neighborhood of $\mathbf{x}_0$ behave "similarly" to those of the linear system near $\mathbf{0}$. "Similarly" means, for example, that if $\mathbf{0}$ is a sink, then in some neighborhood of $\mathbf{x}_0$ all solutions tend toward $\mathbf{x}_0$.

Intuitively, the reason no conclusion can be drawn about the flow near $\mathbf{x}_0$ in case (*b*) is that the linearized system is not structurally stable (see Section 3.5). For example, $\mathbf{0}$ can be a center, in which case the flow of the linearized system resembles Figure 5.1.3(*a*). But this is just an approximation for the flow of the nonlinear system near $\mathbf{x}_0$. A slight perturbation (which necessarily occurs due to the nonlinear terms) may break this configuration into a spiral sink (Figure 5.1.3(*b*)) or a spiral source (Figure 5.1.3(*c*)), or it can keep it as it is. Since any of these situations can occur, we cannot draw a general conclusion. In case (*a*) the equilibrium is hyperbolic, so the flow about it is structurally stable and therefore a slight perturbation does not affect its qualitative properties.

Notice that the set of nonhyperbolic equilibria is "negligible" compared with the set of hyperbolic ones. In the two-dimensional case, for example, nonhyperbolic equilibria are those for which at least one eigenvalue is zero. Thus in the $\lambda_1\lambda_2$-plane of all possible pairs of eigenvalues, nonhyperbolic equilibria are given by points on the lines $\lambda_1 = 0$ and $\lambda_2 = 0$, whereas hyperbolic ones are points in the rest of the plane (see Figure 5.1.4).

Figure 5.1.5. The flow of (*a*) system (10) and (*b*) system (9) near the origin.

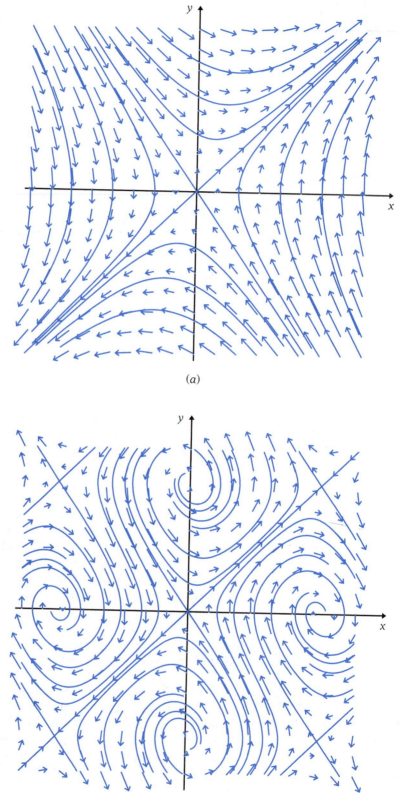

(*a*)

(*b*)

EXAMPLE 7 The nonlinear system (9) has (10) as the linearized system around the origin. The eigenvalues of the matrix that defines (10) are -3 and 2, so the origin of system (10) is a saddle. Thus according to the Hartman-Grobman theorem, the flow near the origin of system (9) is such that two solutions tend to $\mathbf{0}$, two tend away from $\mathbf{0}$, and the others come close to $\mathbf{0}$ but never reach $\mathbf{0}$. With the help of one of the programs in Section 4.6, we have drawn the vector fields of systems (10) (9) and then sketched the flows (see Figures 5.1.5(a) and (b)). Notice that they are similar near the origin.

EXAMPLE 8 The nonlinear system (4) has (8) as the linearized system after the frame is shifted to the equilibrium $(-2, 1)$. The eigenvalues of the matrix defining (8) are $i\sqrt{6}$ and $-i\sqrt{6}$, so the origin of (8) is a center. Thus according to point (b) of the Hartman-Grobman theorem, we cannot draw any conclusion about the flow near the equilibrium $(-2, 1)$ of the nonlinear system (4). So in this case the linearization method fails.

Classification   The Hartman-Grobman theorem suggests the following classification of the equilibria for a nonlinear system.

DEFINITION 5.1.1

An equilibrium of system (1) is called *hyperbolic (source, sink, saddle)* or *nonhyperbolic*, respectively, if the corresponding equilibrium of the linearized system is hyperbolic (source, sink, saddle) or nonhyperbolic.

Notice that there is an important difference between the equilibria of a linear system and those of a nonlinear system. Whereas the former determine the global behavior of solutions, the latter influence the qualitative picture only locally. In Figure 5.1.6 are typical examples of a source, a sink, and a saddle of a nonlinear system in the plane. Each determines the local character of the flow near the equilibrium.

Figure 5.1.6. The flow near (a) a source, (b) a sink, and (c) a saddle of a nonlinear system.

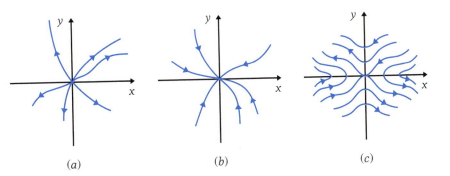

(a)   (b)   (c)

Silver fox.

Applications

Competing species

In an isolated region of the Canadian Northwest Territories, a population of white wolves, x, and one of silver foxes, y, compete for survival. (For each population, one unit represents 100 individuals.) They have a common, limited food supply, which consists mainly of mice. A model that describes the dynamics of these competing species is given by the nonlinear system

$$\begin{cases} x' = x - x^2 - xy \\ y' = \dfrac{3}{4}y - y^2 - \dfrac{1}{2}xy. \end{cases} \tag{11}$$

In other words, each population increases with its size, but decreases with the square of its size and with the product of its own size and the size of the other species. The proportionality factors were obtained statistically. Can the two species survive together?

White wolf.

Solution.   We are interested in the flow for $x, y \geq 0$. First notice that system (11) has four equilibria: $(0, 0)$, $(0, \frac{3}{4})$, $(1, 0)$, and $(\frac{1}{2}, \frac{1}{2})$. Let us first tackle the flow near $(\frac{1}{2}, \frac{1}{2})$. The change of variables $(u, v) = (x - \frac{1}{2}, y - \frac{1}{2})$ leads to the linear system

$$\begin{cases} u' = -\dfrac{1}{2}u - \dfrac{1}{2}v \\ v' = -\dfrac{1}{4}u - \dfrac{1}{2}v, \end{cases}$$

around the equilibrium $(0, 0)$. The eigenvalues are

$$\lambda_1 = -\frac{2 - \sqrt{2}}{4} < 0 \qquad \text{and} \qquad \lambda_2 = -\frac{2 + \sqrt{2}}{4} < 0,$$

so this equilibrium is a sink.

Analogously we can show that for system (11), $(0, 0)$ is a source whereas $(0, \frac{3}{4})$ and $(1, 0)$ are saddles (see Problem 21). Plotting the vector field with the help of one of the programs in Section 4.6 and using the above information, we can sketch the flow in Figure 5.1.7.

The flow shows that except for the case where one of the species is absent from the start, the two species can survive together in spite of the

Figure 5.1.7. The vector field and the flow of system (11).

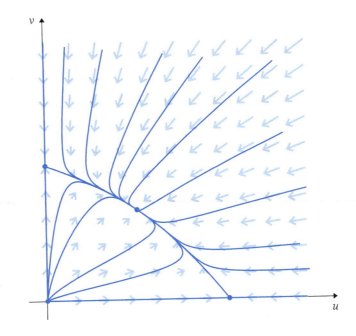

competition. However, this is not necessarily true if the dynamics are governed by similar equations with different coefficients (see Problems 22 and 23).●

PROBLEMS

Find the equilibrium solutions of the nonlinear systems in Problems 1 through 8 and decide which equilibria are isolated.

1. $\begin{cases} x' = 8x^3 - 2xy \\ y' = 4y^2 - \dfrac{x^2}{4} \end{cases}$

2. $\begin{cases} u' = u^2 - v^2 - 1 \\ v' = u - \frac{1}{3} \end{cases}$

3. $\begin{cases} z' = 27w^3 - \dfrac{z^3}{8} \\ w' = 3w - \dfrac{z}{2} \end{cases}$

4. $\begin{cases} r' = \cos\theta - \sin\theta \\ \theta' = r^2 - 3r + 2 \end{cases}$

5. $\begin{cases} z' = z^2 + v + w \\ v' = v^2 + w + z \\ w' = w^2 + z + v \end{cases}$

6. $\begin{cases} x' = \sigma(y - x) \\ y' = \rho x - y - xz \\ z' = -\beta z + xy \end{cases}$

(the Lorenz equations), where $\sigma, \rho, \beta > 0$ are constants

7. $\begin{cases} u' = u + v + \cos w \\ v' = u - v - \sin w \\ w' = 2w \end{cases}$

8. $\begin{cases} r' = r^2 + s^2 + p^2 - 27 \\ s' = r - p \\ p' = r - s \end{cases}$

Linearize the systems in Problems 9 through 16 around the indicated equilibrium solution, and determine whether the equilibrium is hyperbolic or not. In case it is hyperbolic, is it a source, sink, or saddle?

9. $\begin{cases} x' = y^2 - 1 \\ y' = x^2 - 4 \end{cases}$
at $(2, 1)$ and then at $(-2, -1)$

10. $\begin{cases} \theta' = v \\ v' = -g\sin\theta \end{cases}$
at $(-\pi, 0)$ and then at $(\pi, 0)$, where $g > 0$ is a constant

11. $\begin{cases} u' = \sin u \cos v \\ v' = -\cos u \sin v \end{cases}$
at $(0, 0)$ and then at $(\pi/2, \pi/2)$

12. $\begin{cases} z' = z^2 + 3zu \\ u' = 2z + u^3 \end{cases}$

at $(0, 0)$ and then at $(-3, \sqrt{6})$

13. $\begin{cases} x' = 10(y - x) \\ y' = 28x - y - xz \\ z' = -\frac{8}{3}z + xy \end{cases}$

at $(0, 0, 0)$, $(-6\sqrt{2}, -6\sqrt{2}, 27)$, and $(6\sqrt{2}, 6\sqrt{2}, 27)$

14. $\begin{cases} R' = R - R^2 - RS \\ S' = S - S^2 + RS - SP \\ P' = P - P^2 + SP \end{cases}$

at $(0, 0, 1)$, $(0, 1, 0)$, $(1, 0, 0)$, and $(1, 0, 1)$

15. $\begin{cases} u' = \dfrac{v}{4} - v^3 \\ v' = \dfrac{w}{4} - w^3 \\ w' = \dfrac{u}{4} - u^3 \end{cases}$

at $(0, 0, 0)$ and then at $(\frac{1}{2}, \frac{1}{2}, \frac{1}{2})$

16. $\begin{cases} X' = X^2 - Y^2 \\ Y' = X^2 + Y^2 \\ z' = z \end{cases}$

at $(0, 0, 0)$

17. Consider the systems

$\begin{cases} x' = 2x + y - y^2 \\ y' = x - 2y + x^2, \end{cases}$ $\begin{cases} x' = 2x + y + y^2 \\ y' = x - 2y - x^2, \end{cases}$

$\begin{cases} x' = 2x - y - y^2 \\ y' = x + 2y + x^2, \end{cases}$

which have $(0, 0)$ as an equilibrium solution. Determine which of these systems have similar flows near the origin.

18. Consider the system

$\begin{cases} u' = -u + 1 \\ v' = -2u^3 + 3v, \end{cases}$

which has an equilibrium at $(1, \frac{2}{3})$. Solve the first equation of the system, and then use the solution to solve the second equation and write the general solution of the system. Is $(1, \frac{2}{3})$ a singular solution? Can it be recovered from the general solution? Then linearize the system and decide the nature of the equilibrium $(1, \frac{2}{3})$.

19. Consider the system

$\begin{cases} r' = 4r^2 - \alpha \\ s' = -\dfrac{s}{4}(r^2 + 4), \end{cases}$

where α is a real parameter. Find out how the number of equilibria of the system changes when α varies. In each case linearize the system and determine, if possible, the nature of the equilibria.

20. If written as a system, Holmes's version of Düffing's equation for $\omega = 0$ is

$\begin{cases} x' = y \\ y' = -\delta y + x - x^3, \end{cases}$

where $\delta > 0$ is a constant. Find the equilibria and see if you can determine their nature by linearization.

21. For the model that describes the dynamics of the wolf and fox populations in the Canadian Northwest Territories, show that the equilibrium $(0, 0)$ is a source and that the equilibria $(0, \frac{3}{4})$ and $(1, 0)$ are saddles.

22. Assume that the dynamics of two competing species is described by the system

$\begin{cases} x' = x - x^2 - xy \\ y' = \frac{1}{2}y - \frac{1}{4}y^2 - \frac{3}{4}xy. \end{cases}$

Can the two species survive together?

23. Assume that the dynamics of two competing species is described by the system

$\begin{cases} x' = 4x - x^2 - xy \\ y' = 2y - ay^2 - bxy, \end{cases}$

where a and b are positive constants. For what values of a and b can the two species survive together?

24. For the saddles of the model that describes the dynamics of wolf and fox populations, compute the eigenvectors corresponding to the eigenvalues of the linearized system. Draw these eigenvectors in the phase plane. Do you see any connection between the lines plotting these eigenvectors and the curves representing the solutions that tend toward and away from the equilibria? Can you draw a general conclusion?

In addition to equilibria, periodic solutions play an important role in helping us understand the flow of a nonlinear system. But unlike the case for equilibria, there is no general method of finding periodic solutions. In this section we deal with isolated periodic orbits, define limit cycles and use polar coordinates in their study, present Dulac's theorem on the number of limit cycles for a two-dimensional system, and then apply these results to a model describing the change in temperature of an engine and its coolant.

Isolated Periodic Solutions

Let us first define the notion of a *periodic solution* and provide some examples.

DEFINITION 5.2.1

A solution φ of the system

$$\mathbf{x}' = \mathbf{F}(\mathbf{x}) \tag{1}$$

is called *periodic* if there is a $T > 0$ such that $\varphi(t + T) = \varphi(t)$ for all t for which the solution is defined. The smallest T with this property is called the *prime period* of φ.

In phase space, a periodic solution is represented by a closed curve (see Figure 5.2.1). Can a periodic solution intersect itself?

EXAMPLE 1 The $(0, 0)$ equilibrium of the linear system

$$\begin{cases} x' = y \\ y' = -x \end{cases} \tag{2}$$

is a center. Every solution is of the form

$$\begin{pmatrix} x(t) \\ y(t) \end{pmatrix} = c_1 \begin{pmatrix} \sin t \\ \cos t \end{pmatrix} + c_2 \begin{pmatrix} \cos t \\ -\sin t \end{pmatrix},$$

where c_1 and c_2 are constants. These solutions are periodic with period 2π. Since at some point, say at $(0, 1)$, $x' = 1$ and $y' = 0$, it follows that x is increasing and y is constant, so the orientation of all solutions (why all?) is as in Figure 5.2.2(a).

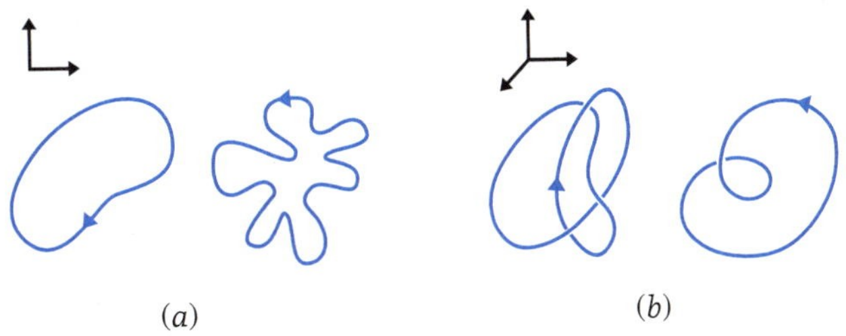

Figure 5.2.1. Possible periodic solutions (a) in the plane and (b) in space.

(a) $\qquad\qquad\qquad\qquad$ (b)

Figure 5.2.2. (*a*) The continuous band of periodic solutions of system (2). (*b*) The isolated periodic solution of system (3).

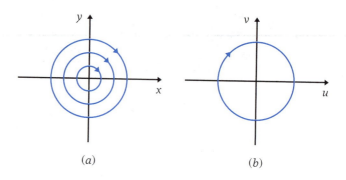

EXAMPLE 2   The nonlinear system

$$\begin{cases} u' = v + u(1 - u^2 - v^2) \\ v' = -u + v(1 - u^2 - v^2) \end{cases} \tag{3}$$

has $(0, 0)$ as an equilibrium solution. Moreover, $(u(t), v(t)) = (\sin t, \cos t)$ is also a solution and it represents a circle (indeed, $u^2 + v^2 = 1$). If we choose a point on the circle, say $(u, v) = (1, 0)$, we see that $u' = 0$ and $v' = -1$; i.e., u remains unchanged while v is decreasing. This means that (u, v) is a periodic solution oriented as in Figure 5.2.2(*b*).

Notice that the periodic solutions of system (2) form a continuous band, whereas the periodic solution of system (3) is *isolated* (this will be proved in Example 4); i.e., there is a neighborhood around it free of other periodic solutions. (A neighborhood of a solution can be imagined as a "strip" in the plane or as a "pipe" in space—see Figure 5.2.3.)

Limit Cycles   We will now deal with two-dimensional systems whose phase space is a plane. A famous result, called *Jordan's curve theorem* after the French mathematician Camille Jordan (1838–1922), states that a closed planar curve with no self-intersections divides the plane into two regions: an exterior and an interior one. This is not as obvious as it may seem. Just think of the curve as resembling a complicated maze. It is then hard to decide if a given point belongs to the interior or to the exterior region.

Figure 5.2.3. Examples of neighborhoods of solutions (*a*) in the plane and (*b*) in space.

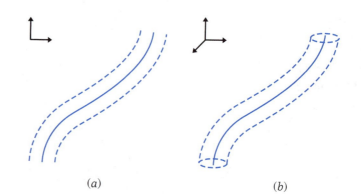

Figure 5.2.4. Limit cycles to which other solutions tend from (*a*) the interior region, (*b*) the exterior region, and (*c*) both regions.

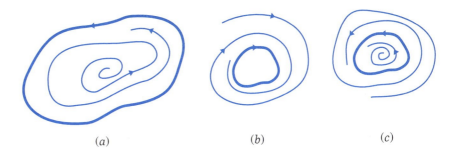

(*a*)   (*b*)   (*c*)

DEFINITION 5.2.2

A periodic orbit φ is called a *limit cycle* if there is a solution that tends to it from the interior region, the exterior region, or both, when $t \to \infty$ or when $t \to -\infty$ (see Figure 5.2.4).

EXAMPLE 3   Consider the system

$$\begin{cases} x' = -y + x(x^2 + y^2)\sin \dfrac{1}{\sqrt{x^2 + y^2}} \\[4mm] y' = x + y(x^2 + y^2)\sin \dfrac{1}{\sqrt{x^2 + y^2}} \end{cases} \tag{4}$$

if $(x, y) \neq (0, 0)$ and $x' = y' = 0$ if $(x, y) = (0, 0)$. The flow in the phase plane is easier to draw if we write the system in polar coordinates, i.e., use the change of variables

$$\begin{cases} x = r \cos \theta \\ y = r \sin \theta. \end{cases} \tag{5}$$

The polar coordinates express a system relative to the distance r from the origin and the angle θ made with the x-axis (see Figure 5.2.5(*a*)). To perform this transformation, differentiate the relations in (5) and obtain

$$\begin{cases} x' = r' \cos \theta - r\theta' \sin \theta \\ y' = r' \sin \theta + r\theta' \cos \theta. \end{cases} \tag{6}$$

Figure 5.2.5. (*a*) Polar coordinates. (*b*) The flow of the system in Example 3.

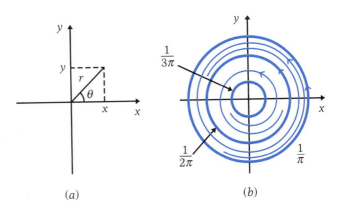

(*a*)   (*b*)

Using the fact that $r^2 = x^2 + y^2$, introducing (5) into (4), and comparing with (6), we are led to the relations

$$\begin{cases} r'\cos\theta - r\theta'\sin\theta = -r\sin\theta + r^3\cos\theta\sin\dfrac{1}{r} \\ r'\sin\theta + r\theta'\cos\theta = r\cos\theta + r^3\sin\theta\sin\dfrac{1}{r}. \end{cases}$$

Solving with respect to r' and θ', we obtain system (4) in polar coordinates:

$$\begin{cases} r' = r^3\sin\dfrac{1}{r} \\ \theta' = 1. \end{cases}$$

At $r = 0$ we have $r' = 0$. Since $\theta' = 1$, the angle increases with constant speed, so all solutions rotate counterclockwise around the origin. Therefore a solution is periodic if the distance from the origin remains constant, i.e., if $r' = 0$, which is equivalent to $\sin(1/r) = 0$. This happens when r takes one of the values $(1/\pi)$, $1/(2\pi)$, ..., $1/(n\pi)$, ..., with n a positive integer. Thus the flow contains an infinite sequence of periodic solutions given by concentric circles whose radii decrease with increasing n (see Figure 5.2.5(b)).

If r belongs to the interval $(1/(2\pi),\ 1/\pi)$ then $r' < 0$, so r decreases. If r belongs to the interval $(1/(3\pi),\ 1/(2\pi))$, then $r' > 0$, so r increases. Continuing this process, we see that the sign of r' alternates, so every periodic orbit is a cycle (see Figure 5.2.5(b)).

Dulac's Theorem

Examples like the ones above have inspired mathematicians to seek more general properties regarding limit cycles of two-dimensional systems. In 1923, the French mathematician Henri Dulac published a paper in which he stated the following result.

Dulac's Theorem

In any bounded region of the plane, a two-dimensional polynomial system has at most a finite number of limit cycles.

This theorem claims that if the vector field is given by a polynomial function (as in Example 2), then—within any bounded region of the plane—there cannot be infinitely many limit cycles. The vector field in Example 3 involves trigonometric functions, so it is not polynomial, and therefore it doesn't fulfill the hypothesis of the theorem.

It is interesting that in the 1980s errors were found in Dulac's original proof, but they were independently corrected in 1988 by a Russian and by several French mathematicians. The proof is beyond the scope of this book.

For specific equations we might be able to conclude more than what Dulac's theorem ensures; for example, we may be able to find the exact number of limit cycles.

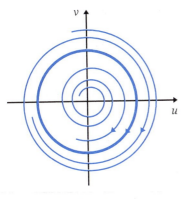

Figure 5.2.6. The flow of system (3).

EXAMPLE 4 In Example 2 we saw that system (3) has a periodic solution. We will now show that there are no other periodic solutions and that this one is a limit cycle. Indeed, using polar coordinates

$$\begin{cases} u = r \cos \theta \\ v = r \sin \theta, \end{cases}$$

system (3) becomes

$$\begin{cases} r' = r(1 - r^2) \\ \theta' = -1. \end{cases}$$

This shows that inside the periodic solution (i.e., for $r < 1$) $r' > 0$, so r is increasing, whereas outside the periodic solution (i.e., for $r > 1$) $r' < 0$, so r is decreasing. Since this occurs while the angle θ rotates clockwise with constant speed ($\theta' = -1$), the flow is as in Figure 5.2.6.

Applications

The engine and the coolant

Worker adding coolant to the engine of a car.

A mechanical engineer measures the respective temperatures x and y of an engine and its coolant and notices that the two values have a slight periodic variation. He would like to understand this phenomenon and comes up with a mathematical model. First he sets a scale in which the ideal temperatures are at level zero, i.e., $(x, y) = (0, 0)$, and considers temperatures above and below the ideal ones as positive and negative, respectively. Then he argues that the rate of change of the engine's temperature, x', increases with the temperature x but decreases with the temperature y of the coolant, so he takes x' proportional to $x - y$. On the other hand, x' decreases the farther x is from 0 and the greater the deviation of the two temperatures is from $(0, 0)$. He takes this deviation to be the "distance" $\sqrt{x^2 + y^2}$ and thus chooses the first equation of his model as

$$x' = x - y - x \sqrt{x^2 + y^2}.$$

Analogously, he considers that the rate of change of the coolant's temperature, y', increases like $x + y$ and decreases like $y \sqrt{x^2 + y^2}$, so the second equation is

$$y' = x + y - y \sqrt{x^2 + y^2}.$$

Can this model explain the periodic change in temperatures?

Solution. The system to study,

$$\begin{cases} x' = x - y - x \sqrt{x^2 + y^2} \\ y' = x + y - y \sqrt{x^2 + y^2}, \end{cases} \tag{7}$$

is obviously nonlinear and has a single equilibrium, $(x, y) = (0, 0)$. Linearization around this equilibrium leads to the system

$$\begin{cases} x' = x - y \\ y' = x + y, \end{cases}$$

which has the eigenvalues $\lambda_1 = 1 + i$ and $\lambda_2 = 1 - i$. This means that the origin is a spiral source, but this indicates no periodic behavior near $(0, 0)$.

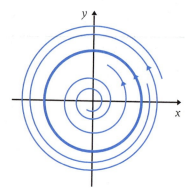

Figure 5.2.7. The flow of the nonlinear system (7) that models the change in temperature for the engine and the coolant.

The use of polar coordinates was previously successful because the studied systems had circular periodic orbits in the phase plane. Let us see whether system (7) exhibits such behavior. Consider again the change of variables

$$\begin{cases} x = r\cos\theta \\ y = r\sin\theta, \end{cases}$$

which transforms system (7) into

$$\begin{cases} r' = r(1 - r) \\ \theta' = 1. \end{cases} \tag{8}$$

Since $\theta' = 1$, the flow rotates counterclockwise. From the first equation in (8) we recover the equilibrium at $r = 0$. For $0 < r < 1$, $r' > 0$, so r is increasing; therefore the origin is a spiral source, a conclusion in agreement with the one obtained from linearization. For $r = 1$, however, $r' = 0$, i.e., $r = $ constant. This reveals the periodic solution in Figure 5.2.7. Moreover, this is a limit cycle and it is the only periodic solution of the system. Indeed, for $r > 1$, $r' < 0$, so the cycle is approached from both the inside and the outside. This proves that unless ideal from the start (a situation as improbable as making a pencil stand on its sharp end), the two temperatures will tend to oscillate periodically around the ideal values as shown in Figure 5.2.7.●

PROBLEMS

Transform the systems in Problems 1 through 4 using polar coordinates.

1. $\begin{cases} x' = x^2 + y^2 - 1 \\ y' = 3x - x^2 - y^2 + 1 \end{cases}$

2. $\begin{cases} u' = 2u - v + u^4 + v^4 \\ v' = v - u^2 + v^2 \end{cases}$

3. $\begin{cases} X' = X\sin\sqrt{X^2 + Y^2} \\ Y' = -Y\cos\sqrt{X^2 + Y^2} \end{cases}$

4. $\begin{cases} s' = 2p - \sqrt{2}s^2 \\ p' = -\frac{1}{3}s + p^2 \end{cases}$

5. For the polar-coordinate transformation $x = r\cos\theta, y = r\sin\theta$, show that

$$r^2\theta' = xy' - x'y.$$

6. Show that the system

$$\begin{cases} x' = x + y + x^3 - y^3 \\ y' = -x + 2y + x^3 + y^3 \end{cases}$$

has no periodic solutions that are circles with the center in the origin of the frame.

7. Show that the linear system

$$\begin{cases} x' = -4y \\ y' = x \end{cases}$$

has a continuous band of periodic solutions. What is the period of each solution?

8. Show that the nonlinear system

$$\begin{cases} x' = -y\sqrt{x^2 + y^2} \\ y' = x\sqrt{x^2 + y^2} \end{cases}$$

has a continuous band of periodic solutions. Then show that these solutions are of the form $(x_\alpha, y_\alpha)(t) = (\alpha\cos\alpha t, \alpha\sin\alpha t)$, for $\alpha > 0$. What is the period of each solution? Show that the function of α describing the period tends to 0 when $\alpha \to \infty$ and tends to infinity when $\alpha \to 0$.

9. Show that the system

$$\begin{cases} x' = -y + x(1 - x^2 - y^2) \\ y' = x + y(1 - x^2 + y^2) \\ z' = z \end{cases}$$

has one isolated limit cycle. Sketch the flow.

10. Using polar coordinates, show that the system

$$\begin{cases} u' = -v + u(1 - u^2 + v^2)(9 - u^2 - v^2) \\ v' = u + v(1 - u^2 - v^2)(9 - u^2 - v^2) \end{cases}$$

has exactly two limit cycles. Sketch the flow.

11. Find a system other than (4) that has infinitely many limit cycles but does not contain a continuous band of periodic solutions. (Can this system be polynomial?) After finding such a system, sketch its flow.

Using the result below, solve Problems 12 through 14. Consider the two-dimensional system

$$\begin{cases} x' = F(x, y) \\ y' = G(x, y), \end{cases} \tag{9}$$

where F and G have continuous partial derivatives in a domain D of the xy-plane. If this system has a periodic solution in D, then the closed curve that represents it must enclose at least one equilibrium solution. Moreover, if it encloses only one equilibrium, then that equilibrium cannot be a saddle.

12. Using the above result, explain why the origin of the engine-coolant system,

$$\begin{cases} x' = x - y - x\sqrt{x^2 + y^2} \\ y' = x + y - y\sqrt{x^2 + y^2}, \end{cases}$$

cannot be a saddle.

13. Using the above result, show that the system

$$\begin{cases} x' = x^2 + y^2 + 1 \\ y' = (x - 1)^2 + 4 \end{cases}$$

has no periodic solutions.

14. Using the above result, show that the system

$$\begin{cases} u' = u - u^2 - uv \\ v' = \frac{1}{2}v - \frac{1}{4}v^2 - \frac{3}{4}uv \end{cases}$$

has no periodic solutions in the first quadrant.

Using the result below, solve Problems 15 and 16. Assume that the functions F and G that define the two-dimensional system (9) have continuous first partial derivatives in a simply connected domain D of the xy-plane, i.e., a domain with no holes (a point is also considered a hole, so a punctured domain is not simply connected). If

$$\frac{\partial F}{\partial x}(x, y) + \frac{\partial G}{\partial y}(x, y)$$

does not change sign in D, there is no periodic solution lying entirely in D.

15. Using the above result, show that the system

$$\begin{cases} x' = x + x^3 - 2y \\ y' = -3x + y^5 \end{cases}$$

has no periodic solutions.

16. Using the above result, show that the system

$$\begin{cases} x' = y \\ y' = x - x^3 - \delta y + x^2 y, \end{cases}$$

where $\delta > 0$ is a constant, has no periodic solutions lying entirely inside any of the three regions delimited by the vertical lines $x = -\delta$ and $x = \delta$. Does this imply that the system has no periodic solutions at all?

5.3 Gradient and Hamiltonian Systems

In this section we will study two kinds of two-dimensional systems: gradient and Hamiltonian. Though they are rather uncommon in the set of systems, many natural phenomena are modeled with their help. We will provide formal definitions, show an easy way to identify them, and present some of their properties—in particular, the orthogonality of their reciprocal flows. Finally, we will use the reciprocal Hamiltonian system to sketch the flow of a gradient system that explains why lobsters only rarely move in straight lines.

Gradient Systems

A two-dimensional system of the form

$$\begin{cases} x' = f(x, y) \\ y' = g(x, y) \end{cases} \tag{1}$$

is called a *gradient system* if there is a real function G of variables x and y, called a *gradient function*, that has continuous partial derivatives and satisfies the relations

$$f(x, y) = \frac{\partial G}{\partial x}(x, y) \quad \text{and} \quad g(x, y) = \frac{\partial G}{\partial y}(x, y). \tag{2}$$

These conditions imply that

$$\frac{\partial f}{\partial y} = \frac{\partial^2 G}{\partial y \partial x} \quad \text{and} \quad \frac{\partial g}{\partial x} = \frac{\partial^2 G}{\partial x \partial y},$$

which means that

$$\frac{\partial f}{\partial y} = \frac{\partial g}{\partial x}. \tag{3}$$

To see if a given two-dimensional system is a gradient system, we can either find a function G that satisfies conditions (2) or check whether relation (3) is satisfied.

EXAMPLE 1   The nonlinear equations

$$\begin{cases} x' = 9x^2 - 10xy^2 \\ y' = 2y - 10x^2 y \end{cases}$$

form a gradient system. Indeed,

$$f(x, y) = 9x^2 - 10xy^2, \quad g(x, y) = 2y - 10x^2 y,$$

and

$$\frac{\partial f}{\partial y}(x, y) = \frac{\partial g}{\partial x}(x, y) = -20xy,$$

so relation (3) is satisfied.

We can also find a function G that satisfies (2). Since

$$f(x, y) = \frac{\partial G}{\partial x}(x, y),$$

integrating $f(x, y) = 9x^2 - 10xy^2$ with respect to x leads to

$$G(x, y) = 3x^3 - 5x^2 y^2 + \alpha(y), \tag{4}$$

where α is a function depending on y alone, i.e., is constant relative to x. To determine α, differentiate relation (4) with respect to y and obtain

$$\frac{\partial G}{\partial y}(x, y) = \frac{d}{dy}\alpha(y) - 10x^2 y,$$

which if compared with $g(x, y) = 2y - 10x^2 y$ yields

$$\frac{d}{dy}\alpha(y) = 2y.$$

Therefore $\alpha(y) = y^2 + c$, where c is a constant, so we can take $G(x, y) = 3x^3 - 5x^2 y^2 + y^2$. (Is the choice of G unique?)

Let us prove the following result, which shows that two-dimensional gradient systems exhibit uncomplicated flows.

Theorem 5.3.1

Gradient systems do not have periodic solutions but can have equilibria. The equilibria, however, cannot be spiral sources, spiral sinks, or centers.

PROOF

Let (x, y) be a solution of the gradient system (1). Then by the chain rule,

$$\frac{d}{dt} G(x(t), y(t)) = \frac{\partial G}{\partial x}(x(t), y(t)) x'(t) + \frac{\partial G}{\partial y}(x(t), y(t)) y'(t)$$

$$= \left(\frac{\partial G}{\partial x}(x(t), y(t))\right)^2 + \left(\frac{\partial G}{\partial y}(x(t), y(t))\right)^2 \geq 0.$$

Notice that the equality is satisfied only for equilibria. This means that G is increasing along any nonequilibrium solution (x, y). If (x, y) were periodic, then G would take some value at least twice, so (x, y) could not be periodic.

To understand the behavior of the flow near isolated equilibria, assume that $(0, 0)$ is an isolated equilibrium solution of system (1) (if not, we can shift the frame's origin into the equilibrium). The linearization technique of Section 5.1 leads us to the linear system

$$\begin{cases} x' = ax + by \\ y' = bx + cy, \end{cases}$$

where

$$a = \frac{\partial^2 G}{\partial x^2}(0, 0), \qquad b = \frac{\partial^2 G}{\partial x \partial y}(0, 0), \qquad \text{and} \qquad c = \frac{\partial^2 G}{\partial y^2}(0, 0).$$

The polynomial equation $\lambda^2 - (a + c)\lambda + ac - b^2 = 0$ yields the eigenvalues

$$\frac{1}{2}(a + c) \pm \frac{1}{2}\sqrt{(a - c)^2 + 4b^2},$$

which are real since $(a - c)^2 + 4b^2 \geq 0$. So the equilibrium cannot be a spiral source, a spiral sink, or a center. This completes the proof.●

Hamiltonian Systems

A two-dimensional system of the form (1) is called *Hamiltonian*—after the Irish mathematician William Rowan Hamilton (1805–1865)—if there is a real function H of variables x and y, called a *Hamiltonian function*, that has continuous partial derivatives and satisfies the relations

$$f(x, y) = \frac{\partial H}{\partial y}(x, y) \qquad \text{and} \qquad g(x, y) = -\frac{\partial H}{\partial x}(x, y). \qquad (5)$$

These conditions imply that

$$\frac{\partial f}{\partial x} = \frac{\partial^2 H}{\partial x \partial y} \qquad \text{and} \qquad \frac{\partial g}{\partial y} = -\frac{\partial^2 H}{\partial y \partial x},$$

which means that

$$\frac{\partial f}{\partial x} = -\frac{\partial g}{\partial y}. \qquad (6)$$

To see whether a given two-dimensional system is Hamiltonian, we can either find a function H that satisfies conditions (5) or check whether relation (6) is satisfied.

EXAMPLE 2   The nonlinear equations

$$\begin{cases} x' = y \\ y' = x - x^2 \end{cases}$$

form a Hamiltonian system. Indeed, $f(x, y) = y$, $g(x, y) = -x + x^2$, and

$$\frac{\partial f}{\partial x}(x, y) = -\frac{\partial g}{\partial y}(x, y) = 0,$$

so relation (6) is satisfied.

We can also find the Hamiltonian function of this system. Using an integration-differentiation technique similar to the one that produced the gradient function in Example 1, we obtain

$$H(x, y) = \frac{1}{2}y^2 - \frac{1}{2}x^2 + \frac{1}{3}x^3.$$

The main property of Hamiltonian systems is that the Hamiltonian function provides an *energy relation*. (This terminology is borrowed from physics, where systems endowed with this property are Hamiltonian.) In other words, H is a conserved quantity. We can formally express this as follows.

Theorem 5.3.2

If H is the Hamiltonian function and (x, y) is a solution of a Hamiltonian system, then

$$H(x(t), y(t)) = h \tag{7}$$

for all t for which the solution is defined, where h is a constant called the *energy constant*.

PROOF
Let (x, y) be a solution of a Hamiltonian system with Hamiltonian function H. Then, by the chain rule and relations (5),

$$\frac{d}{dt}H(x(t), y(t)) = \frac{\partial H}{\partial x}(x(t), y(t))x'(t) + \frac{\partial H}{\partial y}(x(t), y(t))y'(t)$$

$$= \frac{\partial H}{\partial x}(x(t), y(t))\frac{\partial H}{\partial y}(x(t), y(t)) - \frac{\partial H}{\partial x}(x(t), y(t))\frac{\partial H}{\partial y}(x(t), y(t))$$

$$= 0.$$

Equation (7) follows by integrating

$$\frac{d}{dt}H(x(t), y(t)) = 0.$$

This completes the proof.●

EXAMPLE 3   The equation $\phi'' = -(g/L)\phi$ of the linear pendulum (see Section 3.2) can be written as the system

$$\begin{cases} \phi' = \theta \\ \theta' = -\dfrac{g}{L}\phi. \end{cases}$$

But this is a Hamiltonian system with $H(\phi, \theta) = \frac{1}{2}\theta^2 + [g/(2L)]\phi^2$. The energy relation

$$\frac{1}{2}\theta^2 + \frac{g}{2L}\phi^2 = h$$

shows that the sum of the *kinetic energy*, $\frac{1}{2}\theta^2$, and the *potential energy*, $[g/(2L)]\phi^2$, is constant.

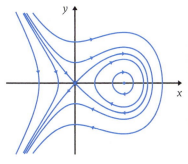

Figure 5.3.1. The flow of the Hamiltonian system in Example 2.

Notice that the energy constant h is in general different for distinct solutions, but it is possible that different solutions correspond to the same constant. In Figure 5.3.1, for example, the three curves that reach or leave the equilibrium $(0, 0)$ represent three distinct solutions that have the same energy. This is because, from the geometric point of view, the energy relation (7) describes so-called level curves and the three above-mentioned curves belong to the same level. The terminology is borrowed from cartography, since these curves can be imagined as level curves on a map (see Figures 5.3.1 and 5.3.3). They can help us sketch the flow.

The following result characterizes the equilibrium solutions of Hamiltonian systems. As for gradient systems, certain types of equilibria may not appear.

Theorem 5.3.3

If a Hamiltonian system has equilibria, they are not sources or sinks.

PROOF

Assume that $(0, 0)$ is an isolated equilibrium of a Hamiltonian system. Then linearization leads to the system

$$\begin{cases} x' = ax + by \\ y' = cx - ay, \end{cases}$$

where

$$a = \frac{\partial^2 H}{\partial x \partial y}(0, 0), \qquad b = \frac{\partial^2 H}{\partial y^2}(0, 0), \qquad \text{and} \qquad c = -\frac{\partial^2 H}{\partial x^2}(0, 0),$$

which has the eigenvalues $\pm\sqrt{a^2 + bc}$. There are three cases to discuss:

(i)   If $a^2 + bc > 0$, the two eigenvalues are real and have opposite signs.
(ii) If $a^2 + bc = 0$, zero is the only eigenvalue.
(iii) If $a^2 + bc < 0$, both eigenvalues are imaginary and have zero real part.

Therefore the equilibria are not sources or sinks. This completes the proof.●

Orthogonality   We will now present an interesting connection between two-dimensional reciprocal systems. But let us first make the notions precise.

DEFINITION 5.3.1

A gradient system of gradient function G and a Hamiltonian system of Hamiltonian function H are called *reciprocal* if $G = H$.

Notice that if

$$\begin{cases} x' = f(x, y) \\ y' = g(x, y) \end{cases} \tag{8}$$

is a gradient system, then the reciprocal Hamiltonian has the form

$$\begin{cases} x' = g(x, y) \\ y' = -f(x, y). \end{cases} \tag{9}$$

DEFINITION 5.3.2

Two planar curves[1] are *orthogonal* at a point of intersection if the tangents to the curves at the point are perpendicular to each other (see Figure 5.3.2(a)). Two families of curves are orthogonal if any two intersecting curves that belong to different families are orthogonal at their intersection points (see Figure 5.3.2(b)).

Let us recall from vector calculus that the conditions in (2) imply that, with respect to the surface $z = G(x, y)$, flows of gradient systems move in the direction of steepest ascent or descent. The conditions in (5) show that flows of Hamiltonian systems move along the level curves of the surface

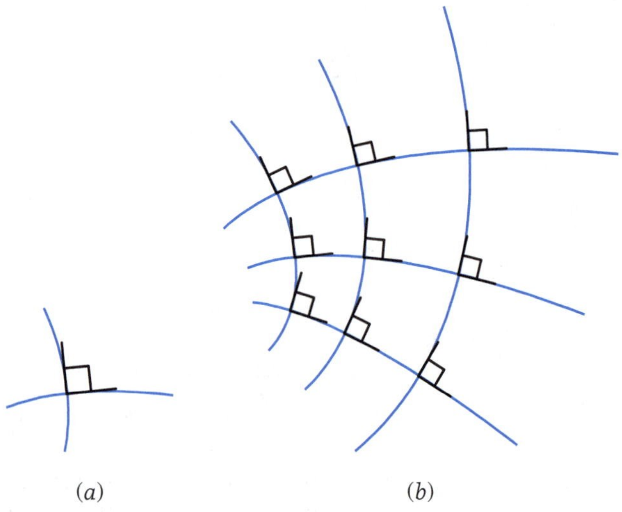

Figure 5.3.2. (a) Two orthogonal curves and (b) orthogonal families of curves.

(a) (b)

---

[1]We assume that the curves are given by differentiable functions, so that they are smooth. This is a natural assumption since the curves we are interested in correspond to solutions of differential equations.

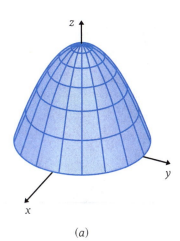

(a)

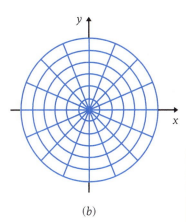

(b)

Figure 5.3.3. (a) The flows of reciprocal systems on the surface $z = G(x, y)$ and (b) their projections in the xy-plane.

Figure 5.3.4. The orthogonal flows of the systems in Example 4.

$z = H(x, y)$. Since reciprocal systems have $G = H$, the corresponding flows can be represented on the same surface (as, for example, in Figure 5.3.3(a)). This suggests that the two families of curves are orthogonal to each other on the surface. Moreover, if projected onto the xy-plane, these families of curves still remain orthogonal (see Figure 5.3.3(b)). Indeed, we can now prove the following result for the flows of reciprocal systems in the plane.

Theorem 5.3.4

The flows of reciprocal gradient and Hamiltonian systems are orthogonal to each other.

PROOF

Let (x_0, y_0) be the intersection point of two curves that represent solutions of the reciprocal gradient and the Hamiltonian system, respectively. Then from (8) and (9), the directions of the tangents to these curves are given by the vectors $(f(x_0, y_0), g(x_0, y_0))$ and $(g(x_0, y_0), -f(x_0, y_0))$, respectively. The dot product of these vectors is $f(x_0, y_0)g(x_0, y_0) - g(x_0, y_0)f(x_0, y_0)$, which is 0, so the two vectors are perpendicular to each other. This completes the proof.●

EXAMPLE 4 Consider the systems

$$\begin{cases} x' = x \\ y' = 2y \end{cases} \quad \text{and} \quad \begin{cases} x' = 2y \\ y' = -x, \end{cases}$$

which are reciprocal gradient and Hamiltonian, respectively. The Hamiltonian is $H(x, y) = \frac{1}{2}x^2 + y^2$, so the energy relation has the form

$$\frac{x^2}{2} + y^2 = h.$$

Thus for $h > 0$ the level curves are ellipses (see Figure 5.3.4), and for $h = 0$ they degenerate to a single point, the origin of the frame. The flow of the gradient system is orthogonal to the ellipses, as Figure 5.3.4 shows.

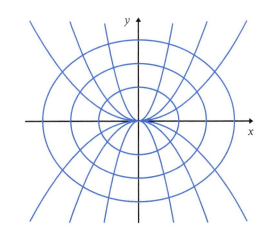

Lobster in its habitat.

Applications

How lobsters scavenge

Practically blind in murky waters, lobsters use their sensitive antennae to detect the concentrations of chemicals signaling dead fish. How can we explain the fact that their scavenging path is only rarely a straight line?

Solution. Let us assume that the lobster moves in a plane, that its position at time t is $(x(t), y(t))$, and that it heads in the direction of the velocity vector $(x'(t), y'(t))$. If $R(x, y)$ is the function that describes the concentration of the chemicals that indicate food in the xy-plane, then we know from vector calculus that the direction in which R increases the fastest is that of the *gradient vector*, $(\partial R/\partial x, \partial R/\partial y)$. Since the lobster moves toward the highest concentration of those chemicals, we can write the equations of motion[2]

$$\begin{cases} x' = \dfrac{\partial R}{\partial x}(x, y) \\ y' = \dfrac{\partial R}{\partial y}(x, y). \end{cases} \tag{10}$$

It is reasonable to think that in perfectly still waters the chemicals indicating food propagate like concentric circles, with each circle having the same concentration of chemicals. In this ideal case the function R has the form $R(x, y) = x^2 + y^2$. On the other hand the reciprocal Hamiltonian system of (10) is

$$\begin{cases} x' = \dfrac{\partial R}{\partial y}(x, y) \\ y' = -\dfrac{\partial R}{\partial x}(x, y), \end{cases} \tag{11}$$

and its Hamiltonian function is R, so the level curves are exactly the circles of equal concentration of chemicals, $x^2 + y^2 = h$, for $h > 0$. The flow of (11) is formed by concentric circles, and then the flow of the gradient system, which is orthogonal to it, is formed by straight rays (see Figure 5.3.5).

The ideal situation, however, is never attained. The waters in which lobsters live are far from still, so the propagation of chemicals in our

---

[2]Though the vectors $(x'(t), y'(t))$ and $(\partial R/\partial x, \partial R/\partial y)$ are not necessarily equal in length, a rescaling transformation still leads to a gradient system like (10).

Figure 5.3.5. The flows of systems (10) and (11) for $R(x, y) = x^2 + y^2$.

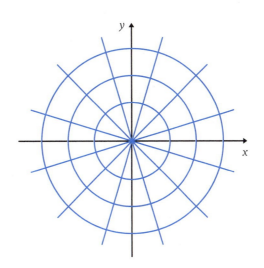

model is not circular. For example, if the propagation of chemicals is elliptic—given, say, by the function in Example 4—the motion of the lobster is described by the gradient system whose flow is sketched in Figure 5.3.4. Therefore the path of the lobster is rarely a straight line.●

PROBLEMS

Determine whether the equations in Problems 1 through 6 are gradient systems and, in case they are, find the gradient function.

1. $\begin{cases} x' = 2\cos x \sin y \\ y' = 2\sin x \cos y \end{cases}$

2. $\begin{cases} u' = 3u - u^2 + 3v^2 \\ v' = u - 3v + u^2 - 3v^2 \end{cases}$

3. $\begin{cases} r' = 10rs - s\cot r \\ s' = 5r^2 - \tan r \end{cases}$

4. $\begin{cases} v' = \sqrt{2}v^3 - w \\ w' = 2\sqrt{2}w^3 - 2v \end{cases}$

5. $\begin{cases} P' = -8P^3 \\ Q' = -8Q^3 \end{cases}$

6. $\begin{cases} v' = \mu + \mu^2 - 2\tan v \\ \mu' = v - 3\mu^2 \end{cases}$

Determine whether the equations in Problems 7 through 12 are Hamiltonian systems and, in case they are, find the Hamiltonian function.

7. $\begin{cases} z' = -w \\ w' = -z \end{cases}$

8. $\begin{cases} \alpha' = 5\sin \alpha \cos \beta \\ \beta' = -5\cos \alpha \sin \beta \end{cases}$

9. $\begin{cases} x' = 14x + y + x^2 - 2xy + 3y^2 \\ y' = x - 14y + x^2 - 2xy + y^2 \end{cases}$

10. $\begin{cases} r' = p \\ p' = r^3 - r \end{cases}$

11. $\begin{cases} u' = -u + 4w - \alpha w^2 \\ w' = u - 4u^2w + \beta, \end{cases}$
where α and β are constants

12. $\begin{cases} X' = aX + Z + Z^2 \\ Z' = bZ + X + X^2, \end{cases}$
where a and b are constants

Explain why the flows in Problems 13 through 16 may not belong to gradient systems.

13.

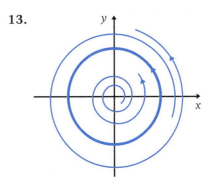

14.

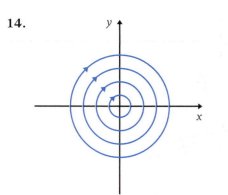

15.

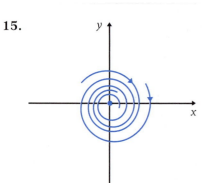

16.

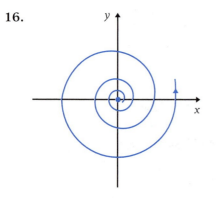

*Explain why the flows in Problems 17 through 20
may not belong to Hamiltonian systems.*

17.

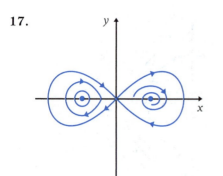

18.

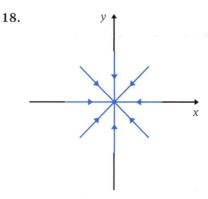

19.

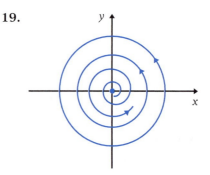

20.

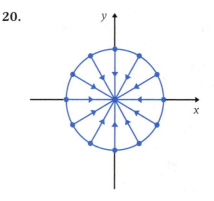

21. Find all two-dimensional systems that are
both gradient and Hamiltonian.

*Sketch the flows of the systems that have the
Hamiltonians given in Problems 22 through
27. For this purpose use your knowledge of
geometry or of plotting functions, or use one of
the computer programs in Section 4.6. Then write
the reciprocal gradient systems and sketch their
flows.*

22. $H(x, y) = x^2 + 4y^2$

23. $H(u, v) = u^2 - v^2$

24. $H(s, \omega) = s \cos \omega$

25. $H(z, w) = 2z \sin w$

26. $H(\rho, r) = \rho - r^2 + r$

27. $H(X, Y) = X^2 - X^4 + 2Y^2$

28. Show that the equilibrium $(\frac{1}{2}, 0)$ of the gradient system given by the gradient function $G(x, y) = x^2 - 2x^3 + x^4 + y^2$ is a saddle.

29. The *trace* of a two-dimensional matrix $\mathbf{A}$, denoted by tr $\mathbf{A}$, is defined as the sum of the elements of its main diagonal (i.e., from the upper left to the lower right element). Let $\mathbf{x}' = \mathbf{F}(\mathbf{x})$ be a system having $\mathbf{0}$ as an equilibrium, and $\mathbf{x}' = \mathbf{A}\mathbf{x}$ be the corresponding linearized system around $\mathbf{0}$. Show that if tr $\mathbf{A} = 0$, then the initial system is Hamiltonian. Is the converse true?

30. Gradient systems can be defined in higher dimensions. What would be the natural form of an n-dimensional gradient system?

31. The natural extension of the definition of Hamiltonian systems to higher dimensions is

$$\begin{cases} \mathbf{x}' = \dfrac{\partial H}{\partial \mathbf{y}}(\mathbf{x}, \mathbf{y}) \\ \mathbf{y}' = -\dfrac{\partial H}{\partial \mathbf{x}}(\mathbf{x}, \mathbf{y}), \end{cases}$$

where $\mathbf{x} = (x_1, \ldots, x_n)$, $\mathbf{y} = (y_1, \ldots, y_n)$, $\partial H / \partial \mathbf{x} = (\partial H / \partial x_1, \ldots, \partial H / \partial x_n)$, and $\partial H / \partial \mathbf{y} = (\partial H / \partial y_1, \ldots, \partial H / \partial y_n)$. This means that the dimension of a Hamiltonian system is always even. Such systems also have the property that the Hamiltonian is constant along solutions. Show that the system

$$\begin{cases} x'' = -\dfrac{Gmx}{(x^2 + y^2)^{3/2}} \\ y'' = -\dfrac{Gmy}{(x^2 + y^2)^{3/2}}, \end{cases}$$

which describes the planar two-body problem (see system (27) in Chapter 1), can be written as a four-dimensional Hamiltonian system. Find the Hamiltonian function.

5.4 Stability

The notion of *stability* has several meanings in mathematics. We already encountered it in Section 3.5 when dealing with the qualitative structure of a parametric flow. In this section we will investigate the concept of *Liapunov stability*, named after the Russian mathematician Aleksandr Mikhailovich Liapunov (1857–1918), who defined and studied it in 1892. Then we will introduce Liapunov's method, suitable for checking the stability of solutions, and finally apply this method to understand the flow of the nonlinear pendulum.

Definitions and Examples

Roughly speaking, a solution is stable (in the sense of Liapunov) if all solutions close to it at some initial point will stay close forever. Let us give a formal definition for equilibria. Consider the system

$$\mathbf{x}' = \mathbf{F}(\mathbf{x}), \tag{1}$$

where $\mathbf{F}$ is the vector field and $\mathbf{x}$ is the vector variable.

DEFINITION 5.4.1

An equilibrium solution $\mathbf{x}_0$ of system (1) is called *stable* if for any neighborhood U of $\mathbf{x}_0$ there is a neighborhood V of $\mathbf{x}_0$ contained in U such that for any solution φ, with $\varphi(0)$ in V, $\varphi(t)$ belongs to U for any $t \geq 0$ (see Figure 5.4.1(a)). The equilibrium $\mathbf{x}_0$ is called *unstable* if it is not stable.

Figure 5.4.1. (*a*) Stability and (*b*) asymptotic stability of equilibrium solutions.

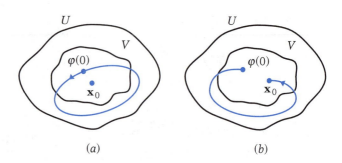

In other words, an equilibrium is stable if all solutions close to it at some initial moment will not depart too far from it later on. If we additionally ask that all solutions initially close to the equilibrium will eventually tend to it, then we have a stronger property, called *asymptotic stability*. We can formally define it as follows.

DEFINITION 5.4.2

An equilibrium solution $\mathbf{x}_0$ of system (1) is called *asymptotically stable* if it is stable and if there is a neighborhood V_0 contained in V such that any solution φ, with $\varphi(0)$ in V_0, tends to $\mathbf{x}_0$ when $t \to \infty$ (see Figure 5.4.1(*b*)).

Aleksandr Mikhailovich Liapunov.

EXAMPLE 1   A physical example to keep in mind is the undamped pendulum. As we will see at the end of this section, its lower and upper positions are equilibria, with the lower stable and the upper unstable. Physically this is pretty clear. If not exactly in the upper equilibrium position, no matter how close, the pendulum will move away from it. If close to the lower equilibrium position, the pendulum will oscillate around it (see Figures 5.4.2(*a*) and (*b*)).

EXAMPLE 2   For two-dimensional systems, every center is stable. In any dimension, sinks are asymptotically stable, whereas sources and saddles are unstable.

EXAMPLE 3   Some nonhyperbolic equilibria are stable (the centers, for example); others are unstable—such as the one in Figure 5.4.3. (Recall that flows near nonhyperbolic equilibria can have any behavior, even saddle-like ones as in Figure 5.4.3.)

Figure 5.4.2. (*a*) All solutions leave the upper equilibrium of the pendulum. (*b*) For zero or small initial velocities, solutions close to the lower equilibrium remain close to it.

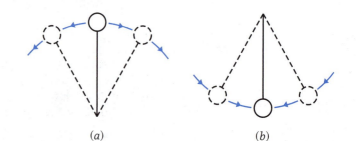

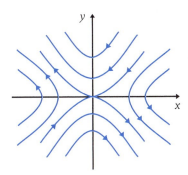

Figure 5.4.3. An example of an unstable nonhyperbolic equilibrium solution.

The definition of asymptotic stability requires that the equilibrium is stable. This seems unnecessary. Intuition suggests that if all the orbits tend to the equilibrium, stability must follow. Deeper reflection, however, reveals that this statement is false. We can construct flows for which solutions pass close to the equilibrium, then go far away, return, and end up at the equilibrium. The shape of these solutions resembles a large loop. Moreover, the flow must be such that the closer a solution's first passage is to the equilibrium, the larger the loop. Such flows, which are easier to imagine than to draw, are not stable in the sense of Definition 5.4.1 (can you explain why?), though all solutions tend to the equilibrium.

A simpler example is the "spider" in Figure 5.4.4. Every solution is a loop from the equilibrium E to itself. The closer the solution is to the horizontal axis, the larger the loop. Notice that except for the point E, the horizontal axis is removed from the phase plane. Can you explain why? Can you prove that E is not stable?

Liapunov's Method

A method that helps us decide whether an equilibrium is stable, asymptotically stable, or unstable was given by Liapunov in 1892. The idea is to find a function whose properties determine the nature of the equilibrium. Unfortunately, there is no algorithm for finding the desired function. Liapunov's method only ensures that if such a function exists, we can draw the right conclusion. But in many cases the function can be guessed. The advantage of this method is that it may succeed when linearization fails.

For simplicity we will state the following result in terms of two-dimensional systems, but the theorem is true for the n-dimensional case. Let (x_0, y_0) be an isolated equilibrium solution of the system

$$\begin{cases} x' = f(x, y) \\ y' = g(x, y). \end{cases}$$

Consider a real function $V(x, y)$ defined in a neighborhood D of (x_0, y_0), differentiable with respect to x and y, and define the function

$$\dot{V}(x, y) = \frac{\partial V}{\partial x}(x, y)x' + \frac{\partial V}{\partial y}(x, y)y'.$$

Notice that

$$\dot{V}(x(t), y(t)) = \frac{d}{dt} V(x(t), y(t)).$$

This means that if $\dot{V}(x, y) < 0$, then V decreases along solution trajectories, which will cross the curves $V(x, y) = c$ (constant) toward the equilibrium (see Figure 5.4.5(a)). If $\dot{V}(x, y) = 0$, then V is constant along solution trajectories, so we expect that these trajectories will follow the curves $V(x, y) = c$ (see Figure 5.4.5(b)). If $\dot{V}(x, y) > 0$, then V increases along solution trajectories, which will cross the curves $V(x, y) = c$ away from the equilibrium (see Figure 5.4.5(c)).

These remarks suggest the following result, whose rigorous proof is beyond our scope.

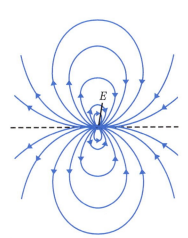

Figure 5.4.4. An example of a flow with an unstable equilibrium for which all solutions tend to this equilibrium.

Figure 5.4.5. The behavior of solutions with respect to the curves $V(x, y) = c$ (constant) in case (a) $\dot{V}(x, y) < 0$, (b) $\dot{V}(x, y) = 0$, and (c) $\dot{V}(x, y) > 0$.

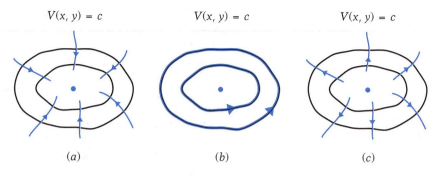

$V(x, y) = c$ $V(x, y) = c$ $V(x, y) = c$

(a) (b) (c)

Liapunov's Theorem

Assume that the function V defined in some neighborhood D of (x_0, y_0) is continuous, V has continuous partial derivatives in D, $V(x_0, y_0) = 0$, $V(x, y) > 0$ for all (x, y) in D with $(x, y) \neq (x_0, y_0)$, and the points (x, y) satisfying the equation $V(x, y) = C$ form a closed curve around (x_0, y_0) for every constant C for which these curves are in D. Then

(i) If $\dot{V}(x, y) \leq 0$ for all (x, y) in D, (x_0, y_0) is stable.
(ii) If $\dot{V}(x, y) < 0$ for all (x, y) in D, (x_0, y_0) is asymptotically stable.
(iii) If $\dot{V}(x, y) > 0$ for all (x, y) in D, (x_0, y_0) is unstable.

A function V with the properties in the hypothesis and that also satisfies (i), (ii), or (iii) is called a *Liapunov function*. If the domain D is the whole space of the variables (x, y), then V is called a *global Liapunov function* and signals that the equilibrium is *globally (asymptotically) stable*; i.e., all solutions around the equilibrium stay close to it, not only those in some sufficiently small neighborhood, as required in Definitions 5.4.1 and 5.4.2.

It is interesting to note that finding a Liapunov function requires no knowledge of the solutions around the equilibrium. But, as mentioned, there is no method of finding such a function. For Hamiltonian systems, however, a good candidate is the Hamiltonian itself.

EXAMPLE 4   Consider the system

$$\begin{cases} x' = -y^3 \\ y' = x^3, \end{cases}$$

whose origin is an isolated equilibrium solution. Since both eigenvalues of the linearized system are 0, the linearization method fails to determine the nature of this equilibrium.

On the other hand, the system is Hamiltonian, given by $H(x, y) = -\frac{1}{4}(x^4 + y^4)$. So let us check whether the function

$$V(x, y) = x^4 + y^4$$

is a Liapunov function. V is obviously continuous and has continuous partial derivatives. The equation $x^4 + y^4 = C$ represents closed curves, one for every $C > 0$, all encircling $(0, 0)$. Moreover, $V(0, 0) = 0$ and $V(x, y) > 0$ for all (x, y) in the plane, except $(0, 0)$. On the other hand,

$$\dot{V}(x, y) = 4x^3 x' + 4y^3 y' = -4x^3 y^3 + 4y^3 x^3 = 0$$

for all (x, y) in the plane. Therefore, V is a (global) Liapunov function, so $(0, 0)$ is (globally) stable. Since $\dot{V} = 0$, $(0, 0)$ is not asymptotically stable, so the solutions do not approach the origin; therefore, the other solutions must be periodic around the equilibrium, so the flow is formed by closed curves around $(0, 0)$.

EXAMPLE 5   Consider the system

$$\begin{cases} x' = -2y + yz - x^3 \\ y' = x - xz - y^3 \\ z' = xy - z^3, \end{cases}$$

whose origin is an isolated equilibrium solution. Since the eigenvalues of the linearized system are $\lambda_1 = 0$, $\lambda_2 = 2i$, $\lambda_3 = -2i$, the linearization method fails. The system is not Hamiltonian either. Sometimes, however, Liapunov functions can be found among functions of the form

$$V(x, y, z) = ax^2 + by^2 + cz^2,$$

where a, b, and c are positive constants to be determined. In our case V is continuous and has continuous partial derivatives for all (x, y, z). The equation $ax^2 + by^2 + cz^2 = C$ (constant) represents ellipsoids around the origin, one for every $C > 0$. Moreover, $V(0, 0, 0) = 0$ and $V(x, y, z) > 0$ for all (x, y, z) except $(0, 0, 0)$. So the constants a, b, and c can be determined from the condition concerning $\dot{V}$. Computing $\dot{V}$, we obtain

$$\dot{V}(x, y, z) = 2(a - b + c)xyz + 2(-2a + b)xy - 2ax^4 - 2by^4 - 2cz^4.$$

This shows that we can choose the constants such that the first two terms vanish and then $\dot{V}$ becomes negative. The choice $a = 1$, $b = 2$, $c = 1$ leads to $\dot{V}(x, y, z) < 0$ for all (x, y, z) except $(0, 0, 0)$, so the function $V(x, y, z) = x^2 + 2y^2 + z^2$ is a (global) Liapunov function. According to point (ii) of Liapunov's theorem, the origin is (globally) asymptotically stable. However, notice that though all solutions tend to it, the origin is not a sink. (Can you explain why?)

EXAMPLE 6   Consider the system

$$\begin{cases} x' = -y + xy^2 + 2x^3 \\ y' = x + x^2 y + 3y^3, \end{cases}$$

whose origin is an isolated equilibrium. The linearization method fails since the corresponding eigenvalues are $\lambda_1 = i$, $\lambda_2 = -i$. The system is not Hamiltonian either. In polar coordinates, the equations look even more complicated. So let us try Liapunov's method and seek a function of the type $V(x, y) = ax^2 + by^2$, with $a, b > 0$ constants to be determined. The equation $ax^2 + by^2 = C$ represents ellipses around $(0, 0)$, one for each $C > 0$. V is continuous and has continuous partial derivatives for all (x, y) in the plane, $V(0, 0) = 0$, and $V(x, y) > 0$ for all $(x, y) \neq (0, 0)$. The computations show that

$$\dot{V}(x, y) = 2(b - a)xy + 2(a + b)x^2 y^2 + 2ax^4 + 6by^4,$$

so if $a = b > 0$, $\dot{V}$ is positive for all $(x, y) \neq (0, 0)$. The choice $a = b = 1$ makes V satisfy condition (iii) of Liapunov's theorem. Therefore, the origin is an unstable equilibrium.

Applications

Galileo's pendulum revisited   In Section 3.2 we obtained the second-order equation that describes the motion of Galileo's pendulum and then solved it for small oscillations, i.e., for what we called the *linear pendulum*. Using the more powerful methods of this chapter, we can now study the general case of the *nonlinear pendulum*. Recall that the equation describing the motion is

$$\phi'' = -\frac{g}{L} \sin \phi,$$

where L is the length of the rod, ϕ is the angle with the vertical, and g is the gravitational acceleration (see Figure 3.2.1). To simplify our presentation we will take $g/L = 1$ and leave the general case as an exercise (see Problem 12). If we denote $\theta = \phi'$, the above equation can be rewritten as a first-order nonlinear system,

$$\begin{cases} \phi' = \theta \\ \theta' = -\sin \phi, \end{cases} \tag{2}$$

which has infinitely many equilibrium solutions, represented in the plane by all points of the form $(\phi, \theta) = (k\pi, 0)$, where k is an integer. The ones with k even correlate to the downward vertical rest position, whereas those with k odd correspond to the upward vertical rest position. To understand the motion of the pendulum, let us sketch the flow of system (2).

Solution. We start with the equilibrium $(0, 0)$. Linearization leads to the system

$$\begin{cases} \phi' = \theta \\ \theta' = -\phi, \end{cases}$$

with eigenvalues $\pm i$, so linearization fails. We can see, however, that system (2) is given by the Hamiltonian $H(\phi, \theta) = \frac{1}{2}\theta^2 + 1 - \cos \phi$. Therefore, a possible candidate for a Liapunov function is

$$V(\phi, \theta) = \frac{1}{2}\theta^2 + 1 - \cos \phi.$$

Indeed, V is continuous, it has continuous partial derivatives, $V(0, 0) = 0$, and $V(\phi, \theta) > 0$ for any value of θ and for any ϕ in the interval $(-\pi, \pi)$, except 0. A direct computation shows that $\dot{V}(\phi, \theta) = 0$. The discussion at point (i) below shows that the equation $\frac{1}{2}\theta^2 + 1 - \cos \phi = h$ represents closed curves around the equilibrium for all h in the interval $(0, 2)$. According to Liapunov's theorem, $(0, 0)$ is stable. Since the system is two-dimensional, $(0, 0)$ must be a center (see Figure 5.4.6).

The same argument applies to any equilibria of the form $(\phi, \theta) = (2n\pi, 0)$, where n is an integer. For $(2\pi, 0)$, for example, $V(2\pi, 0) = 0$, $V(\phi, \theta) > 0$ for any value of θ and for any ϕ in the interval $(\pi, 3\pi)$ except 2π, and $\dot{V}(\phi, \theta) = 0$. So all equilibria $(2n\pi, 0)$, with n an integer, are centers.

Let us see what happens around equilibria of the form $((2n + 1)\pi, 0)$. It is easy to show that all of them behave like the equilibrium $(\pi, 0)$,

Figure 5.4.6. The flow of the nonlinear undamped pendulum.

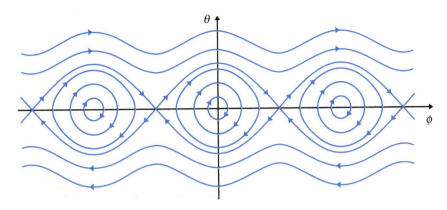

so for simplicity we will deal with this one. To obtain the corresponding linear system, we first shift the coordinate system to $(\pi, 0)$ with the transformation

$$\begin{cases} u = \phi - \pi \\ v = \theta. \end{cases}$$

This leads to the system

$$\begin{cases} u' = v \\ v' = \sin u, \end{cases}$$

whose linear correspondent is

$$\begin{cases} u' = v \\ v' = u. \end{cases}$$

The eigenvalues are ± 1, so according to the Hartman-Grobman theorem, $(\pi, 0)$ is a saddle (see Figure 5.4.6).

To sketch the whole flow we can use the above information about the equilibria and the energy relation

$$\frac{1}{2}\theta^2 + 1 - \cos \phi = h,$$

where h is a constant. We can express θ as a function of ϕ in the upper half of the phase plane (i.e., for $\theta > 0$) as $\theta(\phi) = \sqrt{2(h + \cos \phi - 1)}$ and then in the lower half (i.e., for $\theta < 0$) as $\theta(\phi) = -\sqrt{2(h + \cos \phi - 1)}$. Depending on h, there are three cases to discuss:

(i) For $0 < h < 2$ we obtain the closed curves around each of the equilibria $(2n\pi, 0)$ ($h = 0$ gives the equilibria and for $h < 0$ the graphs are empty). Physically these curves correspond to periodic motions around the lower equilibrium position, similar to the ones of a clock pendulum. We can see from Figure 5.4.6 that the velocity of motion, given by θ, changes: It is largest (in absolute value) when the pendulum passes through the vertical position and 0 when the motion changes direction.

(ii) For $h = 2$ we obtain the curves connecting consecutive equilibria of the form $((2n + 1)\pi, 0)$. Physically they correspond to a rare situation: The pendulum starts from the upper vertical position and performs a rotation until it reaches the same position. The velocity is zero at start, reaches a maximum when the pendulum passes through the downward vertical position, and becomes 0 after a complete rotation. The difference between the upper and the lower curves with $h = 2$ in the phase plane is the direction of rotation. These curves are called *separatrices* of the flow because they are at the border of two different patterns of behavior: those given by (i) above and (iii) below.

(iii) For $h > 2$ we obtain the wave curves in Figure 5.4.6. Physically they correspond to everlasting rotation in one direction or another. The speed is highest when the pendulum passes through the downward vertical position and lowest when it crosses the upward one.•

PROBLEMS

Use Liapunov's method with the function $V(x, y) = x^2 + y^2$ to show what is indicated in Problems 1 to 3.

1. $(0, 0)$ is a stable equilibrium of the system
$$\begin{cases} x' = -y - xy \\ y' = x + x^2. \end{cases}$$

2. $(0, 0)$ is an asymptotically stable equilibrium of the system
$$\begin{cases} x' = -y - x^3 - xy^2 \\ y' = x - y^3 - x^2y. \end{cases}$$

3. $(0, 0)$ is an unstable equilibrium of the system
$$\begin{cases} x' = -y + x^3 + xy^2 \\ y' = x + y^3 + x^2y. \end{cases}$$

Using a function of the form $V(x, y) = ax^2 + by^2$, show what is indicated in Problems 4 to 6.

4. $(0, 0)$ is a stable equilibrium of the system
$$\begin{cases} x' = -x^3 + 2y^3 \\ y' = -2xy^2. \end{cases}$$

5. $(0, 0)$ is an asymptotically stable equilibrium of the system
$$\begin{cases} x' = -x^3 + 2xy^2 \\ y' = -y^3. \end{cases}$$

6. $(0, 0)$ is an unstable equilibrium of the system
$$\begin{cases} x' = x^3 - y^3 \\ y' = xy^2 + 4x^2y + 2y^3. \end{cases}$$

7. Consider the function $F(x, y) = ax^2 + bxy + cy^2$ and prove the following:

   (i) $F(x, y) > 0$ for all (x, y) if and only if $a > 0$ and $b^2 - 4ac < 0$.

   (ii) $F(x, y) < 0$ for all (x, y) if and only if $a < 0$ and $b^2 - 4ac < 0$.

Using this fact and Liapunov's method, show that the system that describes the dynamics of two competing species,
$$\begin{cases} x' = x - xy - x^2 \\ y' = \frac{3}{4}y - \frac{1}{2}xy - y^2, \end{cases}$$
has an asymptotically stable equilibrium at $(\frac{1}{2}, \frac{1}{2})$.

8. Using a Liapunov function of the form $V(x, y, z) = ax^2 + by^2 + cz^2$, show that $(0, 0, 0)$ is an asymptotically stable equilibrium of the system
$$\begin{cases} x' = y + z^3 - x^3 \\ y' = -x - x^2y + z^2 - y^3 \\ z' = -yz - y^2z - xz^2 - z^5. \end{cases}$$

9. Show that $(0, 0, 0)$ is an asymptotically stable equilibrium but not a sink of the system
$$\begin{cases} u' = -2v + vw - u^3 \\ v' = u - uw - v^3 \\ w' = uv - w^3. \end{cases}$$

10. Consider the system
$$\begin{cases} x' = y \\ y' = -g(x), \end{cases}$$
where g is a continuous function such that $xg(x) > 0$ for all $x \neq 0$ in the domain in which g is defined and $g(0) = 0$. For a suitably chosen domain of g, this generalizes the undamped-pendulum model. This system has the Hamiltonian
$$H(x, y) = \frac{1}{2}y^2 + \int_0^x g(s)\, ds.$$
Show that $(0, 0)$ is a stable equilibrium.

11. Consider the system

$$\begin{cases} x' = y - xg(x, y) \\ y' = -x - yg(x, y), \end{cases}$$

where g is a continuously differentiable function. Show that if $g > 0$, then the equilibrium $(0, 0)$ is asymptotically stable. (*Hint:* Find a quadratic Liapunov function.)

12. Consider the general case of the undamped pendulum

$$\begin{cases} \phi' = \theta \\ \theta' = -\dfrac{g}{L} \sin \phi, \end{cases}$$

find the equilibria, discuss their stability, and sketch the flow.

13. Consider the system that describes the motion of the *nonlinear damped pendulum*,

$$\begin{cases} \phi' = \theta \\ \theta' = -\theta - \sin \phi, \end{cases}$$

where the term $-\theta$ models the effect of friction. Find the equilibria, discuss their stability, and sketch the flow. Compare the flows of the nonlinear damped pendulum

and the nonlinear pendulum and give the physical interpretation in the damped case.

14. Consider the system

$$\begin{cases} x' = -x + y^2 \\ y' = -2y + 3x^2, \end{cases}$$

which has the equilibrium $(0, 0)$. Show that $V(x, y) = x^2/2 + y^2/4$ is a Liapunov function. Then prove that the largest ellipse of the form $x^2/2 + y^2/4 = r$ (where $r > 0$ is constant), which contains orbits that tend to the origin, is the one with $r = \frac{1}{9}$.

15. Show that every two-dimensional gradient system has a Liapunov function.

16. Give an example of a two-dimensional system that has a Liapunov function but is not a gradient system.

17. Show that $F(x, y) = x^2 + y^2$ is not a Liapunov function of the system

$$\begin{cases} x' = y \\ y' = -4x - \dfrac{1}{10}y. \end{cases}$$

Does this system have a Liapunov function? If so, find one. If not, prove that it has none.

5.5 Chaos

In this section we will give a mathematical description for the highly unstable character exhibited by the solutions of certain differential equations, a phenomenon called *chaos*. We present some systems that have this property, discuss the concept of *sensitivity of the solutions with respect to initial data*, and then introduce the "language" of *symbolic dynamics*, which expresses chaos in mathematical terms. Most chaos techniques are beyond the scope of this text, but many of the ideas are accessible.

Highly Unstable Solutions

For his work on the three-body problem, Henri Poincaré in 1889 received the prestigious prize established by King Oscar II of Sweden and Norway. One of the phenomena Poincaré encountered was the possibility of extremely irregular and complicated solutions. It is interesting that he was led to this discovery only after correcting a mistake in the work he had submitted to the competition.[1]

Except for a small circle of mathematicians who knew about it, Poincaré's discovery remained in obscurity for over 8 decades. But in the 1960s, when the first computers made their way into scientific research, the complicated behavior of solutions for nonlinear differential equations was recognized in systems other than the three-body problem. Today we know that many nonlinear systems have this property.

---

[1]The story of Poincaré's discovery of chaos can be found in Florin Diacu and Philip Holmes, *Celestial Encounters—The Origins of Chaos and Stability* (Princeton University Press, Princeton, N.J., 1996), a book accessible to anybody who has reached this mathematical level.

Satellite photo showing turbulent weather.

EXAMPLE 1   One of those who rediscovered the chaotic behavior of solutions was an MIT meteorology professor, Edward Lorenz, who in 1963 published a numerical study of the system

$$\begin{cases} x' = \sigma(y - x) \\ y' = \rho x - y - xz \\ z' = -\beta z + xy, \end{cases} \tag{1}$$

where $\sigma, \rho, \beta > 0$ are constants. These equations describe the fluid convection in a two-dimensional layer heated from below, so they are a

Figure 5.5.1. Two graphs of y for the Lorenz system for close but different initial conditions near the saddle point $(0, 0, 0)$.

$\beta = 1.95$
$x_0 = 10^{-8}$
$y_0 = -1$
$z_0 = 0$

$\beta = 1.95$
$x_0 = 1.9999 \times 10^{-8}$
$y_0 = -1$
$z_0 = 0$

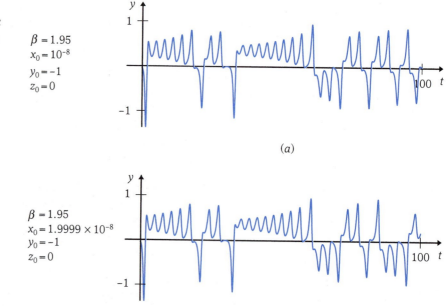

(a)

(b)

simple model for the dynamics of the earth's atmosphere. To match the physical reality, Lorenz chose the constants $\sigma = 10$, $\rho = 28$, and $\beta = \frac{8}{3}$; showed that the equilibrium $(0, 0, 0)$ is a saddle point (see Problem 10); and integrated the system numerically for initial conditions close to $(0, 0, 0)$. For the first 300 iterations he obtained the graph of y shown in Figure 5.5.1(a).

Then he took some initial conditions very close to but different from the previous ones. He was surprised to notice that though the solution stayed close at first, it soon differed from the other. New computer experiments led to the same puzzling behavior: Every new set of initial conditions, though close to $(0, 0, 0)$, offered a different graph.

EXAMPLE 2   Another system with similar behavior is Holmes's version of Düffing's equation. Figure 3.6.3(a) gives the graph of the numerical solution x for initial conditions $(x_0, y_0) = (0.5, 0)$. Figure 5.5.2 shows the numerical results for two different but close sets of initial data.

These numerical endeavors show that in spite of being continuous with respect to initial data (which implies that if they start close, they remain close for a while), the solutions of certain nonlinear systems are highly unstable and part from each other fairly quickly. This property is called *sensitivity of the solutions with respect to initial data*, which is the main feature of mathematical *chaos*.

The main consequence of this phenomenon is disconcerting. In terms of the Lorenz system, it means that accurate long-term weather predictions are hard to make. Numerical work for systems of this type encounters great difficulties because errors add up quickly. So what can we do? How can we understand chaotic behavior? A way to approach this phenomenon was proposed in 1965 by the American mathematician

Figure 5.5.2. Two graphs of x for the Düffing system for close but different initial conditions near the point (0.5, 0).

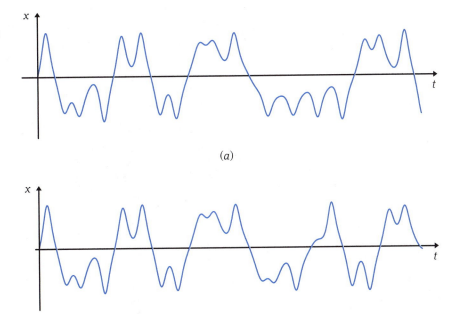

(a)

(b)

and Fields medalist[2] Stephen Smale. Smale's idea was to express chaos in terms of a new "language," that of symbolic dynamics.

Symbolic Dynamics   Symbolic dynamics first appeared in a paper on differential geometry published in 1897 by the French mathematician Jacques Hadamard (1865–1963). In its simplest form, it deals with infinite sequences of two elements (symbols), say 0 and 1, starting from an *origin* and extending to the left and to the right of it—for example,

$$\mathbf{s} = (\ldots, s_{-2}, s_{-1}, s_0, s_1, s_2, \ldots) = (\ldots, 1, 0; 1, 0, 0, \ldots).$$

The semicolon indicates that the first element to its right is s_0 (in this case, $s_0 = 1$).

In the set of all such sequences we define a *distance*. For this let us suppose that

$$\mathbf{s} = (\ldots, s_{-2}, s_{-1}, s_0, s_1, s_2, \ldots) \quad \text{and} \quad \mathbf{r} = (\ldots, r_{-2}, r_{-1}, r_0, r_1, r_2, \ldots)$$

are two sequences whose symbols agree on a central block (i.e., $s_j = r_j$ for all j between $-N$ and N). Then the larger N is, the closer the sequences. Obviously, the distance between two sequences is 0 only if all corresponding elements coincide, i.e., $N = \infty$.

EXAMPLE 3   The sequences $\mathbf{s}$ and $\mathbf{r}$ are closer than the sequences $\mathbf{s}$ and $\mathbf{p}$, where

$$\mathbf{s} = (\ldots, 0, 1, 0, 0, 1, 0; 0, 1, 0, 0, 1, 1, \ldots),$$
$$\mathbf{r} = (\ldots, 1, 0, 0, 0, 1, 0; 0, 1, 0, 0, 1, 0, \ldots),$$
$$\mathbf{p} = (\ldots, 1, 1, 1, 0, 1, 0; 0, 1, 0, 0, 0, 1, \ldots).$$

Indeed, $\mathbf{s}$ and $\mathbf{r}$ agree on a central block of nine elements around the origin, whereas $\mathbf{s}$ and $\mathbf{p}$ agree on a central block of only seven elements.

We further define a function σ, called *shift map*, from the set of sequences to itself, which shifts the elements of any sequence one step to the left, i.e., the next element to the right becomes the origin.

EXAMPLE 4 $\sigma((\ldots, 0, 0, 0; 1, 1, 1, \ldots)) = (\ldots, 0, 0, 0, 1; 1, 1, \ldots).$

If we apply the shift map to a symbolic sequence $\mathbf{S}_0$, we obtain a new symbolic sequence $\mathbf{S}_1 = \sigma(\mathbf{S}_0)$. If we further apply the shift map to the symbolic sequence $\mathbf{S}_1$, we obtain a new symbolic sequence $\mathbf{S}_2 = \sigma(\mathbf{S}_1)$. If this process goes on, we finally obtain a so-called orbit of symbolic sequences: $\mathbf{S}_0, \mathbf{S}_1, \mathbf{S}_2, \ldots, \mathbf{S}_n, \ldots$. We can consider such an orbit as the analogue of the solution of a differential equation and the closeness of orbits as the analogue of the closeness of solutions of

---

[2]Since there is no Nobel Prize for mathematics, the highest distinction is the Fields Medal, named after the Canadian mathematician John Charles Fields (1863–1932), who was a professor at the University of Toronto. The Fields Medal is awarded only to mathematicians under the age of 40, at the International Congress of Mathematicians, which takes place every 4 years.

differential equations. With this analogy in mind, let us show that the shift map has a similar property as the sensitivity of solutions with respect to initial data.

Indeed, let $\mathbf{S}_0$ and $\mathbf{R}_0$ be two close symbolic sequences in the sense that their central $2N + 1$ elements coincide, where N is a large number. If with the help of the shift map σ we construct the corresponding orbits of symbolic sequences,

$$\mathbf{S}_0, \mathbf{S}_1, \mathbf{S}_2, \ldots, \mathbf{S}_n, \ldots$$
$$\mathbf{R}_0, \mathbf{R}_1, \mathbf{R}_2, \ldots, \mathbf{R}_n, \ldots,$$

we see that the larger n is, the farther apart the corresponding symbolic sequences $\mathbf{S}_n$ and $\mathbf{R}_n$ will be. This means that the two orbits part from each other like the curves representing solutions of differential equations that have the property of sensitivity with respect to initial data.

In some cases it is suitable to use sequences with more than two symbols. All the preceding conclusions can be drawn for sequences of three, four, or any other number of symbols, even infinitely many (see Problem 6).

We can often substitute a differential equation with a function that characterizes it (as when applying numerical methods that, instead of dealing with the initial equation, use some suitable formula derived from the equation). Smale's idea was to show the existence of chaotic behavior by proving a one-to-one correspondence between a certain function characterizing the differential equation and the shift map defined on symbolic sequences. Since the shift map has the sensitivity property, the function and therefore the differential equation have it too.

Of course, this is just the idea. There is no general rule for proving this correspondence, so every equation or system is a new problem. The existing methods work only in special cases; they are complicated and beyond the scope of this introductory text. They form an important research area in the qualitative theory of dynamic systems.

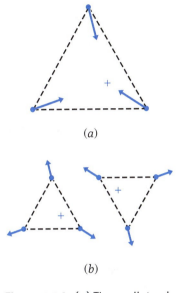

(a)

(b)

Figure 5.5.3. *(a)* The equilateral three-body problem and *(b)* the motion after total collapse.

Applications

We will discuss three examples and a way in which the above symbolic dynamics method may be applied to them. Two of the examples are taken from celestial mechanics and the third is the Lorenz system.

The equilateral three-body problem

A nice property in the three-body problem is that if the three point masses are released with zero initial velocity from the vertices of an equilateral triangle, they move toward a simultaneous triple collapse in their common center of mass (see Figure 5.5.3(*a*)). The idea of proof is based on the fact that the angular momentum is zero, so the bodies will not rotate around the center of mass but will move in straight lines toward it, maintaining the shape of an equilateral triangle.

If the vector field is required to be only continuous but not differentiable at collision, it can be shown that there are two ways to continue the motion beyond collision. One is when the three particles return backward on their collision paths, and the other is when they pass through collision and follow their straightforward motion. In each case they simultaneously stop after some time and proceed again toward collision.

After collision one of two possibilities occurs again. We can thus attach to each solution a symbolic sequence in which 0 and 1 correspond to the two possible continuations of the motion beyond collision. Each solution then has a unique correspondent in the set of sequences. We can thus consider the shift map σ along such a sequence and obtain an orbit, which is the analogue of a solution.

This shows that the uniqueness property is not fulfilled. (Is this in agreement with the results of Section 2.5?) Two solutions that have identical initial conditions can move far apart in terms of the distance defined for symbolic sequences.

The Sitnikov problem

In 1961 the Russian mathematician K. Sitnikov constructed an interesting solution for the three-body problem. He considered two bodies of equal mass, called primaries, that move in a plane on elliptical orbits, and a third body of negligible mass, that moves along a line containing the center of mass of the primaries and is perpendicular to their plane of motion (see Figure 5.5.4). Obviously, the third particle does not influence the motion of the primaries. Sitnikov showed that for suitable initial conditions the third particle oscillates up and down with growing oscillations such that the motion becomes unbounded but does not tend to a definite position when $t \to \infty$.

In 1968 another Russian mathematician, V. M. Alekseev, found a much richer set of solutions. With the help of symbolic dynamics on infinitely many symbols, he showed that any symbolic sequence corresponds to a solution of the Sitnikov problem and vice versa. He attached sequences to solutions by counting how many times the primaries rotate completely between two consecutive passages of the third particle through the plane of the primaries. For example, the sequence

$$(\ldots, 2, 5; 6, 3, \ldots)$$

means that after the initial moment the primaries rotate six times before the third particle passes through their plane of motion, then they rotate three times until the next passage, and so on. The symbol ∞ is also used. For example, the sequence $(\ldots, 3, 7; 5, 4, \infty)$ means that the third body moves to infinity and never returns. The shift map can be further applied to each symbolic sequence to obtain an orbit, which is the analogue of a solution. Again, from the point of view of the distance between sequences of symbols, solutions that are initially close together move far apart after some time.

Figure 5.5.4. The Sitnikov problem.

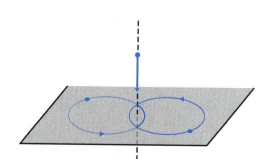

Figure 5.5.5. The Lorenz attractor.

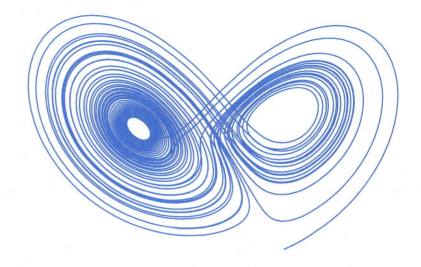

The Lorenz equations It is easy to show that for Lorenz's choice of the parameters, $\sigma = 10$, $\rho = 28$, and $\beta = \frac{8}{3}$, there are three equilibria, all unstable (see Problem 10); however, most of what is known about the Lorenz equations is due to numerical work. If we take some initial conditions close to $(0, 0, 0)$, the solution will go toward one equilibrium, rotate around it several times, then move toward another and rotate around it, and then switch between these two equilibria in an unpredictable manner (see Figure 5.5.5), forming what is called the *Lorenz attractor*.

A natural way to apply symbolic dynamics is to count how many times the solution rotates around one equilibrium, assign the corresponding positive integer, count how many times it moves around the other equilibrium, assign a negative integer, and thus obtain a symbolic sequence like

$$(\ldots 14, -8, 25, -18; 58, -13, 7, -52, \ldots).$$

Unfortunately, all attempts to show that there is a one-to-one correspondence between such sequences and a subset of solutions have failed up to now. Without a sound proof, we cannot exclude the possibility that the Lorenz attractor is, say, a periodic solution of very long period. So though all numerical endeavors indicate that the system leads to chaos, the problem is still open.

Symbolic dynamics is not the only method that detects chaos. Mathematicians have found other ways to describe this phenomenon. None of them, however, is easy to apply to practical problems. As is the case with the preceding examples, the details are often complicated and beyond the scope of this introductory text.

PROBLEMS

1. Use Euler's method and one of the programs in Section 4.6 to draw the graphs of x for the Lorenz equations for initial conditions $(5, 5, 5)$ and $(5.01, 5, 5)$. Use step size 0.01 and 2000 steps. Then use step size 0.02 and 1000 steps. Compare the results.

2. Use the second-order Runge-Kutta method and one of the programs in Section 4.6 to draw the graphs of y for Holmes's version of Düffing's equation with $\delta = 0.26$, $\gamma = 0.3$, $\omega = 1$ and initial conditions $(0.5, 0)$ and $(0.501, 0)$. Use step size 0.1 and 1200 steps.

Then use step size 0.05 and 2400 steps. Compare the results.

3. The Rössler equations are given by the system

$$\begin{cases} x' = -y - z \\ y' = x + ay \\ z' = b + (x - c)z, \end{cases}$$

where a, b, and c are constants.

(a) Take $a = b = 0.2, c = 2.5$ and find a numerical solution for t in $(0, 500)$ for the initial condition $(1, 1, 1)$. (Use various step sizes and methods.) What do you obtain?

(b) Take the same a and b but $c = 3.5$ and repeat the procedure. What do you obtain? Compare the results at (a) and (b).

4. A symbolic sequence is called *periodic* if after some finite number of steps the shift map leads to the initial sequence. The minimum number of steps to reach the same sequence is called the *period*. For example, the sequence

$$(\ldots, 1, 0, 0, 1, 0, 0; 1, 0, 0, 1, 0, 0, \ldots),$$

for which the pattern persists forever, is periodic, and has period 3. Show that any sequence can be approximated by an infinite sequence of symbolic sequences; i.e., if $\mathbf{s}$ is a given symbolic sequence, then there is a sequence $\mathbf{s}^0, \mathbf{s}^1, \ldots, \mathbf{s}^n, \ldots$ such that $\mathbf{s}^n \to \mathbf{s}$ when $n \to \infty$. Explain how this sequence of symbolic sequences can be constructed.

5. Consider the symbolic sequence

$$\mathbf{s} = (\ldots; 0, 1, \quad 0, 0, \ 0, 1, \ 1, 0,$$
$$1, 1, \quad 0, 0, 0, \ 0, 0, 1, \ldots),$$

constructed as follows. The first two entries describe all possible choices of one symbol, i.e., 0 and 1; the next eight entries describe all possible choices of pairs of symbols, i.e., 0, 0, 0, 1, 1, 0, 1, 1; then we take all possible triples of symbols, starting with 0, 0, 0, 0, 0, 1, etc.; and continue this process ad infinitum. Show that if $\mathbf{r}$ is any sequence, then after a finite number of steps of applying the shift map to $\mathbf{s}$, we obtain a sequence whose first 200 entries agree with the first 200 entries of $\mathbf{r}$. Is this property true only for the first 200 entries?

6. Prove the properties in Problems 4 and 5 for symbolic sequences of three symbols.

7. Describe the set of all symbolic sequences of two symbols that have period 3 (i.e., those sequences $\mathbf{s}$ of 0s and 1s with the property that $\sigma(\sigma(\sigma(\mathbf{s}))) = \mathbf{s}$).

8. Let Θ be the set of all sequences of two symbols (0 and 1) that never have two consecutive zeroes. Show that if $\mathbf{s}$ belongs to Θ, then $\sigma(\mathbf{s})$ also belongs to Θ.

9. Let Θ be the set of all sequences of two symbols (0 and 1) that never have two consecutive zeroes. Show that every periodic sequence in Θ (see Problem 4) is the limit of a sequence of symbolic sequences from Θ (i.e., if $\mathbf{s}$ is a given symbolic sequence in Θ, then there is a sequence $\mathbf{s}^0, \mathbf{s}^1, \ldots, \mathbf{s}^n, \ldots$ of elements from Θ such that $\mathbf{s}^n \to \mathbf{s}$ when $n \to \infty$).

10. Find the equilibria of the Lorenz system (1) and determine their nature by linearization, wherever possible. Discuss all cases with respect to the values of the constants σ, ρ, and β.

11. Show that $V(x, y, z) = x^2 + \sigma y^2 + \sigma z^2$ is a Liapunov function for the equilibrium $(0, 0, 0)$ of the Lorenz equations. Use this fact to prove that if $\rho < 1$, the equilibrium $(0, 0, 0)$ is asymptotically stable.

5.6 Modeling Experiments

Before you attempt to work on any of the modeling experiments below, read the introductory paragraph of Section 2.8. This will give you an idea about what is expected from you, how to proceed, and what to emphasize.

Van der Pol's Equation

Recall from Chapter 1 the van der Pol equation

$$x'' + \alpha(x^2 - 1)x' + x = \beta \cos \omega t,$$

where α, β, and ω are constants. This second-order nonlinear equation has many applications. It can model an electric circuit with a triode, the resistive properties of which change with the current, (the negative resistance becomes positive as the current increases). It can be used to model wind-induced oscillations of buildings due to vortex shedding. It is suitable for studying the stability of both rubber-tired and tracked vehicles. It was used in connection with models for radar equipment. It can also model certain chemical reactions.

Write the equation as a two-dimensional system, and fix some values for the constants. Use computer programs of previous chapters to draw the vector field in some region of the phase plane around the origin. Then obtain some solution curves in the phase plane. If a certain region seems to possess some interesting types of solutions, zoom in and study it further. What can you say about the qualitative behavior of these solutions? Do you see any evidence of chaotic behavior? Before drawing any conclusion, take several different initial conditions that are close to each other and compute the solutions numerically in some larger intervals of t. Do they diverge quickly or do they stay close together? Find some references regarding applications of the van der Pol equation and interpret your results in the framework of the model. You can start your library research with the book by Guckenheimer and Holmes (see the Bibliography), which will guide you to other references.

The Düffing Equation

Recall Holmes's version of the Düffing equation mentioned in Chapter 1 and considered in Section 3.6,

$$x'' = -\delta x' + x - x^3 + \gamma \cos \omega t,$$

where δ, γ, and ω are constants, which models the forced vibrations of a cantilever beam in the nonuniform field of two permanent magnets (see Figure 3.6.2). Write the equation as a system, and first take $\delta = \gamma = 0$. Show that the system is Hamiltonian and draw its flow. What does the flow look like if $\gamma = 0$ but $\delta > 0$?

Next take $\delta = 0.25$, $\gamma = 0.3$, and $\omega = 1$ and the initial conditions $x(0) = 0.5$ and $x'(0) = 0$. If you use one of the numerical computer programs in Section 3.7, for the interval $[0, 120]$ you will obtain the graph and the phase-plane representation in Figure 3.6.3. Now use a larger interval and see what you obtain. Vary then the initial conditions just a little. How large is the interval in which the two solutions remain close together? Repeat this experiment for several other nearby initial conditions. Can you draw any conclusions about how quickly the solutions separate from each other?

Repeat the numerical experiment for other values of the constants. What do you obtain? What conclusions can you draw?

The Lorenz System

Recall the Lorenz system mentioned in Chapter 1 and considered in Section 5.5,

$$\begin{cases} x' = \sigma(y - x) \\ y' = \rho x - y - xz \\ z' = -\beta z + xy, \end{cases}$$

where σ, ρ, and $\beta > 0$ are constants, which is a simple theoretical model for weather forecasts. Take $\sigma = 10$, $\beta = \frac{8}{3}$, and $\rho = 28$. Then show that the system has three equilibria, one being $(0, 0, 0)$, which is a saddle. Next consider some initial condition close to $(0, 0, 0)$. Use one of the numerical computer programs in Section 4.6 to draw the graphs of x, y, and z in the interval $[1, 100]$. Then change the initial conditions by only a little bit and see what happens. How large is the interval in which the two solutions remain similar? Now take several close initial conditions and compare the graphs of the corresponding solutions. How fast do solutions separate from each other? What conclusions can you draw?

Next choose different values for the constants σ, ρ, and $\beta > 0$ and repeat the procedure. Do the solutions still present chaotic behavior? Choose other constants. Can you draw any final conclusions about the Lorenz system?

The Three-Body Problem

Consider the nine-dimensional second-order system that describes the motion of three bodies in a plane:

$$\begin{cases} q_{1i}'' = Gm_2 \dfrac{q_{2i} - q_{1i}}{r_{21}^3} + Gm_3 \dfrac{q_{3i} - q_{1i}}{r_{31}^3} \\[2mm] q_{2i}'' = Gm_1 \dfrac{q_{1i} - q_{2i}}{r_{12}^3} + Gm_3 \dfrac{q_{3i} - q_{2i}}{r_{32}^3} \qquad (i = 1, 2, 3), \\[2mm] q_{3i}'' = Gm_1 \dfrac{q_{1i} - q_{3i}}{r_{13}^3} + Gm_2 \dfrac{q_{2i} - q_{3i}}{r_{23}^3} \end{cases}$$

in which G is the gravitational constant, $q_{ji}(j, i = 1, 2, 3)$ represents the ith coordinate of the jth body with respect to a three-dimensional frame, and r_{ji} ($j, i = 1, 2, 3$) is the distance between the jth and the ith body. What will the system become if the motion of the three bodies takes place in a plane? Write these new equations as a first-order system. How many equations do you have?

Next consider three bodies of masses 3, 4, and 5 mass units and place them at the vertices of a right triangle with sides of 3, 4, and 5 length units, such that the side and the mass that bear the same number are opposite each other. Then release the masses from these positions with zero initial velocity. To obtain an approximate solution, adapt to your needs one of the numerical computer programs in Section 4.6. Do the computations for long intervals of time. Do you see any particular pattern occurring? How long does it take to reach this pattern? What does this pattern mean from the physical point of view?

Now change the initial positions by just a little bit and proceed as before. Do you still encounter the same pattern? Does this pattern have any connection with the fact that the bodies come close to a triple collision? What is the physical reason for this phenomenon? Do you see any evidence of chaos? What conclusions can you draw? Compare your conclusions with the ones in the book by Diacu and Holmes (see Bibliography).

6 The Laplace Transform

One could say that application's constant relation to theory is the same as that of the leaf to the tree: one supports the other, but the former feeds the latter.

JACQUES HADAMARD
Gold Medal of the French Academy of Sciences, 1962

6.1 Fundamental Properties

The goal of this chapter is to solve differential equations using the *Laplace transform*, named after the French mathematician Pierre Simon, Marquis de Laplace (1749–1827), who published it in 1782. This method has many applications, especially in physics and engineering. In this introductory section we define the transform, describe its main properties, give an existence theorem, and find the Laplace and the inverse Laplace transforms of several elementary functions. This is a background for Section 6.3, in which we will solve certain types of differential equations—in particular, those given by discontinuous vector fields.

The Laplace Transform

A change of variables, or a coordinate transformation, is similar to expressing a fact in different words. The Laplace transform does more than this. It can be compared with a translator of sentences into another language, in which not only the words differ but the grammar rules too.

DEFINITION 6.1.1

The *Laplace transform* of a function $f(t)$, with $t \geq 0$, is a function usually denoted by $\mathscr{L}[f]$ or $F(s)$, given by the formula

$$\mathscr{L}[f] = F(s) = \int_0^\infty e^{-st} f(t) \, dt.$$

The Laplace transform is defined for all s for which the integral $\int_0^\infty e^{-st} f(t) \, dt$ converges, i.e., for all s for which $\lim_{A \to \infty} \int_0^A e^{-st} f(t) \, dt$ exists and is finite.

In terms of the above language metaphor, the function F is the translation of the function f via the Laplace transform. The goal of this section is to find the functions F for the most common functions f, to determine the "grammar rules" followed by F, (i.e., to see how the properties of f are expressed in terms of F), and to discuss the reverse process—that of finding f when F is given.

Pierre Simon, Marquis de Laplace.

From a different point of view, the Laplace transform is a way of comparing f with the exponential function on the interval $[0, \infty)$, using the integral as a comparison tool. The Laplace transform is part of the larger class of *integral transforms*, which have the general form $F(s) = \int_B^A G(s, t)f(t)\, dt$, where G is a given function.

EXAMPLE 1   The Laplace transform of the constant function $f(t) = 1$, $t \geq 0$, is

$$\mathcal{L}[1] = \int_0^\infty e^{-st}\, dt = \lim_{A \to \infty} \left(-\frac{1}{s} e^{-st} \Big|_0^A \right) = \frac{1}{s}$$

and is defined for all $s > 0$.

EXAMPLE 2   The Laplace transform of the exponential function $f(t) = e^{at}$, $t \geq 0$, where a is a constant, is

$$\mathcal{L}[e^{at}] = \int_0^\infty e^{-st} e^{at}\, dt = \int_0^\infty e^{-(s-a)t}\, dt = \frac{1}{s - a}$$

and is defined for all $s > a$. (Since the above infinite integral is not convergent for $s \leq a$, the Laplace transform of e^{at} is undefined for $s \leq a$.) In particular,

$$\mathcal{L}[e^t] = \frac{1}{s - 1}$$

is defined for $s > 1$.

EXAMPLE 3   The Laplace transform of the function $f(t) = \sin at$, $t \geq 0$, where a is a constant, is

$$\mathcal{L}[\sin at] = F(s) = \int_0^\infty e^{-st} \sin at\, dt.$$

To compute this integral, notice that integration by parts yields

$$F(s) = \lim_{A \to \infty} \left[-\frac{e^{-st} \cos at}{a} \Big|_0^A - \frac{s}{a} \int_0^A e^{-st} \cos at\, dt \right]$$

$$= \frac{1}{a} - \frac{s}{a} \int_0^\infty e^{-st} \cos at\, dt.$$

A new integration by parts leads to

$$F(s) = \frac{1}{a} - \frac{s^2}{a^2} \int_0^\infty e^{-st} \sin at\, dt = \frac{1}{a} - \frac{s^2}{a^2} F(s),$$

so

$$\mathcal{L}[\sin at] = \frac{a}{s^2 + a^2}.$$

Since the infinite integrals are convergent for $s > 0$, this function is defined for $s > 0$.

TABLE 6.1.1

$f(t)$	$\mathscr{L}[f] = F(s)$	Domain of F
1	$\dfrac{1}{s}$	$s > 0$
e^{at}	$\dfrac{1}{s - a}$	$s > a$
$\sin at$	$\dfrac{a}{s^2 + a^2}$	$s > 0$
$\cos at$	$\dfrac{s}{s^2 + a^2}$	$s > 0$
$t^n,\ n = 0, 1, 2, \ldots$	$\dfrac{n!}{s^{n+1}}$	$s > 0$
$t \sin at$	$\dfrac{2as}{(s^2 + a^2)^2}$	$s > 0$
$t \cos at$	$\dfrac{s^2 - a^2}{(s^2 + a^2)^2}$	$s > 0$
$e^{at} \sin bt$	$\dfrac{b}{(s - a)^2 + b^2}$	$s > a$
$e^{at} \cos bt$	$\dfrac{s - a}{(s - a)^2 + b^2}$	$s > a$
$\dfrac{1}{a} \sin at - t \cos at$	$\dfrac{2a^2}{(s^2 + a^2)^2}$	$s > 0$
$\sinh at$	$\dfrac{a}{s^2 - a^2}$	$s > |a|$
$\cosh at$	$\dfrac{s}{s^2 - a^2}$	$s > |a|$
$t^n e^{at}$	$\dfrac{n!}{(s - a)^{n+1}}$	$s > a$
$u_a(t)$	$\dfrac{e^{-as}}{s}$	$s > 0$
$\delta(t - t_0),\ t \geq t_0$	e^{-st_0}	
$f(at)$	$\dfrac{1}{a} F\left(\dfrac{s}{a}\right),\ a > 0$	Same as F
$e^{at} f(t)$	$F(s - a)$	Depends on F
$f'(t)$	$sF(s) - f(0)$	Same as F
$f''(t)$	$s^2 F(s) - sf(0) - f'(0)$	Same as F
$f(t - a)u_a(t)$	$e^{-as} F(s)$	Same as F

Similarly, we can compute the Laplace transforms of other elementary functions (see Problems 15 to 25 and Table 6.1.1).

Existence   One of the problems that arises in dealing with the Laplace transform is similar to that of finding the exact correspondent of a word in another language. Sometimes such a correspondent does not exist. The Laplace transform $F(s)$ of $f(t)$ might be undefined at every s, in which case it

Figure 6.1.1. (*a*) The graph of a piecewise continuous function. (*b*) The graph of a function that is not piecewise continuous since the limit at the end of one subinterval is ∞.

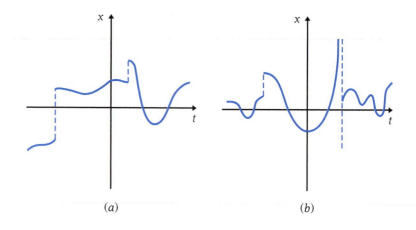

(*a*)   (*b*)

makes no sense at all. We will now provide a result that ensures the existence of the Laplace transform and estimates its domain. For this we first need to define the notions of piecewise continuity and exponential order.

DEFINITION 6.1.2

A function *f* is called *piecewise continuous* if its domain can be divided into several intervals such that *f* is continuous on each of them and has a finite limit at the ends of every such interval, except possibly at ±∞ (see Figure 6.1.1).

DEFINITION 6.1.3

A function *f* defined for $t \geq 0$ is said to be of *exponential order* if there is a real constant, a, and two positive constants, c and M, such that

$$|f(t)| \leq Me^{at} \qquad \text{for } t \geq c.$$

This definition puts into evidence a class of functions which, for *t* large enough, are bounded above and below by some exponential function and by its negative, respectively, each multiplied by the same positive constant (see Figure 6.1.2).

The following theorem ensures the existence of the Laplace transform for a large class of functions. Most functions encountered in applications are of this type.

Existence Theorem

If a function *f* defined on $[0, \infty)$ is piecewise continuous on $[0, \infty)$ and of exponential order with respect to e^{at}, then the Laplace transform $\mathcal{L}[f] = F(s)$ exists for all $s > a$.

Figure 6.1.2. The graph of a function bounded above by e^t and below by $-e^t$.

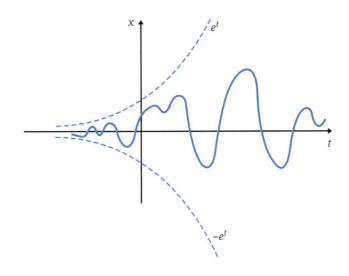

The proof is left as an exercise (see Problem 37). Notice that the functions in Examples 1–3 above satisfy the hypotheses of the theorem.

Properties   Let us now take a look at the "grammar rules" of the Laplace transform. Its main property is *linearity*, which is extensively used in applications. The proof of the following result follows directly from the definition of the transform (see Problem 38).

Linearity Theorem

Let f and g be two functions whose Laplace transforms $\mathscr{L}[f] = F(s)$ and $\mathscr{L}[g] = G(s)$ exist for $s > a$ and $s > b$, respectively, where a and b are constants. Then for any $s > \max\{a, b\}$ and for any real constants c_1 and c_2,

$$\mathscr{L}[c_1 f + c_2 g] = c_1 \mathscr{L}[f] + c_2 \mathscr{L}[g].$$

EXAMPLE 4   Since $\mathscr{L}[e^{2t}] = 1/(s - 2)$ for $s > 2$ (see Example 2) and $\mathscr{L}[\sin 3t] = 3/(s^2 + 9)$ for $s > 0$ (see Example 3), by the linearity theorem,

$$\mathscr{L}[e^{2t} + \sin 3t] = \mathscr{L}[e^{2t}] + \mathscr{L}[\sin 3t] = \frac{1}{s - 2} + \frac{3}{s^2 + 9} \qquad \text{for } s > 2.$$

EXAMPLE 5   To compute the Laplace transform of the function $3 - 2e^{t/3}$, use the fact that $\mathscr{L}[1] = 1/s$ for $s > 0$ and that $\mathscr{L}[e^{t/3}] = 3/(3s - 1)$ for $s > \frac{1}{3}$. The linearity theorem yields

$$\mathscr{L}[3 - 2e^{t/3}] = 3\mathscr{L}[1] - 2\mathscr{L}[e^{t/3}] = \frac{3}{s} - \frac{6}{3s - 1} \qquad \text{for } s > \frac{1}{3}.$$

The second important property concerns the Laplace transform of the derivative. This leads to a formula that is essential for solving differential equations. The proof can be completed following the steps in Problem 39.

Derivative Theorem

Assume that f is differentiable, that its Laplace transform $\mathcal{L}[f] = F(s)$ exists for $s > a$, where a is a constant, and that the derivative f' is piecewise continuous on $[0, \infty)$. Then the Laplace transform of the derivative exists for $s > a$ and

$$\mathcal{L}[f'] = s\mathcal{L}[f] - f(0). \tag{1}$$

Notice that the above result can be easily generalized to higher derivatives. For example, assuming that the Laplace transforms of f and f' exist for $s > a$ and that f'' is piecewise continuous on $[0, \infty)$, we can apply the derivative theorem and obtain that $\mathcal{L}[f'']$ exists for $s > a$ and is given by

$$\mathcal{L}[f''] = s^2\mathcal{L}[f] - sf(0) - f'(0). \tag{2}$$

The generalization of this formula to higher derivatives is left as an exercise (Problem 39).

EXAMPLE 6 Since $\mathcal{L}[\sin t] = 1/(s^2 + 1)$ for $s > 0$ and since both sine and cosine satisfy the conditions of the derivative theorem, we can draw the conclusion that

$$\mathcal{L}[\cos t] = s\mathcal{L}[\sin t] - \sin 0 = \frac{s}{s^2 + 1} \qquad \text{for } s > 0.$$

This agrees with the computation of $\mathcal{L}[\cos t]$ using the definition.

EXAMPLE 7 To compute the Laplace transform of the function $f(t) = t^2$ for $t \geq 0$, we can use the fact that $f''(t) = 2$. Then by (2),

$$\mathcal{L}[2] = s^2\mathcal{L}[t^2] - sf(0) - f'(0).$$

Since $\mathcal{L}[2] = 2/s$ and $f(0) = f'(0) = 0$, it follows that $\mathcal{L}[t^2] = 2/s^3$ for $s > 0$ (see Problem 15 and Table 6.1.1).

Another important rule is given by a shift formula in which the variable s is changed to $s - c$. Its proof follows directly from the definition of the Laplace transform and is left as an exercise (see Problem 40).

Shift Theorem

Let f be a function whose Laplace transform $F(s)$ exists for $s > a$. Then

$$\mathcal{L}[e^{ct}f(t)] = F(s - c) \qquad \text{for } s > a + c.$$

EXAMPLE 8 To obtain the Laplace transform of the function $e^{5t} \sin 2t$, first compute $\mathcal{L}[\sin 2t] = F(s) = 2/(s^2 + 4)$. The shift theorem for $c = 5$ leads to

$$\mathcal{L}[e^{5t} \sin 2t] = F(s - 5) = \frac{2}{(s - 5)^2 + 4} = \frac{2}{s^2 - 10s + 29} \qquad \text{for } s > 5.$$

EXAMPLE 9 To compute the Laplace transform of the function $f(t) = 2t^2 \cosh t$, first notice that $f(t) = 2t^2(e^t + e^{-t})/2 = e^t t^2 + e^{-t} t^2$. This means that according to the shift theorem, $\mathcal{L}[f] = F(s - 1) + F(s + 1)$, where $F(s) = \mathcal{L}[t^2] = 2/s^3$. Thus

$$\mathcal{L}[2t^2 \cosh t] = \frac{2}{(s - 1)^3} + \frac{2}{(s + 1)^3} \qquad \text{for } s > 2.$$

The Inverse Laplace Transform

Like any one-to-one change of variables, the Laplace transform has an inverse. The *inverse Laplace transform* of a function F is a function $f = \mathcal{L}^{-1}[F]$ whose Laplace transform is F.

From Examples 1–3 we can see that $\mathcal{L}^{-1}[1/s] = 1$, $\mathcal{L}^{-1}[1/(s - a)] = e^{at}$, and $\mathcal{L}^{-1}[a/(s^2 + a^2)] = \sin at$, all valid for $t \geq 0$. The inverse Laplace transforms for some elementary functions can be found in Table 6.1.1 by reading from right to left (see also Problems 28 to 36).

Notice that there is a one-to-one correspondence between the set of functions f and the set of functions F via the Laplace transform and its inverse. In other words, starting from f and using the Laplace transform we obtain F, and applying the inverse Laplace transform to this F we recover f. This is unlike differentiation and integration, because differentiating a function and then integrating it leads to the given function plus a real constant.

The inverse Laplace transform is also linear. The proof follows from the linearity theorem and the one-to-one correspondence (see Problem 41).

Linearity of the Inverse Transform

If F and G are defined for $s > a$, then

$$\mathcal{L}^{-1}[k_1 F + k_2 G] = k_1 \mathcal{L}^{-1}[F] + k_2 \mathcal{L}^{-1}[G],$$

which is a relation between functions of variable t that takes place for $t \geq 0$.

EXAMPLE 10 Since $\mathcal{L}^{-1}[4/(4s - 1)] = e^{t/4}$ and $\mathcal{L}^{-1}[2/s^3] = t^2$, both for $t \geq 0$, using the linearity of the inverse Laplace transform we can compute that

$$\mathcal{L}^{-1}\left[\frac{12}{4s - 1} - \frac{8}{s^3}\right] = 3\mathcal{L}^{-1}\left[\frac{4}{4s - 1}\right] - 4\mathcal{L}^{-1}\left[\frac{2}{s^3}\right]$$

$$= 3e^{t/4} - 4t^2 \quad \text{for } t \geq 0.$$

EXAMPLE 11   Since $\mathcal{L}^{-1}[a/(s^2 + a^2)] = \sin at$ and $\mathcal{L}^{-1}[s/(s^2 + a^2)] = \cos at$, using the linearity of the inverse Laplace transform we can compute that

$$\mathcal{L}^{-1}\left[\frac{s + a}{s^2 + a^2}\right] = \mathcal{L}^{-1}\left[\frac{a}{s^2 + a^2}\right] + \mathcal{L}^{-1}\left[\frac{s}{s^2 + a^2}\right]$$

$$= \sin at + \cos at \qquad \text{for } t \geq 0.$$

PROBLEMS

Determine whether the functions in Problems 1 to 12 are of exponential order.

1. $f(t) = 3t^2$

2. $f(t) = 3^{2t}$

3. $f(t) = 3^{3^t}$

4. $f(t) = \ln t$

5. $f(t) = 2 \ln t^3$

6. $f(t) = 5 \sin 3t$

7. $f(t) = \cos 2^t$

8. $f(t) = \sin t \cos t$

9. $f(t) = 2t^{10} + 10^{2t}$

10. $f(t) = 2^{t^2} + 2^{2^t}$

11. $f(t) = \begin{cases} \dfrac{\sin t}{t}, & t > 0 \\ 1, & t = 0 \end{cases}$

12. $f(t) = \begin{cases} e^t, & 0 < t < 1 \\ \sin t, & t \geq 1 \end{cases}$

13. If two functions are of exponential order, is their sum of exponential order? What about their product?

14. Is every bounded function of exponential order? Is every function of exponential order bounded?

Compute the Laplace transform for the functions in Problems 15 to 25.

15. $f(t) = t^2$

16. $f(t) = t^3$

17. $f(t) = t^n, \qquad n = 0, 1, 2, \ldots$

18. $f(t) = e^{5t}$

19. $f(t) = \dfrac{3}{3^t}$

20. $f(t) = 3^t + 1$

21. $f(t) = t \sin at$

22. $f(t) = t \cos at$

23. $f(t) = \dfrac{1}{a} \sin at - t \cos at$

24. $f(t) = \sinh t$, where $\sinh t = \dfrac{e^t - e^{-t}}{2}$

25. $f(t) = \cosh t$, where $\cosh t = \dfrac{e^t + e^{-t}}{2}$

26. Does the Laplace transform of the function $f(t) = 1/t$ exist? What about $f(t) = 1/(t + a), a > 0$?

27. Is the Laplace transform of a product of functions equal to the product of the Laplace transforms of the respective functions?

Compute the inverse Laplace transform of the functions in Problems 28 to 36.

28. $F(s) = \dfrac{1}{s(s + 1)}$

29. $F(s) = \dfrac{2}{s(s + 5)}$

30. $F(s) = \dfrac{1}{s^2 + 7s + 12}$

31. $F(s) = \dfrac{s + 4}{(s + 1)(s + 2)(s + 3)}$

32. $F(s) = \dfrac{2s - 3}{s^2 - 4}$

33. $F(s) = \dfrac{8s^2 - 4s + 12}{s^3 + 4s}$

34. $F(s) = \dfrac{2s + 1}{(s - 1)^2 + 1}$

35. $F(s) = -\dfrac{2s - 1}{s^2 + 4s + 5}$

36. $F(s) = \dfrac{3 - 2s}{s^2 + 2s + 10}$

37. Prove the existence theorem for the Laplace transform.

38. Prove the linearity theorem for the Laplace transform.

39. Prove the derivative theorem for the Laplace transform. (*Hint:* For some $A > 0$, consider $\int_0^A e^{-st} f'(t)\, dt$, take the points $t_1, \dots, t_n$ at

which f' is discontinuous, write the integral as a sum of integrals, and use integration by parts.) Then state a generalization of the derivative theorem for higher-order derivatives and prove it.

40. Prove the shift theorem for the Laplace transform.

41. Prove the linearity theorem for the inverse Laplace transform.

6.2 Step Functions

In many applications we encounter "jumps" of the vector field, which translate into discontinuities of the functions involved. One such example is the Glover-Lazer-McKenna model of the Tacoma Narrows Bridge, studied in Section 3.4. The mathematical base for describing such discontinuities is given by *step functions*, whose graphs resemble the steps of a staircase. In this section we will learn how to represent discontinuous functions with the help of step functions and we will compute their Laplace transforms.

Step Functions

The simplest step function is the *unit step function* (also called the *Heaviside function*, after the English mathematician Oliver Heaviside (1850–1925)), usually denoted by u_c (see Figure 6.2.1):

$$u_c(t) = \begin{cases} 0, & \text{if } t < c \\ 1, & \text{if } t \geq c, \end{cases} \qquad c \geq 0.$$

The Laplace transform of u_c is

$$\mathcal{L}[u_c] = U_c(s) = \int_0^\infty e^{-st} u_c(t)\, dt$$

$$= \int_0^c 0\, dt + \int_c^\infty e^{-st}\, dt = \frac{e^{-cs}}{s} \qquad \text{for } s > 0.$$

(1)

Figure 6.2.1. The graph of the unit step function.

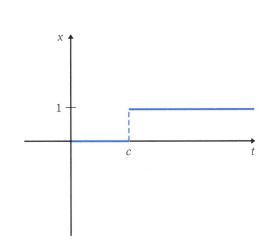

The property that allows us to compute the Laplace transform of any step function is that step functions can be expressed as linear combinations of unit step functions. Indeed, if

$$f(t) = \begin{cases} a_1, & \text{for } t \text{ in } [0, c_1) \\ a_2, & \text{for } t \text{ in } [c_1, c_2) \\ a_3, & \text{for } t \text{ in } [c_2, c_3) \\ \vdots \end{cases}$$

is a step function, then

$$f(t) = a_1 + (a_2 - a_1)u_{c_1}(t) + (a_3 - a_2)u_{c_2}(t) + \cdots. \tag{2}$$

EXAMPLE 1 By formula (2) the function

$$f(t) = \begin{cases} 0, & \text{for } t \text{ in } [0, 3) \text{ or } [4, \infty) \\ 2, & \text{for } t \text{ in } [3, 4) \end{cases}$$

(see Figure 6.2.2(a)) can be written as

$$f(t) = 2u_3(t) - 2u_4(t).$$

We can obtain this linear combination without formula (2) by observing that $f = 2u_3$ on $[0, 3)$ and then determining what function to subtract from $2u_3$ to obtain 0 on $[4, \infty)$. The answer is $2u_4$, so f can be written as above.

Figure 6.2.2. The graphs of the functions in (a) Example 1 and (b) Example 2.

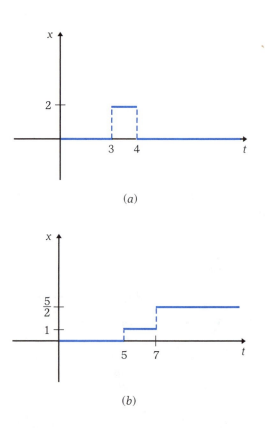

(a)

(b)

The function f is similar to a discontinuous pulse that models, say, an electric circuit connected to a 2-volt battery, switched on for one second (between $t = 3$ and $t = 4$).

Using (1) and the linearity of the Laplace transform, we compute that

$$\mathcal{L}[f] = 2\mathcal{L}[u_3] - 2\mathcal{L}[u_4] = \frac{2(e^{-3s} - e^{-4s})}{s} \qquad \text{for } s > 0.$$

EXAMPLE 2 By formula (2) the function

$$f(t) = \begin{cases} 0, & \text{for } t \text{ in } [0, 5) \\ 1, & \text{for } t \text{ in } [5, 7) \\ \frac{5}{2}, & \text{for } t \text{ in } [7, \infty) \end{cases}$$

(see Figure 6.2.2(b)) can be written as

$$f(t) = u_5(t) + \frac{3}{2}u_7(t).$$

Indeed, from the graph of f we see that $f = u_5$ on $[0, 5)$. Asking what to add to u_5 to obtain $\frac{5}{2}$ on $[7, \infty)$, we see that the answer is $\frac{3}{2}u_7$.

This function could model an electric circuit with a double switch that is first turned on to 1 volt for 2 seconds (between $t = 5$ and $t = 7$) and then switched on to $\frac{5}{2}$ volts and left in this position.

The Laplace transform of f is

$$\mathcal{L}[f] = \mathcal{L}[u_5] + \frac{3}{2}\mathcal{L}[u_7] = \frac{2e^{-5s} + 3e^{-7s}}{2s} \qquad \text{for } s > 0.$$

EXAMPLE 3 By formula (2) the function

$$f(t) = \begin{cases} 3, & \text{for } t \text{ in } [0, 1.2) \\ 2, & \text{for } t \text{ in } [1.2, \pi) \\ 6, & \text{for } t \text{ in } [\pi, 5) \\ \frac{3}{2}, & \text{for } t \text{ in } [5, \infty) \end{cases}$$

Simple electrical systems are fundamental for understanding complex ones, such as the one operated here.

Figure 6.2.3. The graph of the step function in Example 3.

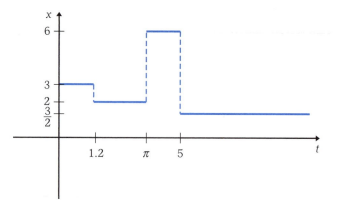

(see Figure 6.2.3) can be written as

$$f(t) = 3 - u_{1.2}(t) + 4u_\pi(t) - \frac{9}{2}u_5(t).$$

The Laplace transform of f is

$$\mathscr{L}[f] = \mathscr{L}[3] - \mathscr{L}[u_{1.2}] + 4\mathscr{L}[u_\pi] - \frac{9}{2}\mathscr{L}[u_5]$$

$$= \frac{3 - e^{-1.2s} + 4e^{-\pi s} - \frac{9}{2}e^{-5s}}{s} \qquad \text{for } s > 0.$$

The Second Shift Formula   We now introduce a formula, similar to the one for the shift theorem of the preceding section, which will help us compute the Laplace transform of some discontinuous functions. For this, notice that if f is a function defined for $t \geq 0$, then the function

$$f_c(t) = \begin{cases} 0, & \text{for } t \text{ in } [0, c) \\ f(t - c), & \text{for } t \text{ in } [c, \infty), \end{cases}$$

where c is a positive constant, is a shift of f to the right of distance c (see Figure 6.2.4). The function f_c, however, can be expressed with the help of the unit step function as

$$f_c(t) = u_c(t)f(t - c).$$

Figure 6.2.4. The shift of the function f to the right.

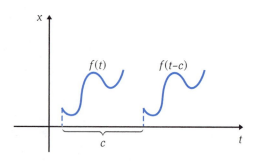

Second Shift Theorem

Let f be a function defined for $t \geq 0$ whose Laplace transform is $F(s) = \mathcal{L}[f]$. Then

$$\mathcal{L}[u_c(t)f(t - c)] = e^{-cs}F(s). \qquad (3)$$

PROOF

To prove the second shift formula (3), denote $u = t - c$ and notice that

$$\mathcal{L}[f_c(t)] = \int_0^\infty u_c(t)f(t - c)e^{-st}\, dt = \int_c^\infty f(t - c)e^{-s(t-c)-sc}\, dt$$

$$= e^{-cs}\int_0^\infty f(u)e^{-su}\, du = e^{-cs}F(s).$$

This completes the proof.●

EXAMPLE 4 Since the Laplace transform of the function $f(t) = t^8$ is $F(s) = 8!/s^9$ for $s > 0$ (see Table 6.1.1), using the second shift formula (3), we can compute the Laplace transform of the function $f_5(t) = (t - 5)^8 u_5(t)$. Indeed,

$$\mathcal{L}[(t - 5)^8 u_5(t)] = \frac{8!e^{-5s}}{s^9} \qquad \text{for } s > 0.$$

EXAMPLE 5 Since the Laplace transform of the function $f(t) = \sin t$ is $F(s) = 1/(s^2 + 1)$ for $s > 0$, using the second shift formula (3), we can compute the Laplace transform of the function $f_{\pi/12}(t) = \sin(t - \pi/12)u_{\pi/12}(t)$. Indeed,

$$\mathcal{L}\left[\sin\left(t - \frac{\pi}{12}\right)u_{\pi/12}(t)\right] = \frac{e^{-(\pi/12)s}}{s^2 + 1} \qquad \text{for } s > 0.$$

EXAMPLE 6 Since the Laplace transform of the function $f(t) = u_6(t)$ is $F(s) = e^{-6s}/s$ for $s > 0$ (see Table 6.1.1), using the second shift formula (3), we can compute the Laplace transform of the function $f_1(t) = u_1(t)u_6(t - 1)$. Indeed,

$$\mathcal{L}[u_1(t)u_6(t - 1)] = \frac{e^{-s}e^{-6s}}{s} = \frac{e^{-7s}}{s} \qquad \text{for } s > 0.$$

This should not be surprising because $u_6(t - 1) = u_7(t)$, or in general, $u_c(t - k) = u_{c+k}(t)$. (Can you check this formula?)

Discontinuous Functions

We can now compute the Laplace transforms of some discontinuous functions by combining the second shift formula with the property that every step function is a linear combination of unit step functions. Though the class of functions whose Laplace transforms can be computed this way is fairly small, they are encountered often in applications.

Figure 6.2.5. The graph of the function in Example 7.

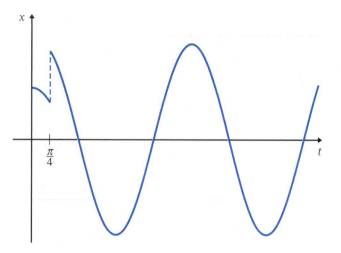

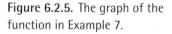

EXAMPLE 7 To compute the Laplace transform of the function

$$f(t) = \begin{cases} \cos t, & \text{for } t \text{ in } [0, \pi/4) \\ \cos t + \cos(t - \pi/4), & \text{for } t \text{ in } [\pi/4, \infty) \end{cases}$$

(see Figure 6.2.5), we can write f using formula (2) but with the functions $\cos t$ and $\cos t + \cos(t - \pi/4)$ instead of the constants a_1 and a_2. We thus obtain

$$f(t) = \cos t + \cos\left(t - \frac{\pi}{4}\right)u_{\pi/4}(t),$$

so using the linearity of the Laplace transform, we have

$$\mathscr{L}[f(t)] = \mathscr{L}[\cos t] + \mathscr{L}\left[\cos\left(t - \frac{\pi}{4}\right)u_{\pi/4}(t)\right].$$

With the help of Table 6.1.1 and the second shift formula, we obtain

$$\mathscr{L}[f(t)] = \frac{s}{s^2 + 1} + e^{-(\pi/4)s}\frac{s}{s^2 + 1} = \frac{s(e^{-(\pi/4)s} + 1)}{s^2 + 1}.$$

EXAMPLE 8 To compute the inverse Laplace transform of the function $F(s) = (1 - e^{-4s})/s^2$, first use the linearity of the transform and find that

$$f(t) = \mathscr{L}^{-1}[F] = \mathscr{L}^{-1}\left[\frac{1}{s^2}\right] - \mathscr{L}^{-1}\left[\frac{e^{-4s}}{s^2}\right].$$

With the help of Table 6.1.1 and the second shift formula applied to the identity function, we obtain

$$f(t) = t - u_4(t)(t - 4) = \begin{cases} t, & \text{for } 0 \leq t < 4 \\ 2, & \text{for } t \geq 4, \end{cases}$$

which is discontinuous. What does its graph look like?

Pulse Functions   Another class of functions encountered in applications is that of *pulse* functions, which are "switched on" and then "switched off"; in other words, they are zero on some interval $[0, c_1)$, nonzero on $[c_1, c_2)$, and zero again on $[c_2, \infty)$, like the step function in Example 1. The computation of the Laplace transform is outlined in the following example for the sine pulse function.

EXAMPLE 9   To compute the Laplace transform of the *sine pulse* function S, whose graph is represented in Figure 6.2.6(*b*), we must find its algebraic expression. For this, recall the step function in Example 1 (see Figure 6.2.2(*a*)), which has something in common with the sine pulse function—namely, it is 0 everywhere except on some embedded interval. Since the function in Example 1 is written as the difference between two unit step functions, it is worth trying something similar here. The pulse is "switched on" at 1 and "switched off" at 2, so the difference $u_1 - u_2$ may be useful. Indeed, $u_1 - u_2$ is 1 in the interval $[1, 2)$ and 0 everywhere else (see Figure 6.2.6(*a*)). To obtain the sine pulse in the interval $[1, 2)$, we need to multiply $u_1 - u_2$ by $\sin 2\pi t$, so the expression of the sine pulse function is

$$S(t) = [u_1(t) - u_2(t)] \sin 2\pi t.$$

Figure 6.2.6. The graphs of the functions (*a*) $u_1 - u_2$ and (*b*) sine pulse.

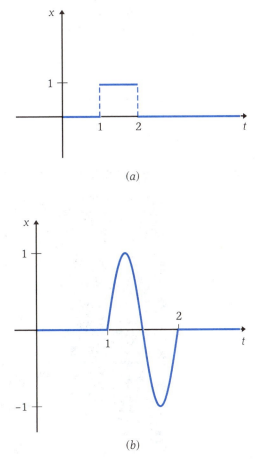

(*a*)

(*b*)

Using the second shift formula, we obtain

$$\mathcal{L}[u_1(t)\sin 2(\pi - 1)t] = \frac{2\pi e^{-s}}{s^2 + 4\pi^2}$$

and

$$\mathcal{L}[u_2(t)\sin 2(\pi - 2)t] = \frac{2\pi e^{-2s}}{s^2 + 4\pi^2}.$$

Since $\sin 2\pi(t-1) = \sin 2\pi(t-2) = \sin 2\pi t$, the linearity of the Laplace transform implies that

$$\mathcal{L}[S(t)] = \mathcal{L}[u_1(t)\sin 2\pi t] - \mathcal{L}[u_2(t)\sin 2\pi t] = \frac{2\pi(e^{-s} - e^{-2s})}{s^2 + 4\pi^2}.$$

Impulsive Functions

In some applications the pulse functions are of an impulsive nature; they act swiftly and with considerable effect, like the blow of a karate master. Such an *impulsive function* can be approximated by

$$d_h(t) = \begin{cases} \dfrac{1}{2h}, & \text{for } t \text{ in } (-h, h) \\ 0, & \text{for } t \text{ in } (-\infty, -h) \text{ or } (h, \infty), \end{cases}$$

whose graph is drawn in Figure 6.2.7(a). The smaller h is, the better the approximation.

The area under the graph of the function d_h is obviously 1, so

$$\int_{-\infty}^{\infty} d_h(t)\, dt = \int_{-h}^{h} d_h(t)\, dt = 1,$$

independently on h as long as $h \neq 0$. When $h \to 0$, $d_h(t) \to 0$ for $t \neq 0$ (see Figure 6.2.7(b)). These facts suggest a definition for the *unit impulsive function* δ, also called the *Dirac delta function*, after the English

Focused power of a karate master.

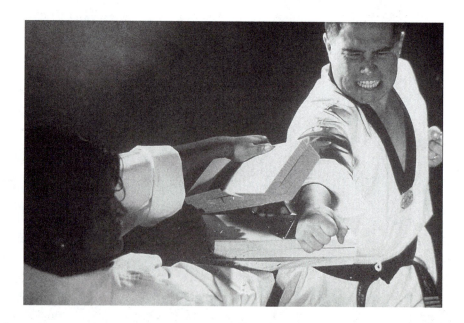

Figure 6.2.7. (*a*) The graph of the function d_h. (*b*) Several graphs of d_h as $h \to 0$.

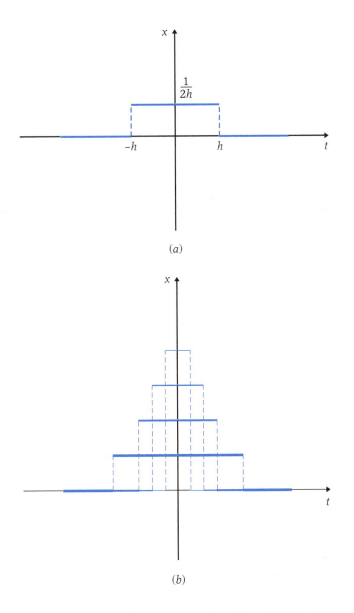

(*a*)

(*b*)

mathematical physicist Paul Adrien Maurice Dirac (1902–1984). This is defined as having the properties[1]

$$\delta(t) = 0 \quad \text{for } t \neq 0 \quad \text{and} \quad \int_{-\infty}^{\infty} \delta(t)\, dt = 1. \qquad (4)$$

Since $\delta(t)$ corresponds to a unit impulse at $t = 0$, we can obtain a unit impulse at $t = t_0$ by taking $\delta(t - t_0)$, which according to (4) must satisfy the conditions

$$\delta(t - t_0) = 0 \quad \text{for } t \neq t_0 \quad \text{and} \quad \int_{-\infty}^{\infty} \delta(t - t_0)\, dt = 1. \qquad (5)$$

---

[1]The Dirac delta function is an example of a *distribution* (also called a *generalized function*), which is not a function in the usual sense of the term. A rigorous treatment of distributions is beyond our scope, so we will rely here on intuition.

Though δ is not a function in the classical sense of the term, and therefore does not satisfy the hypotheses of the existence theorem in Section 6.1, we can still compute its Laplace transform. Assuming that $t_0 > 0$ and defining the Laplace transform of δ as the limit of the Laplace transform of d_h when $h \to 0$, we obtain

$$\mathscr{L}[\delta(t - t_0)] = \lim_{h \to 0} \mathscr{L}[d_h(t - t_0)]$$

$$= \int_0^\infty e^{-st} d_h(t - t_0) \, dt = \int_{t_0-h}^{t_0+h} e^{-st} d_h(t - t_0) \, dt.$$

From the definition of the function d_h, it follows that

$$\mathscr{L}[d_h(t - t_0)] = \frac{e^{sh} - e^{-sh}}{2hs} e^{-st_0}.$$

According to l'Hospital's rule, $\lim_{h \to 0}(e^{sh} - e^{-sh})/(2hs) = 1$, so the Laplace transform of the Dirac delta function is

$$\mathscr{L}[\delta(t - t_0)] = e^{-st_0}.$$

In the next section we will apply this result to see what happens to a pendulum moving inside a frame that is suddenly hit by a karate master.

PROBLEMS

In Problems 1 to 6 sketch the graphs of the given functions.

1. $f(t) = 2u_1(t) + 3u_2(t)$

2. $f(t) = u_3(t) - u_4(t) + \pi u_5(t)$

3. $f(t) = 2tu_4(t)$

4. $f(t) = t^3 u_{1/2}(t)$

5. $f(t) = u_{3\pi}(t)\cos t$

6. $f(t) = (t - 3)u_1(t) + (t - 2)u_2(t) + (t - 1)u_3(t)$

Find an analytical representation for each of the functions whose graphs are drawn in Problems 7 to 12.

7.
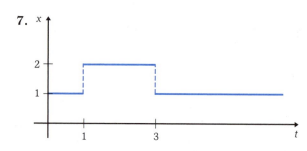

8.
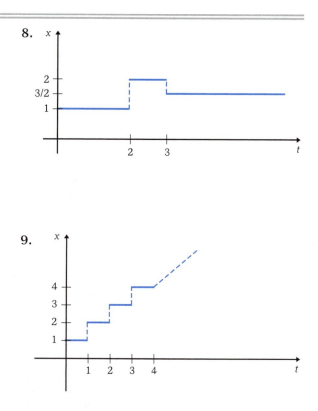

9.

10.

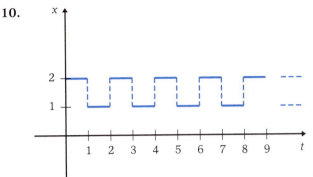

11.

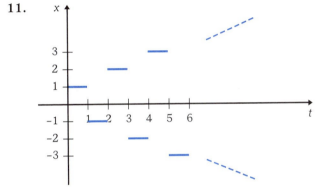

12.

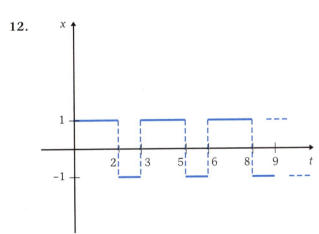

Write the functions in Problems 13 to 16 using unit step functions (see formula (2)).

13. $f(t) = \begin{cases} 1, & \text{for } t \text{ in } [0, 3) \\ 2, & \text{for } t \text{ in } [3, \infty) \end{cases}$

14. $f(t) = \begin{cases} 0, & \text{for } t \text{ in } [0, 1.5) \\ 3, & \text{for } t \text{ in } [1.5, \infty) \end{cases}$

15. $f(t) = \begin{cases} \pi, & \text{for } t \text{ in } [0, 1) \\ 2, & \text{for } t \text{ in } [1, 1.8) \\ \frac{5}{3}, & \text{for } t \text{ in } [1.8, \infty) \end{cases}$

16. $f(t) = \begin{cases} 0, & \text{for t in } [0,2) \\ 1, & \text{for } t \text{ } [2, 4) \\ 0, & \text{for } t \text{ in } [4, \infty) \end{cases}$

In each of Problems 17 to 24 find the Laplace transform of the given functions.

17. $f(t) = \begin{cases} 0, & \text{for } t \text{ in } [0, 3) \\ (t-3)^2, & \text{for } t \text{ in } [3, \infty) \end{cases}$

18. $f(t) = \begin{cases} 0, & \text{for } t \text{ in } [0, 1) \\ (t-1)^2 + 1, & \text{for } t \text{ in } [1, \infty) \end{cases}$

19. $f(t) = \begin{cases} 0, & \text{for } t \text{ in } [0, \pi) \\ t - \pi, & \text{for } t \text{ in } [\pi, 2\pi) \\ 0, & \text{for } t \text{ in } [2\pi, \infty) \end{cases}$

20. $f(t) = \begin{cases} \sin t, & \text{for } t \text{ in } [0, \pi/4) \\ \sin t + \cos(t - \pi/4), & \text{for } t \text{ in } [\pi/4, \infty) \end{cases}$

21. $f(t) = 2u_1(t) - 3u_2(t) + 7u_7(t)$

22. $f(t) = u_{0.3}(t) - (t-1)u_1(t)$

23. $f(t) = (t-3)u_2(t) + (t-2)u_3(t)$

24. $f(t) = 4(t-1)^2 u_1(t) + 3(t-3)^2 u_3(t)$

Compute the inverse Laplace transform of the functions in Problems 25 to 30.

25. $F(s) = (s-2)^{-4}$

26. $F(s) = \dfrac{2e^{-2s}}{(s+1)^2 - s - 3}$

27. $F(s) = \dfrac{(s-1)e^{-2s}}{(s-1)^2 + 1}$

28. $F(s) = \dfrac{e^{-3s}}{(s-3)(s+3)}$

29. $F(s) = \dfrac{e^{-s} + e^{-2s}}{s}$

30. $F(s) = \dfrac{2e^{-s} + 3e^{-2s} + 4e^{-3s}}{5s}$

31. Define the integral of the product between a continuous function g and the Dirac delta function as

$$\int_{-\infty}^{\infty} g(t)\delta(t - t_0)\, dt = \lim_{h \to 0} \int_{-\infty}^{\infty} g(t)d_h(t - t_0)\, dt.$$

Then, using the mean value theorem for integrals, show that

$$\int_{-\infty}^{\infty} g(t)\delta(t - t_0)\, dt = g(t_0).$$

32. Can you apply the result in Problem 31 to compute

$$\int_{-\infty}^{\infty} \delta(t - t_0)\delta(t - t_1)\, dt,$$

where $t_0 > t_1 > 0$?

6.3 Initial Value Problems

In this section we will introduce a three-step method that uses the Laplace transform to solve initial value problems for single differential equations as well as for systems. We will present several examples and then apply the method to determine the motion of a discontinuously forced harmonic oscillator and to find the solution of the equation for an LC circuit with ramped forcing function.

The Three-Step Method

As for linear systems with constant coefficients, which can be solved with the help of linear algebra, the idea of the three-step method is to use the Laplace transform to reduce an initial value problem to an algebraic equation. This can be done as follows.

Step 1. Apply the Laplace transform to both sides of the differential equation (system) and obtain an algebraic equation (system).

Step 2. Substitute the initial data and solve the algebraic equation (system).

Step 3. Apply the inverse Laplace transform to translate the solution of the algebraic equation (system) into the language of the differential equation (system).

> **EXAMPLE 1**   For the sake of outlining the ideas, let us use the three-step method to solve the initial value problem
>
> $$x' = -4x - 1, \qquad x(0) = 3,$$
>
> whose solution can be more directly obtained if the problem is regarded as a separable equation. We will denote by $X(s)$ the Laplace transform $\mathcal{L}[x]$.
>
> *Step 1.* Apply the Laplace transform to both sides of the equation and obtain
>
> $$\mathcal{L}[x'] = \mathcal{L}[-4x - 1],$$
>
> which, if the derivative theorem is used on the left-hand side and the linearity of the Laplace transform on the right-hand side, reduces to
>
> $$sX(s) - x(0) = -4X(s) - \frac{1}{s}.$$
>
> *Step 2.* Substitute $x(0) = 3$ into the above equation and obtain
>
> $$(s + 4)X(s) = 3 - \frac{1}{s},$$

which if solved for X yields

$$X(s) = \frac{3}{s+4} - \frac{1}{s(s+4)}.$$

Step 3. Apply the inverse Laplace transform to both sides of this equation and, using linearity on the right-hand side, obtain

$$\mathcal{L}^{-1}[X(s)] = 3\mathcal{L}^{-1}\left[\frac{1}{s+4}\right] - \mathcal{L}^{-1}\left[\frac{1}{s(s+4)}\right].$$

This leads to

$$x(t) = 3e^{-4t} - \mathcal{L}^{-1}\left[\frac{1}{s(s+4)}\right], \tag{1}$$

so the only thing left is to compute the inverse Laplace transform of $1/[s(s+4)]$. Although the function is not in Table 6.1.1, it can be decomposed into partial to fractions:

$$\frac{1}{s(s+4)} = \frac{1}{4s} - \frac{1}{4(s+4)}.$$

The inverse Laplace transform of the function is thus

$$\mathcal{L}^{-1}\left[\frac{1}{s(s+4)}\right] = \mathcal{L}^{-1}\left[\frac{1}{4s}\right] - \mathcal{L}^{-1}\left[\frac{1}{4(s+4)}\right] = \frac{1}{4} - \frac{1}{4}e^{-4t}.$$

Substitute this into (1) and obtain the solution:

$$x(t) = \frac{13}{4}e^{-4t} - \frac{1}{4}.$$

EXAMPLE 2 Let us now apply the three-step method to the initial value problem

$$\begin{cases} x' = x + y \\ y' = -x + y, \end{cases} \qquad x(0) = y(0) = 1,$$

which can be solved using the method of Section 4.3.

Step 1. Denoting $\mathcal{L}[x] = X(s)$, $\mathcal{L}[y] = Y(s)$ and applying the Laplace transform to each equation, obtain the algebraic system

$$\begin{cases} sX(s) - x(0) = X(s) + Y(s) \\ sY(s) - y(0) = -X(s) + Y(s). \end{cases}$$

Step 2. Substituting the initial data and solving for X and Y yields

$$\begin{cases} X(s) = \dfrac{s}{(s-1)^2 + 1} \\ Y(s) = \dfrac{s-2}{(s-1)^2 + 1}. \end{cases}$$

Step 3. Application of the inverse Laplace transform and the first shift theorem leads to

$$\begin{cases} x(t) = \mathscr{L}^{-1}\left[\dfrac{(s-1)+1}{(s-1)^2+1}\right] = e^t\cos t + e^t\sin t \\[4mm] y(t) = \mathscr{L}^{-1}\left[\dfrac{(s-1)-1}{(s-1)^2+1}\right] = e^t\cos t - e^t\sin t, \end{cases}$$

which represents the solution of the initial value problem.

Discontinuities

In the above examples we used the Laplace transform as an alternative method, since we could have solved the given initial value problems in different ways. Let us now apply the three-step method to differential equations defined by vector fields that are discontinuous or have discontinuous partial derivatives. For such equations or systems, the methods studied in previous chapters may fail.

EXAMPLE 3   Let

$$f(t) = \begin{cases} 0, & \text{for } t \text{ in } [0,3) \text{ or } [4,\infty) \\ 2, & \text{for } t \text{ in } [3,4) \end{cases}$$

and consider the initial value problem

$$x' = x + f, \qquad x(0) = 0.$$

Recall from Example 1 of Section 6.2 that $f = 2u_3 - 2u_4$, so we can write the equation as

$$x' = x + 2u_3 - 2u_4.$$

Applying the Laplace transform and denoting $\mathscr{L}[x] = X(s)$, we obtain the algebraic equation

$$sX(s) - x(0) = X(s) + \frac{2(e^{-3s} - e^{-4s})}{s},$$

which, since $x(0) = 0$, has the solution

$$X(s) = \frac{2(e^{-3s} - e^{-4s})}{s(s-1)}.$$

Decomposing the right-hand side into partial fractions, applying the inverse Laplace transform, and using the linearity property, we obtain

$$x(t) = \mathscr{L}^{-1}\left[\frac{2e^{-3s}}{s-1}\right] - \mathscr{L}^{-1}\left[\frac{2e^{-3s}}{s}\right] - \mathscr{L}^{-1}\left[\frac{2e^{-4s}}{s-1}\right] + \mathscr{L}^{-1}\left[\frac{2e^{-4s}}{s}\right].$$

With the help of Table 6.1.1 and the second shift theorem, we compute that

$$x(t) = 2u_3(t)e^{t-3} - 2u_3(t) - 2u_4(t)e^{t-4} + 2u_4(t),$$

which can be explicitly written as

$$x(t) = \begin{cases} 0, & \text{for } t \text{ in } [0, 3) \\ 2e^{t-3} - 2, & \text{for } t \text{ in } [3, 4) \\ 2e^{t-3} - 2e^{t-4}, & \text{for } t \text{ in } [4, \infty). \end{cases}$$

Indeed, it is easy to verify that this is a solution of the initial value problem. However, since the vector field is discontinuous, it is natural to ask whether the solution is unique for $t \geq 0$. The answer is negative. The function

$$\varphi(t) = \begin{cases} 0, & \text{for } t \text{ in } [0, 3) \\ e^t - 2, & \text{for } t \text{ in } [3, 4) \\ e^t, & \text{for } t \text{ in } [4, \infty), \end{cases}$$

for example, is another solution of the same initial value problem. (Can you show that there are infinitely many solutions?)

EXAMPLE 4   Consider now the function

$$g(t) = \begin{cases} 4t, & \text{for } t \text{ in } [0, 1) \\ 4, & \text{for } t \text{ in } [1, \infty) \end{cases}$$

and the initial value problem for a second-order linear equation,

$$x'' = -4x + g, \qquad x(0) = 1, \qquad x'(0) = 0.$$

The function g is continuous but has discontinuous partial derivatives at $t = 1$ (see Figure 6.3.1), so this makes the application of methods other than the Laplace transform more difficult. First notice that g can be written as $g(t) = 4t - 4(t - 1)u_1(t)$, so the equation to solve is

$$x'' = -4x + 4t - 4(t - 1)u_1(t).$$

Denoting $\mathcal{L}[x] = X(s)$ and applying the Laplace transform, we obtain

$$s^2 X(s) - sx(0) - x'(0) = -4X(s) + \frac{4}{s^2} - \frac{4e^{-s}}{s^2},$$

which if solved for X and using the conditions $x(0) = 1, x'(0) = 0$ leads to the solution

$$X(s) = \frac{s}{s^2 + 4} + \frac{4}{s^2(s^2 + 4)} - \frac{4e^{-s}}{s^2(s^2 + 4)}.$$

Figure 6.3.1. The graph of the function g in Example 4.

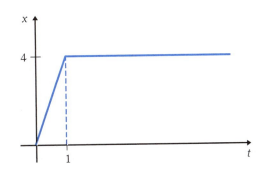

The decomposition of the last two terms into partial fractions yields

$$X(s) = \frac{s}{s^2 + 4} + \frac{1}{s^2} - \frac{1}{s^2 + 4} - \frac{e^{-s}}{s^2} + \frac{e^{-s}}{s^2 + 4}.$$

Applying the inverse Laplace transform and using its linearity, we obtain the solution

$$x(t) = \cos 2t + t - \frac{1}{2}\sin 2t - \left[t - 1 - \frac{1}{2}\sin 2(t - 1)\right]u_1(t) \qquad \text{for } t \geq 0,$$

which can be explicitly written as

$$x(t) = \begin{cases} \cos 2t + t - \dfrac{1}{2}\sin 2t, & \text{for } t \text{ in } [0, 1) \\[2mm] \cos 2t - \dfrac{1}{2}[\sin 2t - \sin 2(t - 1)] + 1, & \text{for } t \text{ in } [1, \infty). \end{cases}$$

Is this solution unique?

Applications

A spring in a decelerating elevator

A problem arising in engineering concerns springs on which discontinuous forces are applied. For example, assume that the initial value problem

$$x'' = -2x' - 5x + f(t), \qquad x(0) = x'(0) = 0$$

describes such a problem, where

$$f(t) = \begin{cases} 1, & \text{for } t \text{ in } [0, 10) \\ 0, & \text{for } t \text{ in } [10, \infty). \end{cases}$$

In other words, a unit force acts during the first 10 seconds on an initially-in-equilibrium spring, and then the force is removed. (This can be the case for, say, a vertical spring in an elevator, which after moving downward with constant speed decelerates for 10 seconds and stops.) What is the motion of the spring?

Solution. The three-step method allows us to treat this problem directly. First notice that the above second-order equation can be written as

$$x'' = -2x' - 5x + 1 - u_{10}(t).$$

Denoting $\mathcal{L}[x] = X(s)$, applying the Laplace transform, and using the linearity and derivative theorems, we obtain

$$s^2 X(s) - sx(0) - x'(0) = -2sX(s) + 2x(0) - 5X(s) + \frac{1}{s} - \frac{e^{-10s}}{s}.$$

Substituting the values of the initial data and solving for X leads to

$$X(s) = \frac{1 - e^{-10s}}{s(s^2 + 2s + 5)}.$$

Since

$$\frac{1}{s(s^2 + 2s + 5)} = \frac{1}{5s} - \frac{s + 2}{5[(s + 1)^2 + 4]},$$

Figure 6.3.2. The graph of the solution to the discontinuously forced harmonic oscillator.

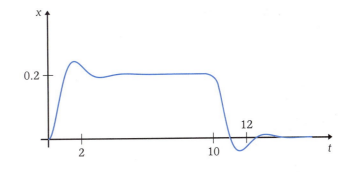

$X(s)$ can be more conveniently written as

$$X(s) = \frac{1 - e^{-10s}}{5s} - \frac{1 - e^{-10s}}{5}\left[\frac{s+1}{(s+1)^2+4} + \frac{1}{(s+1)^2+4}\right].$$

(Notice that we chose to write $X(s)$ in this way in order to have $s + 1$ in both the numerator and the denominator of the first fraction in the brackets. This will help us compute the inverse Laplace transform.)

Applying the inverse Laplace transform, we obtain the solution

$$x(t) = \frac{1 - u_{10}(t)}{5} - e^{-t}\left(\frac{\cos 2t}{5} + \frac{\sin 2t}{10}\right)$$

$$+ u_{10}(t)e^{10-t}\left[\frac{\cos 2(t-10)}{5} + \frac{\sin 2(t-10)}{10}\right],$$

which can be explicitly written as

$$x(t) = \begin{cases} \dfrac{1}{5} - e^{-t}\left(\dfrac{\cos 2t}{5} + \dfrac{\sin 2t}{10}\right), & \text{for } t \text{ in } [0, 10) \\[2mm] -e^{-t}\left(\dfrac{\cos 2t}{5} + \dfrac{\sin 2t}{10}\right) + e^{10-t}\left[\dfrac{\cos 2(t-10)}{5} + \dfrac{\sin 2(t-10)}{10}\right], \\[2mm] \text{for } t \text{ in } [10, \infty). \end{cases}$$

This is a continuous function whose graph is shown in Figure 6.3.2. •

An *LC* circuit with ramped forcing   Consider the initial value problem

$$q'' = -q + \begin{cases} t, & \text{for } t \text{ in } [0, 9) \\ 9, & \text{for } t \text{ in } [9, \infty) \end{cases}, \qquad x(0) = x'(0) = 0,$$

which describes the change in time of the charge in plates q of a simple *LC* circuit (i.e., no resistor included) connected to a potential that increases linearly with time from 0 to 9 volts in the first 9 seconds and remains constant afterward. The graph of the forcing function looks like a ramp (see Figure 6.3.3(*a*)). Let us determine the variation of the charge q.

Solution. Using unit step functions, we can write this equation as

$$q'' = -q + t + (9 - t)u_9(t).$$

Figure 6.3.3. The graphs of
(a) the ramped forcing function
and (b) the solution for the
initial value problem associated
with the LC circuit with ramped
forcing.

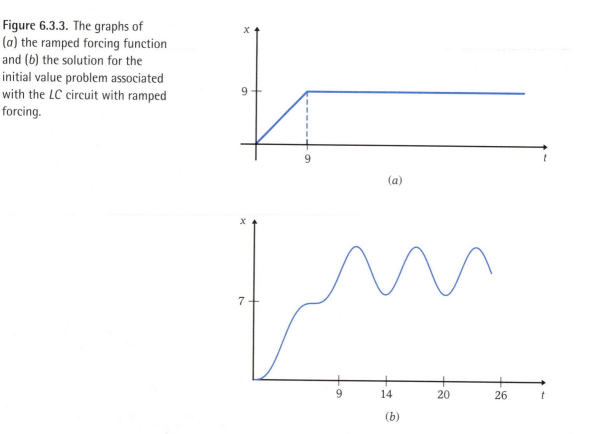

(a)

(b)

Applying the three-step method, we obtain the solution

$$s^2\mathcal{L}[q] - sq(0) - q'(0) = -\mathcal{L}[q] + \mathcal{L}[t] + \mathcal{L}[(9 - t)u_9(t)],$$

which leads to

$$\mathcal{L}[q] = \frac{1}{s^2(s^2 + 1)} - \frac{e^{-9s}}{s^2(s^2 + 1)}.$$

The decomposition in partial fractions yields

$$\mathcal{L}[q] = \frac{1}{s^2} - \frac{1}{s^2 + 1} + \frac{e^{-9s}}{s^2 + 1} - \frac{e^{-9s}}{s^2}.$$

Applying the inverse Laplace transform, we obtain the solution

$$q(t) = t - \sin t - [t - 9 - \sin(t - 9)]u_9(t),$$

whose graph is represented in Figure 6.3.3(b). Notice that in this case the
vector field is continuous but not differentiable at $t = 9$.●

A pendulum hit by a karate blow   A pendulum inside a rigid frame starts swinging with small amplitude
(see Figure 6.3.4). Five seconds later a karate master hits the frame from
above. Since frictions are negligible for short time intervals, we can
model the pendulum's motion with the second-order equation

$$x'' + x = \delta(t - 5)$$

Figure 6.3.4. A pendulum moving inside a frame that suffers a karate blow.

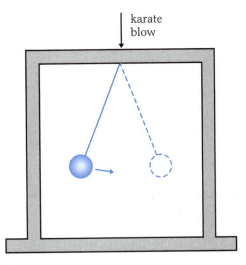

karate blow

(see Sections 3.2 and 3.4), where δ is the Dirac delta function (see Section 6.2). Assuming that the blow does not break the frame, determine the motion of the pendulum for the initial conditions $x(0) = 1, x'(0) = 0$.

Solution. Applying the Laplace transform to the above equation and denoting $X(s) = \mathcal{L}[x]$, we obtain the algebraic equation[1]

$$s^2 X(s) - sx(0) - x'(0) + X(s) = e^{-5s}.$$

After substitution of the values of the initial conditions, this equation yields the solution

$$X(s) = \frac{s + e^{-5s}}{s^2 + 1}.$$

The second shift formula provides the solution of the initial value problem,

$$x(t) = \cos t + u_5(t) \sin(t - 5),$$

whose graph is drawn in Figure 6.3.5. The solution is continuous, but its graph suggests that it is not differentiable at $t = 5$. Indeed, we can check

Figure 6.3.5. The graph of the function $x(t) = \cos t + u_5(t) \sin(t - 5)$ in the interval $[0, 20]$. Notice that x is not differentiable at $t = 5$.

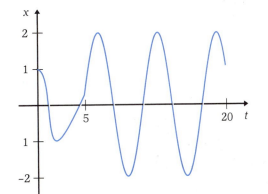

<hr>

[1]As we mentioned in Section 6.2, this is only a formal way to operate with the Dirac delta function. A rigorous analysis, which justifies the limiting operations, is beyond our scope.

that $\lim_{t \to 5, t < 5} x'(t) = -\sin 5$ and $\lim_{t \to 5, t > 5} x'(t) = 1 - \sin 5$, which proves the lack of differentiability at $t = 5$. Physically, this means that at the instant of the karate blow, the velocity of the pendulum experiences a sudden jump.•

PROBLEMS

Solve the initial value problems in Problems 1 to 8 using the three-step method of the Laplace transform.

1. $x' = 5x + 6, \qquad x(0) = 1$

2. $y' = y + 12t, \qquad y(0) = 2$

3. $v' = \frac{1}{2}v, \qquad v(0) = 1$

4. $r' = r + t^3, \qquad r(0) = 3$

5. $x'' = 4x + 10e^{-t}\sin t, \qquad x(0) = x'(0) = 1$

6. $u'' = -u - 10te^{-t}, \qquad u(0) = 1, u'(0) = 2$

7. $z'' = -z' - z - \sin t, \qquad z(0) = z'(0) = 1$

8. $w'' = -2w' - 7w + e^{-t}\cos t,$
$\quad w(0) = 0, w'(0) = 1$

Solve the initial value problems for the systems in Problems 9 to 12 using the three-step method of the Laplace transform.

9. $\begin{cases} x' = x + 2y \\ y' = x - 3y \end{cases} \qquad x(0) = y(0) = 1$

10. $\begin{cases} u' = 1.8u - v \\ v' = 2.9u - 1.6v \end{cases} \qquad u(0) = 1, v(0) = 0.5$

11. $\begin{cases} r' = 2r - 2\theta + e^{-t} \\ \theta' = 6r - 5\theta + e^{-2t}, \end{cases}$
$\quad r(0) = 1, \theta(0) = -1$

12. $\begin{cases} x' = 3x - 2y + t \\ y' = 5x - 2y - t^2 \end{cases} \qquad x(0) = y(0) = 2$

Solve the initial value problems for the discontinuous vector fields given in Problems 13 to 18 using the three-step method of the Laplace transform.

13. $x'' = -x - \begin{cases} 1, & \text{for } t \text{ in } [0, \pi) \\ 0, & \text{for } t \text{ in } [\pi, \infty), \end{cases}$
$\quad x(0) = 0, \ x'(0) = 1$

14. $u'' = 2u - \begin{cases} t, & \text{for } t \text{ in } [0, 1) \\ 0, & \text{for } t \text{ in } [1, \infty), \end{cases}$
$\quad u(0) = 1, \ u'(0) = -1$

15. $v'' = \frac{1}{3}v - \begin{cases} \sin t, & \text{for } t \text{ in } [0, \pi/2) \\ 0, & \text{for } t \text{ in } [\pi/2, \infty), \end{cases}$
$\quad v(0) = v'(0) = 0$

16. $x'' = -x' - 4x + \begin{cases} 2, & \text{for } t \text{ in } [0, 5) \\ 1, & \text{for } t \text{ in } [5, \infty), \end{cases}$
$\quad x(0) = -1, \ x'(0) = 3$

17. $z'' = -z' - \begin{cases} \cos t, & \text{for } t \text{ in } [0, \pi) \\ 0, & \text{for } t \text{ in } [\pi, \infty), \end{cases}$
$\quad z(0) = z'(0) = 0$

18. $w'' = w' - w - \begin{cases} 0, & \text{for } t \text{ in } [0, 1) \\ 1, & \text{for } t \text{ in } [1, \infty), \end{cases}$
$\quad w(0) = w'(0) = 0$

19. Consider an LC circuit connected through a switch to a signal generator, with the whole system being modeled by the initial value problem

$$q'' = -2q - u_{\pi/4}(t)\cos\left(t - \frac{\pi}{4}\right),$$

$$q(0) = 2, \qquad q'(0) = 4.$$

Use the three-step method to find the solution and then draw its graph.

20. Consider an RLC circuit connected at time $t = 5$ to a battery with an electric source of 220 V, which has a forcing function with a jump discontinuity. The system is modeled by the initial value problem

$$x'' = -0.22x' - 4.84x - 220u_5(t),$$

$$x(0) = x'(0) = 0.$$

Use the three-step method to find the solution and then draw its graph.

21. The motion of a spring-mass system is modeled by the initial value problem

$$x'' = -x - \alpha[u_{2.5}(t) - u_{3.5}(t)],$$

$$x(0) = 0, \qquad x'(0) = 10,$$

where α is a parameter.

(i) For $\alpha = 1$ draw the graph of the forcing function and see that it is a pulse.

(ii) Use the three-step method to solve the initial value problem.

(iii) Draw the graphs of the solutions for $\alpha = 0$, $\alpha = 10$, and $\alpha = 20$ and compare the three cases. Which solution exhibits more oscillations in a given time unit?

22. The motion of a spring-mass system is modeled by the initial value problem

$$x'' = -x1 - \alpha u_5(t) + \beta u_6(t),$$

$$x(0) = 10, \qquad x'(0) = 0,$$

where α and β are positive parameters.

(i) For $\alpha = \beta = 1$ draw the graph of the forcing function and see that it is a pulse.

(ii) Use the three-step method to solve the initial value problem.

(iii) Draw the graphs of the solutions for $(\alpha, \beta) = (0, 0)$, $(\alpha, \beta) = (0, 30)$, $(\alpha, \beta) = (30, 0)$, and $(\alpha, \beta) = (\frac{1}{2}, \frac{1}{2})$. Compare the four cases. Which solution exhibits more oscillations in a given time unit?

In Problems 23 to 30, use the Laplace transform to obtain the solutions of the corresponding initial value problems. Discuss the behavior of the solution before and after the impulse. Is x' continuous everywhere?

23. $x'' + 4x = \delta(t - \pi)$, $\qquad x(0) = 0$, $x'(0) = 1$

24. $x'' + 4x = \delta(t - \pi) + \delta(t - 2\pi)$,
$\quad x(0) = 0$, $\qquad x'(0) = 0$

25. $x'' + 4x = \delta(t - \pi) + \delta(t - 2\pi) + \delta(t - 3\pi)$, $\qquad x(0) = 0$, $x'(0) = 1$

26. $x'' - x = 2\delta(t - 1)$, $\qquad x(0) = 1$, $x'(0) = 0$

27. $x'' + 2x' + x = \delta(t) + u_{2\pi}$,
$\quad x(0) = 0$, $x'(0) = 1$

28. $x'' + 2x' + 2x = \delta(t - \pi)$,
$\quad x(0) = 1$, $x'(0) = 0$

29. $x'' + 2x' + 2x = \delta(t - \pi/2) + \cos t$,
$\quad x(0) = 0$, $x'(0) = 0$

30. $x'' + x = \delta(t - \pi)\cos t$, $\qquad x(0) = 0$, $x'(0) = 1$
(*Hint*: Use Problem 31 in Section 6.2.)

6.4 Computer Applications

In this section we will use Maple, *Mathematica*, and MATLAB to solve differential equations and systems with the help of the Laplace transform.

Maple

The Laplace transform is part of a package that can be introduced with the command with(inttrans). Within this package, Maple can directly solve simple initial value problems for equations and systems using the command dsolve with the specification method=laplace. The solution is displayed, but the application of the Laplace transform and of its inverse remain hidden.

EXAMPLE 1   To solve the initial value problem $x'' - 3x' + 2x = 0$, $x(0) = 0$, $x'(0) = 1$ with the help of the Laplace transform, we proceed as follows:

```
> with(inttrans):
> eq := diff(x(t),t$2)-3*diff(x(t),t) + 2*x(t) = 0;
```

$$eq := \left(\frac{\partial^2}{\partial t^2}x(t)\right) - 3\left(\frac{\partial}{\partial t}x(t)\right) + 2x(t) = 0$$

```
> dsolve({eq,x(0)=0,D(x)(0)=1},x(t),
  method=laplace);
```

and Maple displays the answer,

$$x(t) = e^{(2t)} - e^t$$

EXAMPLE 2 To solve the initial value problem for the linear nonhomogeneous system

$$\begin{cases} x' = x - 2y - t \\ y' = 3x + y \end{cases}, \qquad x(0) = y(0) = 0$$

with the help of the Laplace transform, we proceed as follows:

```
> with(inttrans):
> eq := diff(x(t),t)=x(t)-2*y(t)-t,
  diff(y(t),t)=3*x(t)+y(t);
```

$$eq := \frac{\partial}{\partial t}x(t) = x(t) - 2y(t) - t, \ \frac{\partial}{\partial t}y(t) = 3x(t) + y(t)$$

```
> dsolve({eq,x(0)=0,y(0)=0}, {x(t),y(t)},
  method=laplace);
```

and Maple displays the answer,

$$\{y(t) = -\frac{3}{7}t - \frac{6}{49} + \frac{6}{49}e^t \cos(\sqrt{6}t) + \frac{5}{98}e^t \sqrt{6}\sin(\sqrt{6}t),$$

$$x(t) = \frac{1}{7}t - \frac{5}{49} - \frac{2}{49}e^t \sqrt{6}\sin(\sqrt{6}t) + \frac{5}{49}e^t \cos(\sqrt{6}t)\}$$

Maple is also capable of directly solving initial value problems for differential equations and systems for certain discontinuous vector fields. In Maple's language the unit step function is called Heaviside, and $u_c(t)$ is denoted by Heaviside(t-c).

EXAMPLE 3 We can solve the initial value problem $x'' = 3x' - 2x + u_2(t)$, $x(0) = x'(0) = 0$ as follows:

```
> with(inttrans):
> eq := diff(x(t),t$2)-3*diff(x(t),t)+2*x(t)
  -Heaviside(t-2) = 0:
> dsolve({eq,x(0)=0,D(x)(0)=0},x(t),
  method=laplace);
```

and Maple displays the answer,

$$x(t) = \frac{1}{2}\text{Heaviside}(t - 2)(1 - 2e^{(t-2)} + e^{(2t-4)})$$

where Heaviside(t-2) means $u_2(t)$.

Remark. If Maple provides an answer involving e to a complex power, use *Euler's formula*,

$$e^{it} = \cos t + i \sin t,$$

to express the solution in terms of trigonometric functions.

In many cases Maple is unable to solve the equation directly, so you have to combine analytic methods with those provided by the computer. In such cases you can use Maple to compute the Laplace transforms of certain functions and equations and the inverse Laplace transforms of certain functions.

EXAMPLE 4 To compute the Laplace transform of the function $x(t) = t^2 + \sin t$, proceed as follows:

```
> with(inttrans):
> laplace(t^2+sin(t)=x(t), t, s);
```

and Maple displays the answer,

$$\frac{2}{s^3} + \frac{1}{s^2 + 1} = \text{laplace}(x(t), t, s)$$

EXAMPLE 5 To compute the Laplace transform of the equation $x'' - x = \sin at$, proceed as follows:

```
> with(inttrans):
> laplace(diff(x(t), t$2)-x(t)=sin(a*t), t, s);
```

and Maple displays the answer

$$s(\text{laplace}(x(t), t, s) - x(0)) - D(x)(0) - \text{laplace}(x(t), t, s) = \frac{a}{a^2 + a^2}$$

EXAMPLE 6 To compute the inverse Laplace transform of the function

$$\frac{1}{s - a} + \frac{s^2 - a^2}{(s^2 + a^2)^2}$$

proceed as follows:

```
> with(inttrans):
> invlaplace(1/(s-a)+(s^2-a^2)/(s^2+a^2)^2, s, t);
```

and Maple displays the answer

$$e^{(at)} + t \cos(at)$$

All these computer techniques are helpful in dealing with initial value problems that can be approached with the Laplace transform.

Mathematica   Unlike Maple, *Mathematica* has no direct way of solving differential equations and systems using the Laplace transform. Let us therefore present the transform commands and use the three-step method to solve second-order equations and systems. The commands `LaplaceTransform` and `InverseLaplaceTransform` are contained in a package that can be called with the command `<< Calculus`LaplaceTransform'`.

EXAMPLE 7 To compute the Laplace transform of the function $x(t) = t^2 + \sin t$, proceed as follows:

```
<< Calculus`LaplaceTransform'
x[t]=t^2+Sin[t]
```

$$t^2 + Sin[t]$$

```
LaplaceTransform[x[t], t, s]
```

and *Mathematica* displays the answer,

$$\frac{2}{s^3} + \frac{1}{1 + s^2}$$

EXAMPLE 8 To compute the Laplace transform of the equation $x'' - x = \sin at$, proceed as follows:

```
<< Calculus`LaplaceTransform'
y[t]=x''[t]-x[t]==Sin[t]
```

$$-x[t] + x''[t] = \mathrm{Sin}[t]$$

```
LaplaceTransform[y[t],t,s]
```

and *Mathematica* displays the answer,

$$- \mathrm{LaplaceTransform}[x[t], t, s] + s^2 \mathrm{LaplaceTransform}[x[t], t, s]$$
$$- sx[0] - x'[0] = = \frac{a}{a^2 + a^2}$$

EXAMPLE 9 To compute the inverse Laplace transform of the function

$$\frac{1}{s - a} + \frac{s^2 - a^2}{(s^2 + a^2)^2}$$

proceed as follows:

```
<< Calculus`LaplaceTransform'
Y[s]=1/(s-a)+(s^2-a^2)/(s^2+a^2)^2
```

$$\frac{1}{-a + s} + \frac{-a^2 + s^2}{(a^2 + s^2)^2}$$

```
InverseLaplaceTransform[Y[s], s, t]
```

and *Mathematica* displays the answer,

$$E^{at} + t\,\mathrm{Cos}[at]$$

Since *Mathematica* is not endowed with a direct method of solving equations or systems, we can follow the three-step method of Section 6.3. Of course, we could write a program following this method, but a general one that includes all possible situations (first-order equations, second-order equations, systems of different dimensions) is too complicated. Therefore it seems reasonable to simply follow the three steps every time we need to solve differential equations with the help of the Laplace transform.

Step 1. Use `LaplaceTransform` to compute the Laplace transform of the equation or system.

Step 2. Use `Solve` to compute the solution of the algebraic equation or system.

Step 3. Use `InverseLaplaceTransform` to obtain the solution of the initial differential equation or system.

EXAMPLE 10   To solve the initial value problem $x'' - 3x' + 2x = 0$, $x(0) = 0$, $x'(0) = 1$ with the three-step method of the Laplace transform and with *Mathematica*, proceed as follows:

Step 1.

```
<< Calculus`LaplaceTransform`
y[t]=x''[t]-3x'[t]+2x[t]==0
```

$$2x[t] - 3x'[t] + x''[t] = = 0$$

```
LaplaceTransform[y[t],t,s]
```

and *Mathematica* displays the answer,

$$2\text{LaplaceTransform}[x[t], t, s] + s^2\text{LaplaceTransform}[x[t], t, s] -$$
$$3(s\,\text{LaplaceTransform}[x[t], t, s] - x[0]) - sx[0] - x'[0] = = 0$$

Step 2. Using the fact that $x(0) = 0$ and $x'(0) = 1$, we now want to solve for X the algebraic equation $2X + s^2X - 3sX - 1 = 0$, and for this we use the command Solve as follows:

```
X[s]=Solve[2X+s^2*X-3s*X-1==0, X]
```

Mathematica displays the answer,

$$\left\{\left\{X-> -\frac{1}{-2 + 3s - s^2}\right\}\right\}$$

Step 3. We can now obtain the solution of the differential equation x by computing the inverse Laplace transform of the solution of the algebraic equation at Step 2.

```
InverseLaplaceTransform[X[s],s,t]
```

and *Mathematica* displays the answer,

$$-E^t + E^{2t}$$

EXAMPLE 11   To solve the initial value problem for the linear nonhomogeneous system

$$\begin{cases} x' = x - 2y - t \\ y' = 3x + y, \end{cases} \qquad x(0) = y(0) = 0$$

with the three-step method of the Laplace transform, proceed as follows:

Step 1.

```
<< Calculus`LaplaceTransform`
u[t]=x'[t]-x[t]-2y[t]-t==0
```

$$-t - x[t] - 2y[t] + x'[t] = = 0$$

```
v[t]=y'[t]-3x[t]-3x[t]+y[t]==0
```

$$-3x[t] + y[t] + y'[t] = = 0$$

```
LaplaceTransform[u[t],t,s]
```

$$-\frac{1}{s^2} - \text{LaplaceTransform}[x[t], t, s] + s\,\text{LaplaceTransform}[x[t], t, s] -$$
$$- 2\text{LaplaceTransform}[y[t], t, s] - x[0] == 0$$

```
LaplaceTransform[v[t],t,s]
```

$$- 3\text{LaplaceTransform}[x[t], t, s] + \text{LaplaceTransform}[y[t], t, s]$$
$$+ s\,\text{LaplaceTransform}[y[t], t, s] - y[0] == 0$$

Step 2. After taking into account the initial conditions $x(0) = y(0) = 0$, we solve the system of unknowns X and Y given by the equations $-1/s^2 - X + sX - 2Y = 0$ and $-3X + Y + sX = 0$. We can do this with the command Solve:

```
Solve[{-1/s^2-X+s*X-2Y==0,  -3X+Y+s*X==0},  {X,Y}]
```

$$\left\{\left\{X -> -\frac{-1-s}{s^2(-7+s^2)},\quad Y -> \frac{3}{s^2(-7+s^2)}\right\}\right\}$$

Step 3. Then we define $X[s]$ and $Y[s]$,

```
X[s]=(1+s)/(s^2(-7+s^2))
```

$$\frac{1+s}{s^2(-7+s^2)}$$

```
Y[s]=3/(s^2(-7+s^2))
```

$$\frac{3}{s^2(-7+s^2)}$$

and compute the inverse Laplace transforms of these functions to obtain the solutions of the initial value problem:

```
InverseLaplaceTransform[X[s],s,t]
```

$$-\frac{1}{7} - \frac{t}{7} + \frac{1}{7}\left(\text{Cosh}[\sqrt{7}t] + \frac{\text{Sinh}[\sqrt{7}t]}{\sqrt{7}}\right)$$

```
InverseLaplaceTransform[Y[s],s,t]
```

$$3\left(-\frac{t}{7} + \frac{\text{Sinh}[\sqrt{7}t]}{7\sqrt{7}}\right)$$

MATLAB Unlike Maple, MATLAB has no direct way of solving differential equations and systems using the Laplace transform. Let us therefore present the transform commands and use the three-step method to solve second-order equations and systems. The commands laplace and ilaplace are contained in the Symbolic Toolbox package.

EXAMPLE 12 To compute the Laplace transform of the function $x(t) = t^2 + \sin t$, first use the command syms to define the symbolic object t and then apply laplace as follows:

```
>> syms t
>> laplace(t^2+sin(t))
```

MATLAB displays the answer,

```
ans=
2/s^3+1/(s^2+1)
```

EXAMPLE 13 To compute the Laplace transform of the equation $x'' - x = \sin at$, we will compute the Laplace transform of the expression $x'' - x - \sin at$. For this we will use the command syms to define the symbolic objects t and a, and then the command laplace as follows:

```
>> syms t a
>> laplace(diff(diff(sym('x(t)')))-sym('x(t)')-sin(a*t))
```

MATLAB displays the answer,

```
ans=
s*(s*laplace(x(t),t,s)-x(0)-D(x)(0)
-laplace(x(t),t,s)-a/(s^2+a^2)
```

Notice that when introducing $x(t)$ we used sym to let MATLAB know that it should treat this function as a symbol.

EXAMPLE 14 To compute the inverse Laplace transform of the function

$$\frac{1}{s-a} + \frac{s^2 - a^2}{(s^2 + a^2)^2}$$

first use the command sym to define the symbolic objects s and a, and then apply the command ilaplace as follows:

```
>> syms s a
>> ilaplace(1/(s-a)+a^2/(s^2+a^2)^2)
```

MATLAB displays the answer,

```
ans=
exp(a*t) + t*cos(a*t)
```

Since MATLAB is not endowed with a direct method of solving equations or systems, we can follow the three-step method of Section 6.3. Of course, we could write a program following this method, but a general one that includes all possible situations (first-order equations, second-order equations, systems of different dimensions) is too complicated. Therefore it seems reasonable to follow the three steps every time we need to solve differential equations with the help of the Laplace transform.

Step 1. Use laplace to compute the Laplace transform of the equation or system.

Step 2. Use solve to compute the solution of the algebraic equation or system.

Step 3. Use ilaplace to obtain the solution of the initial differential equation or system.

EXAMPLE 15   To solve the initial value problem $x'' - 3x' + 2x = 0$, $x(0) = 0$, $x'(0) = 1$ with the three-step method of the Laplace transform and with MATLAB, proceed as follows:

Step 1.

```
>> syms t s
>> laplace(diff(diff(sym('x(t)')))-3*diff(sym('x(t)'))+
   2*sym('x(t)'))
```

MATLAB displays the answer,

```
ans =
s*(s*laplace(x(t),t,s)-x(0))-D(x)(0)
-3*s*laplace(x(t),t,s)+3*x(0)+2*laplace(x(t),t,s)
```

Step 2. Using the fact that $x(0) = 0$ and $x'(0) = 1$, we now want to solve for X the algebraic equation $2X + s^2X - 3sX - 1 = 0$, and for this we use the command solve as follows:

```
>> solve('2*X+s^2*X-3*s*X-1=0','X')
```

MATLAB displays the answer,

```
X =
1/(2+s^2-3*s)
```

Step 3. We can now obtain the solution of the differential equation x by computing the inverse Laplace transform of the solution of the algebraic equation at Step 2.

```
>> ilaplace(1/(2+s^2-3*s))
```

MATLAB displays the answer,

```
ans =
-exp(t)+exp(2*t)
```

EXAMPLE 16   To solve the initial value problem for the linear nonhomogeneous system

$$\begin{cases} x' = x - 2y - t \\ y' = 3x + y \end{cases}, \qquad x(0) = y(0) = 0$$

with the three-step method of the Laplace transform, proceed as follows:

Step 1.

```
>> syms t s
>> laplace(diff(sym('x(t)'))-sym('x(t)')
   +2*sym('y(t)')-t)
ans =
s*laplace(x(t),t,s)-x(0)-laplace(x(t),t,s)
+2*laplace(y(t),t,s)-1/s^2
>> laplace(diff(sym('y(t)'))-3*sym('x(t)')
   +sym('y(t)'))
```

```
ans =
  s*laplace(y(t),t,s)-y(0)-3*laplace(x(t),t,s)
  +laplace(y(t),t,s)
```

Step 2. After taking into account the initial conditions $x(0) = y(0) = 0$, we solve the system of unknowns X and Y given by the equations $-1/s^2 - X + sX - 2Y = 0$ and $-3X + Y + sX = 0$. We can do this with the command `solve`,

```
>> solve('-1/s^2-X+s*X-2*Y=0','-3*X+Y+s*X=0','X','Y')
```

and obtain the solution

```
X=
-(1+s)/s^2(s^2-7)
Y=
3/s^2(s^2-7)
```

Step 3. We now compute the inverse Laplace transforms of X and Y and obtain the solution of the initial value problem:

```
>> ilaplace(-(1+s)/s^2*(s^2-7))
ans =
-Dirac(1,t)-Dirac(t)+7*t+7
>> ilaplace(3/s^2*(s^2-7))
ans =
3*Dirac(t)-21*t
```

Here `Dirac` denotes the Dirac delta function (see Section 6.2). Is this result in agreement with the ones obtained using Maple and *Mathematica*?

PROBLEMS

Using Maple, Mathematica, or MATLAB, compute the Laplace transforms of the functions in Problems 1 to 12

1. $f(t) = \cos 4t + te^{2t}$

2. $f(t) = \cosh 3t - \sinh 5t$

3. $f(t) = t^2 \cos t + t^3 \sin t$

4. $f(t) = t(\sinh 2t - 5 \cosh 4t)$

5. $f(t) = \sinh t \cos t$

6. $f(t) = \cosh 2t \sin 2t$

7. $f(t) = \cos at - at \sin at$,
where a is a nonzero constant

8. $f(t) = (\sin at - at \cos at)e^{bt}$,
where a and b are nonzero constants

9. $f(t) = \begin{cases} t, & \text{for } t \text{ in } [0, 4) \\ 0, & \text{for } t \text{ in } [4, \infty) \end{cases}$

10. $f(t) = \begin{cases} \sin t, & \text{for } t \text{ in } [0, 4\pi) \\ 0, & \text{for } t \text{ in } [4\pi, \infty) \end{cases}$

11. $f(t) = \begin{cases} 0, & \text{for } t \text{ in } [0, 1) \\ t, & \text{for } t \text{ in } [1, 2) \\ t^2 & \text{for } t \text{ in } [2, \infty) \end{cases}$

12. $f(t) = \begin{cases} \sin 2t, & \text{for } t \text{ in } [0, \pi) \\ \cos 3t, & \text{for } t \text{ in } [\pi, 3\pi) \\ 2 \cos 2t + 3 \sin 3t, & \text{for } t \text{ in } [3\pi, \infty) \end{cases}$

Using Maple, Mathematica, or MATLAB, compute the inverse Laplace transforms of the functions in Problems 13 to 24.

13. $F(s) = \dfrac{e^{-s}}{s + 1}$

14. $F(s) = \dfrac{3}{s^3} + \dfrac{2}{s^2}$

15. $F(s) = \dfrac{5}{s^2 + 2s}$

16. $F(s) = \dfrac{12s + 4}{s^2 - 4s - 5}$

17. $F(s) = \dfrac{2s + \frac{3}{2}}{s^2 + 4s + 13}$

18. $F(s) = -\dfrac{2}{4s + s^3}$

19. $F(s) = \dfrac{1}{2(s^3 - 2s^2)}$

20. $F(s) = \dfrac{1}{s^2(s^2 - 16)}$

21. $F(s) = \dfrac{se^{-2\pi s}}{s^2 + 2s + 5}$

22. $F(s) = \dfrac{2e^{-5s}}{3s(s + 1)^2}$

23. $F(s) = -\dfrac{1}{s^3 + s}$

24. $F(s) = \dfrac{s}{s^2 + 4s + 4}$

Using Maple, Mathematica, or MATLAB and the Laplace transform method, compute the solution of the initial value problems in Problems 25 to 34.

25. $x'' = -4x' - 8x - 10\cos 2t,$
    $x(0) = x'(0) = 0$

26. $x'' = -x' + 2x - 15e^t \sin t,$
    $x(0) = x'(0) = 0$

27. $x'' = 2x' + x - t + u_{\pi/2}(t)(t - \pi/2),$
    $x(0) = x'(0) = 0$

28. $x'' = -4x - u_\pi - u_{3\pi},$
    $x(0) = x'(0) = 0$

29. $x'' = -3x' - 2x + f(t),$
    $x(0) = x'(0) = 0,$

where $f(t) = \begin{cases} \sin t, & \text{for } t \text{ in } [0, \pi) \\ 0, & \text{for } t \text{ in } [\pi, \infty) \end{cases}$

30. $x'' = 2x' + x + f(t), \quad x(0) = x'(0) = 0,$

where $f(t) = \begin{cases} 3, & \text{for } t \text{ in } [0, 5) \\ t, & \text{for } t \text{ in } [5, \infty) \end{cases}$

31. $\begin{cases} x' = x + y + e^{2t} \\ y' = -x + y, \\ x(0) = y(0) = 0 \end{cases}$

32. $\begin{cases} x' = 2y + \sin t \\ y' = x - y + \sin t, \\ x(0) = 1, \ y(0) = 0 \end{cases}$

33. $\begin{cases} x' = 2x + y + e^t \\ y' = 2y + 1 \\ z' = -z, \end{cases}$
    $x(0) = 1, \ y(0) = z(0) = 0$

34. $\begin{cases} x' = -x + 2z + 1 \\ y' = x + y + z \\ z' = 2x - z, \end{cases}$
    $x(0) = y(0) = z(0) = 0$

6.5 Modeling Experiments

Before attempting to work on any of the modeling experiments below, read the introductory paragraph of Section 2.8. This will give you an idea about what is expected from you, how to proceed, and what to emphasize.

Car Suspensions

As mentioned in Section 6.3, a problem arising in engineering is that of discontinuously forced springs. Such forces act, for example, on car suspensions, which are arrangements of springs designed to support the body of an automobile, railway car, etc. External discontinuous forces act on the springs of an automobile on bumpy roads and on the suspension of a railway car as it passes over junctions. Consider a simplified model in which the suspension acts like a spring, i.e., its motion is given by the initial value problem

$$x'' = ax' + bx + f(t),$$
$$x(0) = x'(0) = 0,$$

where a and b are constants and f is a discontinuous forcing function. Choose some negative values for a and b, and choose a discontinuous function that approximates the external forces acting on the spring. What values for a and b are realistic? Use the three-step method of the Laplace transform to solve the equation. What is the physical interpretation of the result? Keep the same forcing function but slightly vary the values of a and b. Determine the solution and compare it with the previous one. Repeat this procedure several times and compare the results. What conclusions can you draw? For what values of a and b is the impact of the external force less stressful for the suspension? For what values of a and b is the ride in the car smoother? What properties of the spring change when you change a and b? In your opinion, what properties should an ideal suspension have?

Electrical Circuits with Ramped Forcing

Consider electrical circuits like the one studied in Section 6.3, given by initial value problems of the form

$$q'' = -aq + r(t), \qquad q(0) = q'(0) = 0,$$

where q is the charge in plates, r is a ramped function, and a is a constant. Using the three-step method, solve the equation for $a = 1$ and for several ramped functions, e.g.,

$$r_1(t) = (1 - t)u_1(t), \qquad r_2(t) = (2 - t)u_2(t), \qquad \ldots,$$

$$r_{10}(t) = (10 - t)u_{10}(t).$$

Graph the solution in each case and compare the graphs. What conclusion can you draw? Then repeat the procedure for $a = 2$. Compare the results you obtain for various ramped functions and then compare the outcomes for $a = 1$ and $a = 2$. What conclusions can you draw? What can you say about the charge? What happens physically when you vary a? Add a resistor to the circuit and proceed as before. What are your conclusions now? Can you state a theorem? Can you prove it?

External Shocks on Harmonic Oscillators

Imagine a vertically oscillating spring that undergoes a vertical external shock (for example, a hammer blow or an instant earthquake shake). The motion of the spring is well approximated by the initial value problem

$$x'' = ax' + bx + \delta(t - t_0),$$

where $x > 0$ describes the length of the spring below the equilibrium position, a and b are constants, δ is the Dirac delta function, and $t_0 > 0$ is the instant at which the shock takes place.

First fix some realistic values for the constants a and b, and solve the equation for various values of t_0. In each case graph the solution and its derivative. Is the solution a continuous function? Is it differentiable everywhere? What about the derivative? What is the physical interpretation of the solution in each case? Compare the various solutions. Can you draw any conclusions? Now change the values of the constants a and b. Repeat the procedure and see what you obtain. What conclusions can you draw about the motion of a spring under external shocks? What is the general effect of shocks on springs? Can you extend your conclusions to other physical systems—the pendulum, for example? What about electrical circuits? What would be a "shock" for an electrical circuit? How can it appear? How would you model the motion of harmonic oscillators under repeated external shocks? Can you solve the corresponding equations? What general conclusions can you draw?

7 Power Series Solutions

Do not worry about your difficulties in mathematics. I can assure you that mine are still greater.

ALBERT EINSTEIN
Nobel Prize for Physics, 1921

7.1 The Power Series Method

In this chapter we will introduce the power series method, which is especially useful for obtaining exact or approximate solutions of linear equations and systems with variable coefficients. We will consider second-order linear equations of the form

$$a(t)x'' + b(t)x' + c(t)x = 0 \qquad (1)$$

where a, b, and c are polynomial functions. The values of t for which $a(t) = 0$ are called *singular points*, and those with $a(t) \neq 0$ are called *ordinary points*. After briefly reviewing the main properties of power series, we will present in this section a method for solving equation (1) near ordinary points. In Section 7.2 we will deal with approximate solutions and with some applications of the method. We will address the case of singular points in Section 7.3.

Power Series

Recall from calculus that a power series $\sum_{n=0}^{\infty} a_n t^n$ converges either at $t = 0$ alone or at least for all t in some interval $(-r, r)$, where $r > 0$ (finite or infinite) is the *radius of convergence*. In the latter case the series represents a function in the interval of convergence; i.e., there is a function f such that

$$f(t) = \sum_{n=0}^{\infty} a_n t^n, \qquad \text{for all } t \text{ in } (-r, r).$$

This function is differentiable and integrable, and its derivative and integral are given by

$$f'(t) = \sum_{n=0}^{\infty} n a_n t^{n-1}$$

and

$$\int_0^t f(s)\, ds = \sum_{n=0}^{\infty} \frac{a_n}{n+1} t^{n+1} \qquad \text{for all } t \text{ in } (-r, r).$$

Also recall that power series can be added and multiplied like polynomials. If $f(t) = \sum_{n=0}^{\infty} a_n t^n$ and $g(t) = \sum_{n=0}^{\infty} b_n t^n$ are convergent in the interval $(-r, r)$, then $f + g$ and fg are also convergent in $(-r, r)$ and are given by

$$f(t) + g(t) = \sum_{n=0}^{\infty}(a_n + b_n)t^n,$$

$$f(t)g(t) = a_0 b_0 + (a_0 b_1 + a_1 b_0)t + (a_0 b_2 + a_1 b_1 + a_2 b_0)t^2 + \cdots.$$

The radius of convergence r of a series $\sum_{n=0}^{\infty} a_n t^n$ can often be determined with the *ratio test:*

$$r = \lim_{n \to \infty} \left| \frac{a_n}{a_{n+1}} \right|.$$

EXAMPLE 1   The series $\sum_{n=0}^{\infty} t^n/(n\, 3^n)$ has the radius of convergence

$$r = \lim_{n \to \infty} \frac{1/(n\, 3^n)}{1/[(n+1)3^{n+1}]} = 3,$$

so it converges in the interval $(-3, 3)$, at least.

EXAMPLE 2   The series $\sum_{n=0}^{\infty} t^n/n!$ has the radius of convergence

$$r = \lim_{n \to \infty} \frac{1/n!}{1/(n+1)!} = \infty,$$

i.e., it converges on $(-\infty, \infty)$. This was to be expected since $\sum_{n=0}^{\infty} t^n/n!$ is the power series representation of the function e^t.

Solutions Near Ordinary Points

The idea of the power series method is to check in equation (1) a solution of the form $\sum_{n=0}^{\infty} a_n t^n$ and to determine the coefficients $a_1, a_2, \ldots, a_n, \ldots$. If this succeeds, we subsequently must find the radius of convergence $r > 0$. If $r = 0$, the series is not a valid solution, so the method fails.

Let us first illustrate the method for a simple equation, which can easily be solved by the direct method of Section 3.2. The following example, however, will help us describe the main steps of the power series technique and fix the ideas.

EXAMPLE 3   To apply the power series method to the equation

$$x'' + x = 0,$$

we will check a solution of the form $x(t) = \sum_{n=0}^{\infty} a_n t^n$. Then $x'(t) = \sum_{n=0}^{\infty} n a_n t^{n-1} = \sum_{n=1}^{\infty} n a_n t^{n-1}$ and $x''(t) = \sum_{n=0}^{\infty} n(n-1)a_n t^{n-2} = \sum_{n=2}^{\infty} n(n-1)a_n t^{n-2}$. Substituting the expressions for x and x'' into the equation, we obtain

$$\sum_{n=2}^{\infty} n(n-1)a_n t^{n-2} + \sum_{n=0}^{\infty} a_n t^n = 0.$$

To add the two series, we need to match the powers. For this we make $n = k + 2$ in the first series and $n = k$ in the second. This leads to the equation

$$\sum_{k=0}^{\infty}(k + 2)(k + 1)a_{k+2}t^k + \sum_{k=0}^{\infty}a_k t^k = 0,$$

which is equivalent to

$$\sum_{k=0}^{\infty}[(k + 2)(k + 1)a_{k+2} + a_k]t^k = 0,$$

an equation that must be satisfied for all t for which the series is convergent. Therefore all coefficients must vanish, so

$$(k + 2)(k + 1)a_{k+2} + a_k = 0, \qquad k = 0, 1, 2, \ldots.$$

This *recurrence relation* yields the coefficients

$$a_2 = -\frac{a_0}{2 \cdot 1} = -\frac{a_0}{2!}, \qquad a_3 = -\frac{a_1}{3 \cdot 2} = -\frac{a_1}{3!},$$
$$a_4 = -\frac{a_2}{4 \cdot 3} = \frac{a_0}{4!}, \qquad a_5 = -\frac{a_3}{5 \cdot 4} = \frac{a_1}{5!},$$
$$a_6 = -\frac{a_4}{6 \cdot 5} = -\frac{a_0}{6!}, \qquad a_7 = -\frac{a_5}{7 \cdot 6} = -\frac{a_1}{7!},$$

and so on, following an obvious pattern. We can thus write the solution as

$$x(t) = a_0 + a_1 t - \frac{a_0}{2!}t^2 - \frac{a_1}{3!}t^3 + \frac{a_0}{4!}t^4 + \frac{a_1}{5!}t^5 - \frac{a_0}{6!}t^6 - \frac{a_1}{7!}t^7 + \cdots$$

$$= a_0\left(1 - \frac{1}{2!}t^2 + \frac{1}{4!}t^4 - \frac{1}{6!}t^6 + \cdots\right)$$

$$\quad + a_1\left(t - \frac{1}{3!}t^3 + \frac{1}{5!}t^5 - \frac{1}{7!}t^7 + \cdots\right)$$

$$= a_0\sum_{k=0}^{\infty}\frac{(-1)^k}{(2k)!}t^{2k} + a_1\sum_{k=0}^{\infty}\frac{(-1)^k}{(2k + 1)!}t^{2k+1}.$$

For the first series the ratio test shows that $r = \lim_{k\to\infty}|[(-1)^k(2k)!]/[(-1)^{k+1}(2k + 1)!]| = \lim_{k\to\infty}|-(2k + 1)| = \infty$. Since the same happens with the second series, we see that each series is convergent for all real values of t. In fact, they are the Taylor series of $\cos t$ and $\sin t$, respectively, so we can write the general solution as $x(t) = a_0\cos t + a_1\sin t$, a result that could have been directly obtained using the method of Section 3.2.

Notice that the truncation of the series offers a polynomial approximation of the solution. For example, the initial value problem $x'' + x = 0$, $x(0) = 1$, $x'(0) = 0$ has the solution $\varphi(t) = \cos t = \sum_{k=0}^{\infty}[(-1)^k/(2k)!]t^{2k}$. The polynomial approximations of degree 2, 4, and 6 are $\varphi_2(t) = 1 - t^2/2!$, $\varphi_4(t) = 1 - t^2/2! + t^4/4!$, and $\varphi_6 = 1 - t^2/2! + t^4/4! - t^6/6!$, respectively. The higher the degree, the better the approximation (see Figure 7.1.1).

Figure 7.1.1. Polynomial approximations of degree n for the solution $\varphi(t) = \cos t$ of the equation $x'' + x = 0$.

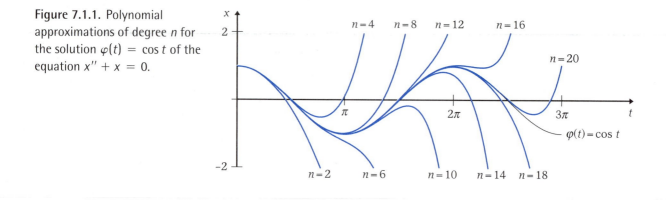

EXAMPLE 4   Let us solve the equation

$$(1 - t^2)x'' - 6tx' - 4x = 0 \tag{2}$$

near the ordinary point $t = 0$. Unlike the equation in Example 3, this one cannot be solved with the exact methods of Chapter 3. To use the power series method, we will check a solution of the form $x(t) = \sum_{n=0}^{\infty} a_n t^n$, so according to the above rules of differentiation $x'(t) = \sum_{n=0}^{\infty} n a_n t^{n-1}$ and $x''(t) = \sum_{n=0}^{\infty} n(n-1)a_n t^{n-2}$. Substituting x, x', and x'' into (2) yields

$$(1 - t^2)\sum_{n=0}^{\infty} n(n-1)a_n t^{n-2} - 6t\sum_{n=0}^{\infty} n a_n t^{n-1} - 4\sum_{n=0}^{\infty} a_n t^n = 0,$$

an equation equivalent to

$$\sum_{n=0}^{\infty} n(n-1)a_n t^{n-2} - \sum_{n=0}^{\infty} n(n-1)a_n t^n - \sum_{n=0}^{\infty} 6n a_n t^n - \sum_{n=0}^{\infty} 4a_n t^n = 0.$$

Adding the last three series, we obtain

$$\sum_{n=0}^{\infty} n(n-1)a_n t^{n-2} - \sum_{n=0}^{\infty} (n+1)(n+4)a_n t^n = 0.$$

To add the two remaining series we make $n = k$ in the first and $n = k - 2$ in the second. This leads to

$$\sum_{k=0}^{\infty} k(k-1)a_k t^{k-2} - \sum_{k=2}^{\infty} (k-1)(k+2)a_{k-2} t^{k-2} = 0,$$

which, since the first two coefficients of the first series are 0, is equivalent to

$$\sum_{k=2}^{\infty} k(k-1)a_n t^{k-2} - \sum_{k=2}^{\infty} (k-1)(k+2)a_{k-2} t^{k-2} = 0.$$

We can now write the equation as

$$\sum_{k=2}^{\infty} [k(k-1)a_k - (k-1)(k+2)a_{k-2}]t^{k-2} = 0.$$

Since this relation has to be satisfied for all t for which the series is convergent, each coefficient must cancel, so we are led to the recurrence relation

$$k(k-1)a_k - (k-1)(k+2)a_{k-2} = 0, \quad k = 2, 3, 4, \dots.$$

We thus successively obtain

$$a_2 = \frac{4}{2}a_0, \qquad a_3 = \frac{5}{3}a_1,$$

$$a_4 = \frac{6}{4}a_2 = 3a_0, \qquad a_5 = \frac{7}{5}a_3 = \frac{7}{3}a_1,$$

$$a_6 = \frac{8}{6}a_4 = 4a_0, \qquad a_7 = \frac{9}{7}a_5 = 3a_1,$$

or, in general,

$$a_{2k} = \frac{2k+2}{2k}a_{2k-2} = (k+1)a_0,$$

$$a_{2k+1} = \frac{2k+3}{2k+1}a_{2k-1} = \frac{2k+3}{3}a_1.$$

Since the even and odd coefficients are given by different formulas, we can write the series $\sum_{k=0}^{\infty} a_k t^k$ as a sum of two series,

$$\left(a_0 + \sum_{k=1}^{\infty} a_{2k}t^{2k}\right) + \left(a_1 t + \sum_{k=1}^{\infty} a_{2k+1}t^{2k+1}\right).$$

Then the general solution x of equation (2) has the form

$$x(t) = a_0\left(1 + \sum_{k=1}^{\infty}(k+1)t^{2k}\right) + a_1\left(t + \sum_{k=1}^{\infty}\frac{2k+3}{3}t^{2k+1}\right),$$

which can be written as

$$x(t) = a_0 \sum_{k=0}^{\infty}(k+1)t^{2k} + a_1 \sum_{k=0}^{\infty}\frac{2k+3}{3}t^{2k+1}.$$

Using the ratio test, we see that for the first series

$$\lim_{k \to \infty}\frac{k+1}{k+2} = 1$$

and for the second

$$\lim_{k \to \infty}\frac{(2k+3)/3}{(2k+5)/3} = 1,$$

so each series has radius of convergence 1. Thus the solution is convergent in the interval $(-1, 1)$. In fact, it can be shown (see Problem 33) that the first series converges to $1/(t^2-1)^2$ and the second to $(3t - t^3)/[3(t^2 - 1)^2]$, which are defined in the interval $(-1, 1)$.

EXAMPLE 5 To find a power series solution $x(t) = \sum_{n=0}^{\infty} a_n t^n$ of the equation

$$x'' = (t + 1)x, \tag{3}$$

we substitute $x(t)$ and $x''(t) = \sum_{n=0}^{\infty} n(n-1)a_n t^{n-2}$ into (3), and after performing computations similar to the ones in the previous examples, we are led to the recurrence relation

$$(n + 1)(n + 2)a_{n+2} = a_n + a_{n-1}.$$

Since we are interested in two linearly independent solutions of equation (3), we will find one by choosing $a_0 = 1, a_1 = 0$ and the other by taking $a_0 = 0, a_1 = 1$. (Why do these choices ensure linear independence?) For $a_0 = 1, a_1 = 0$ we recursively obtain

$$a_2 = \frac{1}{2}a_0 = \frac{1}{2},$$

$$a_3 = \frac{a_1 + a_0}{6} = \frac{1}{6},$$

$$a_4 = \frac{a_2 + a_1}{12} = \frac{1}{24},$$

$$a_5 = \frac{a_3 + a_2}{20} = \frac{1}{30},$$

$$a_6 = \frac{a_4 + a_3}{30} = \frac{1}{144},$$

so a solution of equation (3), linearly independent with x_1, has the form

$$x_1(t) = 1 + \frac{1}{2}t^2 + \frac{1}{6}t^3 + \frac{1}{24}t^4 + \frac{1}{30}t^5 + \frac{1}{144}t^6 + \cdots.$$

For $a_0 = 1, a_1 = 0$ we recursively obtain

$$a_2 = \frac{1}{2}a_0 = 0,$$

$$a_3 = \frac{a_1 + a_0}{6} = \frac{1}{6},$$

$$a_4 = \frac{a_2 + a_1}{12} = \frac{1}{12},$$

$$a_5 = \frac{a_3 + a_2}{20} = \frac{1}{120},$$

$$a_6 = \frac{a_4 + a_3}{30} = \frac{1}{120},$$

so another solution of (3), linearly independent with x_1, has the form

$$x_2(t) = t + \frac{1}{6}t^3 + \frac{1}{12}t^4 + \frac{1}{120}t^5 + \frac{1}{120}t^6 + \cdots.$$

Unlike in previous examples, we failed to express the coefficients a_n with a formula depending on n alone. Though we can recursively compute many more coefficients, we could never obtain them all. So instead of having a power series solution, we must content ourselves with a polynomial approximation. However, even this approximation is questionable in the absence of a proof that the power series is convergent in some interval about 0. But how do we check convergence when the series is not entirely known? We will answer this question in the next section.

PROBLEMS

In each of Problems 1 to 6 add the given series.

1. $\displaystyle\sum_{n=0}^{\infty} \frac{t^n}{(2n)!}, \quad \sum_{n=1}^{\infty} \frac{t^n}{n!}$

2. $\displaystyle\sum_{n=1}^{\infty} \frac{(n+1)t^n}{n^2}, \quad \sum_{n=0}^{\infty} \frac{n^2 t^n}{n+1}$

3. $\displaystyle\sum_{n=1}^{\infty} \frac{t^{2n}}{2n}, \quad \sum_{n=1}^{\infty} \frac{t^n}{n}$

4. $\displaystyle\sum_{n=0}^{\infty} \frac{2t^n}{(n+1)^2}, \quad \sum_{n=1}^{\infty} \frac{t^{3n}}{n!}$

5. $\displaystyle\sum_{n=0}^{\infty} \frac{t^n}{(2n)!}, \quad \sum_{n=0}^{\infty} \frac{t^{n+1}}{2^n}$

6. $\displaystyle\sum_{n=2}^{\infty} \frac{t^{n-1}}{n-1}, \quad \sum_{n=1}^{\infty} (-1)^n \frac{t^n}{n(n+1)}$

In each of Problems 7 to 12 multiply the given functions.

7. $f(t) = t, \quad g(t) = \displaystyle\sum_{n=0}^{\infty} \frac{t^n}{n!}$

8. $f(t) = \dfrac{1}{t}, \quad g(t) = \displaystyle\sum_{n=1}^{\infty} \frac{nt^n}{(n+1)(2n+1)}$

9. $f(t) = t^2, \quad g(t) = \displaystyle\sum_{n=0}^{\infty} \frac{2^n t^n}{(n+1)(n+2)}$

10. $f(t) = 2t + 1, \quad g(t) = \displaystyle\sum_{n=1}^{\infty} \frac{t^n}{n}$

11. $f(t) = 1 + t + t^2, \quad g(t) = \displaystyle\sum_{n=0}^{\infty} \frac{(-1)^n t^{2n}}{(2n)!}$

12. $f(t) = \displaystyle\sum_{n=1}^{\infty} (-1)^{n+1} \frac{t^{2n+1}}{(n+1)!},$

$g(t) = \displaystyle\sum_{n=0}^{\infty} \frac{(-1)^n t^{2n}}{(2n)!}$

In Problems 13 and 14 determine the coefficients a_n for all indicated values of n such that the corresponding identity takes place for all t.

13. $\displaystyle\sum_{n=0}^{\infty} na_n t^n + \sum_{n=1}^{\infty} a_n t^{n-1} = 0$

14. $\displaystyle\sum_{n=0}^{\infty} (n+1)a_n t^n + \sum_{n=2}^{\infty} a_n t^{n-2} = 0$

Determine the radius of convergence of each of the series in Problems 15 to 20.

15. $\displaystyle\sum_{n=0}^{\infty} (t-1)^n$

16. $\displaystyle\sum_{n=0}^{\infty} t^{3n}$

17. $\displaystyle\sum_{n=0}^{\infty} \frac{n}{3^n} t^n$

18. $\displaystyle\sum_{n=1}^{\infty} \frac{1}{n^2} (t+1)^n$

19. $\displaystyle\sum_{n=1}^{\infty} \frac{n!}{n^n} t^n$

20. $\displaystyle\sum_{n=0}^{\infty} \frac{(-1)^n n^3}{2^n} (2t+1)^n$

In Problems 21 to 26 obtain power series solutions of the given second-order equations near the ordinary point $t = 0$. Try to obtain the general term. In case you fail, find a polynomial approximation of degree 5.

21. $x'' = x$

22. $x'' + tx' + x = 0$

23. $x'' + tx' - x = 0$

24. $x'' + tx' - 2x = 0$

25. $x'' + tx' + 2x = 0$

26. $(1 + t)x'' + x = 0$

In Problems 27 to 32 obtain power series solutions for the given initial value problems near the ordinary point $t = 0$. Try to obtain the general term. In case you fail, find a polynomial approximation of degree 5.

27. $x'' = 4x, \quad x(0) = 1, \quad x'(0) = 2$

28. $x'' + x = 0, \quad x(0) = 1, \quad x'(0) = 0$

29. $x'' + tx' - 3x = 0, \qquad x(0) = 1, \; x'(0) = \frac{1}{2}$

30. $x'' + 2tx' + \frac{1}{2}x = 0, \qquad x(0) = x'(0) = 1$

31. $(t - 2)x'' + x = 0, \quad x(0) = 2, \; x'(0) = -\frac{1}{3}$

32. $(t - 1)x'' + 2x' + tx = 0, \quad x(0) = 0, \; x'(0) = 2$

33. Write the Taylor series of the functions
$f(t) = (t^2 - 1)^{-2}$ and $g(t) = \frac{1}{3}t(3 - t^2)(t^2 - 1)^{-2}$
and then draw the conclusion that

$$f(t) = \sum_{k=0}^{\infty} (k + 1)t^{2k}$$

and $g(t) = \displaystyle\sum_{k=0}^{\infty} \frac{2k + 3}{3} t^{2k+1}.$

34. Using the fact that $(\sin t)/t = \sum_{n=0}^{\infty} (-1)^n$ $[t^{2n}/(2n + 1)!]$ for all real t, find a power series solution of the equation

$$tx'' + (\sin t)x = 0.$$

35. Find the general power series solution of the equation

$$x'' = tx,$$

called *Airy's equation*, after the English mathematician and astronomer George Airy (1801–1892). Determine the radius

of convergence. Then show that for $t < 0$ the solutions are oscillatory and that for $t > 0$ they are monotonic.

36. Consider *Chebyshev's equation*,

$$(1 - t^2)x'' = tx' - \alpha^2 x,$$

named after the Russian mathematician Pafnuti Lvovich Chebyshev (1821–1894), where α is a constant. This equation is connected to the *Chebyshev polynomials*, which the Russian mathematician introduced in his investigations regarding the approximation of functions by polynomials. Find two linearly independent power series solutions that converge for $|t| < 1$.

37. Consider *Legendre's equation*,

$$(1 - t^2)x'' = 2tx' - \lambda(\lambda + 1)x,$$

named after the French mathematician Adrien-Marie Legendre (1752–1833), where $\lambda > 0$ is a constant. This equation is encountered in modeling certain physical phenomena in spherical coordinates. Find the polynomial approximation of degree 5 of two linearly independent solutions. Then take the particular cases $\lambda = 0, 1, 2, 3, 4, 5$ and see what you obtain.

7.2
Approximations

In this section we will find approximate power series solutions for equations whose exact solutions are difficult or impossible to obtain. We will first show how to determine the radius of convergence before finding the solution and present some examples in which this goal can be achieved. Then we will apply the power series method to analyze the bending of a flexible pole in pole vaulting, to estimate the expenditures of a foreign country in case of an arms race, to find the general solution of Hermite's equation, and to determine the Hermite polynomials, which have applications in physics.

Radius of Convergence

A formula to compute all the coefficients $a_1, a_2, \ldots, a_n, \ldots$ of a power series solution is usually hard to obtain because the recurrence relation is in most cases too complicated. So we must often content ourselves with the first few terms of the series, which offer only an approximate solution. In a way, this is similar to the method of successive approximations, studied in Section 2.5, for which every new step provides a better approximation of the solution to an initial value problem. The advantage of a power series approximation is that the method is not restricted to initial value problems. An initial difficulty, however, is that we cannot use the ratio test to check whether the series has a positive radius of convergence—a necessary condition for relying on the computed approximation. Fortunately, there is a way of finding this radius without knowing the solution. This is given by the following theorem, whose proof is beyond our scope.

Convergence Theorem

The radius of convergence of a power series solution $\sum_{n=0}^{\infty} a_n t^n$ of the equation $a(t)x'' + b(t)x' + c(t)x = 0$ around the ordinary point $t = 0$, where $a(t), b(t), c(t)$ are polynomials and $a(t)$ is nonconstant, is given by the distance in the complex plane from 0 to the closest real or complex root of the polynomial equation $a(t) = 0$. If this equation has no roots (when $a(t)$ is a nonzero constant, for example) then the radius of convergence is infinite.

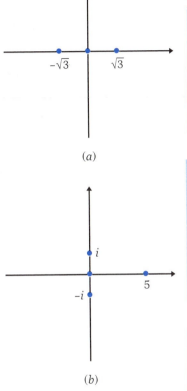

(a)

EXAMPLE 1 The radius of convergence of the power series solution of the equation

$$(t^2 - 3)x'' - 4tx' + x = 0$$

is given by the real or complex solution closest to 0 for the equation $t^2 - 3 = 0$. This equation has the roots $t_1 = -\sqrt{3}$ and $t_2 = \sqrt{3}$, both at distance $\sqrt{3}$ from 0 in the complex plane (see Figure 7.2.1(a)). So the radius of convergence is $r = \sqrt{3}$.

EXAMPLE 2 The radius of convergence of the power series solution of the equation

$$(t - 5)(t^2 + 1)x'' - (5t - 2)x' + 2tx = 0$$

is given by the real or complex solution closest to 0 for the equation $(t - 5)(t^2 + 1) = 0$. This equation has the roots $t_1 = 5, t_2 = -i$, and $t_3 = i$. Since the first root is at distance 5 from 0 and the second and the third are at distance 1, the radius of convergence is $r = 1$ (see Figure 7.2.1(b)).

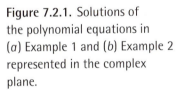

(b)

Figure 7.2.1. Solutions of the polynomial equations in (a) Example 1 and (b) Example 2 represented in the complex plane.

EXAMPLE 3 Recall the equation

$$x'' = (t + 1)x,$$

discussed in Example 5 of Section 7.1. Now we can conclude that the polynomial approximation of the solution found there is valid for all real t. This follows from the convergence theorem since $a(t) = 1$ in the above equation.

Applications

The pole vault   A model for analyzing the bending of a flexible pole in pole vaulting is given by the equation

$$\theta'' = k(L - x)\theta, \tag{1}$$

Pole vaulter.

where k is a positive constant that depends on certain physical constants of the pole (Young's modulus, linear density, and cross-sectional moment of inertia), on the gravitational acceleration, and on the the mass of the jumper; L is the length of the pole; x is the distance measured from the bottom along the pole; and $\theta(x)$ is the angular deflection from the vertical at the point of distance x (see Figure 7.2.2).

Equation (1) is linear and has a variable coefficient, so the power series method is suitable for solving it. Assume that a particular pole has the constants $k = 1$ and $L = 5$, so the equation to solve is

$$\theta'' = (5 - x)\theta. \tag{2}$$

First notice that since the coefficient of θ'' is 1, according to the convergence theorem a power series solution is convergent for every x. Then substitute $\theta(x) = \sum_{n=0}^{\infty} a_n x^n$ and $\theta''(x) = \sum_{n=2}^{\infty} n(n-1)a_n x^{n-2}$ into equation (2) and obtain

$$\sum_{n=2}^{\infty} n(n-1)a_n x^{n-2} = (5-x)\sum_{n=0}^{\infty} a_n x^n,$$

which is equivalent to

$$\sum_{n=2}^{\infty} n(n-1)a_n x^{n-2} = \sum_{n=0}^{\infty} 5a_n x^n - \sum_{n=0}^{\infty} a_n x^{n+1}.$$

Making $n = k+2$ in the first series, $n = k$ in the second, and $n = k-1$ in the third, we obtain

$$\sum_{k=0}^{\infty} (k+2)(k+1)a_{k+2} x^k = \sum_{k=0}^{\infty} 5a_k x^k - \sum_{k=1}^{\infty} a_{k-1} x^k,$$

which is equivalent to

$$2a_2 + \sum_{k=1}^{\infty} (k+2)(k+1)a_{k+2} x^k = 5a_0 + \sum_{k=1}^{\infty} (5a_k - a_{k-1})x^k.$$

This leads to the recurrence relation

$$a_2 = \frac{5}{2}a_0, \quad a_{k+2} = \frac{5a_k - a_{k-1}}{(k+1)(k+2)}, \quad k = 1, 2, 3, \ldots.$$

To obtain two linearly independent solutions of equation (2), we first choose $a_0 = 0, a_1 = 1$ and then $a_0 = 1, a_1 = 0$. The first choice leads to the coefficients $a_2 = 0$, $a_3 = \frac{5}{6}$, $a_4 = -\frac{1}{12}$, $a_5 = \frac{5}{24}$, $a_6 = -\frac{1}{24}$, $a_7 = \frac{3}{112}$, $a_8 = -\frac{5}{672}, \ldots$; and the second choice yields the coefficients $a_2 = \frac{5}{2}$, $a_3 = -\frac{1}{6}$, $a_4 = \frac{25}{24}$, $a_5 = -\frac{1}{6}$, $a_6 = \frac{43}{240}$, $a_7 = -\frac{5}{112}$, $a_8 = \frac{17}{896}, \ldots$. Thus the general solution of equation (2) has the form $c_1\theta_1(x) + c_2\theta_2(x)$, where c_1, c_2 are constants and

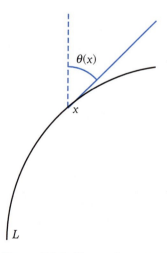

Figure 7.2.2. The mathematical elements needed for measuring the bending of the pole in pole vaulting.

$$\theta_1(x) = x + \frac{5}{6}x^3 - \frac{1}{12}x^4 + \frac{5}{24}x^5 - \frac{1}{24}x^6 + \frac{3}{112}x^7 - \frac{5}{672}x^8 + \cdots,$$

$$\theta_2(x) = 1 + \frac{5}{2}x^2 - \frac{1}{6}x^3 + \frac{25}{24}x^4 - \frac{1}{6}x^5 + \frac{43}{240}x^6 - \frac{5}{112}x^7 + \frac{17}{896}x^8 + \cdots.$$

Military exercises during an arms race.

This offers an approximation of the exact solution, good enough for measuring the bending of the pole at any desired point and for understanding its elasticity properties.

An arms race   The political conflict between two neighboring countries A and B escalates, and an arms race begins. Let x and y denote the military expenditures in tens of billions of dollars by A and B, respectively. The initial expenditure by each country was $10 billion, i.e., $x(0) = y(0) = 1$. Through information received from its spies, B knows that A started to increase its expenditures with a rate of change that equals the present expenditure of country B, i.e., $x' = y$. B would like to increase its expenditures proportionally in time with those of A, but knows that it will be unable to sustain this for long, so it will have to decrease them in proportion to its own expenditures. An equation that models their intention is $y' = 2tx - y + 1$. Upon receiving information from its intelligence service, A will adjust its expenditure according to the changes made by B. B is aware of this and would like to determine the expenditures of A for the next 6 months. The answer can be obtained by finding the x-solution in the interval $[0, \frac{1}{2}]$ for the initial value problem

$$\begin{cases} x' = y \\ y' = 2tx - y + 1 \end{cases}, \quad x(0) = y(0) = 1.$$

This is equivalent to

$$x'' = -x' + 2tx + 1, \quad x(0) = x'(0) = 1. \tag{3}$$

One coefficient is variable, so an exact solution seems difficult to obtain. A way of finding an approximate solution is to use power series. The advantage of this method is that, unlike numerical solutions, it offers a formula that can be used with any initial conditions.

Notice that since the coefficient of x'' is 1, the convergence theorem ensures the convergence of the power series solution $x(t) = \sum_{n=0}^{\infty} a_n t^n$ for all t. Substituting the corresponding $x, x',$ and x'' into equation (3), we obtain

$$\sum_{n=2}^{\infty} n(n-1)a_n t^{n-2} = -\sum_{n=1}^{\infty} na_n t^{n-1} + \sum_{n=0}^{\infty} 2a_n t^{n+1} + 1.$$

If we make the substitution $n = k + 2$ in the first series, $n = k + 1$ in the second, and $n = k - 1$ in the third, the equation becomes

$$\sum_{k=0}^{\infty} (k+2)(k+1)a_{k+2} t^k = 1 - \sum_{k=0}^{\infty}(k+1)a_{k+1} t^k + \sum_{k=1}^{\infty} 2a_{k-1} t^k.$$

This is equivalent to

$$2a_2 + \sum_{k=1}^{\infty}(k+2)(k+1)a_{k+2} t^k = 1 - a_1 + \sum_{k=1}^{\infty}[-(k+1)a_{k+1} + 2a_{k-1}]t^k,$$

which leads to the recurrence relation

$$a_2 = \frac{1-a_1}{2}, \quad a_{k+2} = \frac{2a_{k-1} - (k+1)a_{k+1}}{(k+1)(k+2)}, \quad k = 1, 2, 3\ldots.$$

To obtain two linearly independent solutions, first make the choice $a_0 = 0, a_1 = 1$ and then choose $a_0 = 1, a_1 = 0$. The first choice leads to the coefficients $a_2 = 0,$ $a_3 = 0,$ $a_4 = \frac{1}{6},$ $a_5 = -\frac{1}{30},$ $a_6 = \frac{1}{180},$ $a_7 = \frac{1}{140}, \ldots;$ and the second choice leads to $a_2 = \frac{1}{2},$ $a_3 = \frac{1}{6},$ $a_4 = -\frac{1}{24},$ $a_5 = \frac{7}{120},$ $a_6 = \frac{1}{720},$ $a_7 = -\frac{11}{5040}, \ldots.$ These provide the general solution $x(t) = c_1 x_1(t) + c_2 x_2(t),$ where c_1, c_2 are constants and

$$x_1(t) = t + \frac{1}{6}t^4 - \frac{1}{30}t^5 + \frac{1}{180}t^6 + \frac{1}{140}t^7 + \cdots,$$

$$x_2(t) = 1 + \frac{1}{2}t^2 + \frac{1}{6}t^3 - \frac{1}{24}t^4 + \frac{7}{120}t^5 + \frac{1}{720}t^6 - \frac{11}{5040}t^7 + \cdots.$$

Notice that $x(0) = c_2$ and $x'(0) = c_1$, so $c_1 = c_2 = 1$; thus the solution of the initial value problem (3) is $\phi(t) = x_1(t) + x_2(t)$, i.e.,

$$\phi(t) = 1 + x + \frac{1}{2}t^2 + \frac{1}{6}t^3 + \frac{1}{8}t^4 + \frac{1}{40}t^5 + \frac{1}{144}t^6 + \frac{5}{1008}t^7 + \cdots.$$

To find an approximate figure for the probable expenditures of country A after 6 months, we compute $\phi(\frac{1}{2}) \approx 1 + 0.5 + 0.125 + 0.0208333 + 0.0078125 + \cdots \approx 1.65$, which means that country A will spend at least \$6.5 billion over the initial \$10 billion. (Could the amount be much bigger than this lower estimate? See Problem 21.)

Hermite's equation   The second-order equation

$$x'' = 2tx' - 2\nu x \tag{4}$$

(where ν is a constant), called *Hermite's equation* after the French mathematician Charles Hermite (1822–1901), appears in quantum mechanics in connection with Schrödinger's equation—a partial differential equation for the harmonic oscillator. The physical reasons behind this connection are beyond our scope. The methods of Section 3.2 do not apply to equation (4), but we can solve it using power series solutions. According to the convergence theorem, a power series solution is convergent for all t. The usual substitution leads to the equation

$$\sum_{k=0}^{\infty}[(k+2)(k+1)a_{k+2} - 2(k-\nu)a_k]t^k = 0,$$

from which we find the recurrence relation

$$a_{k+2} = \frac{2(k-\nu)a_k}{(k+1)(k+2)}, \qquad k = 0, 1, 2, \ldots.$$

We thus successively obtain

$$a_2 = -\frac{2\nu}{1\cdot 2}a_0, \qquad\qquad a_3 = \frac{2(1-\nu)}{2\cdot 3}a_1,$$

$$a_4 = \frac{2(2-\nu)}{3\cdot 4}a_2, \qquad\qquad a_5 = \frac{2(3-\nu)}{4\cdot 5}a_3,$$

$$\vdots \qquad\qquad\qquad\qquad \vdots$$

$$a_{2r} = \frac{2(2r-2-\nu)}{2r(2r-1)}a_{2r-2}, \qquad a_{2r+1} = \frac{2(2r-1-\nu)}{2r(2r+1)}a_{2r-1},$$

which means that the coefficients are given by two formulas, one for even subscripts and another for odd ones:

$$a_{2r} = \frac{2^r(0-\nu)(2-\nu)\cdots(2r-2-\nu)}{(2r)!}a_0, \qquad r = 1, 2, 3, \ldots,$$

$$a_{2r+1} = \frac{2^r(1-\nu)(3-\nu)\cdots(2r-1-\nu)}{(2r+1)!}a_1, \qquad r = 1, 2, 3, \cdots.$$

Thus the general solution of Hermite's equation has the form $x(t) = c_1x_1(t) + c_2x_2(t)$, where $c_1 = a_0, c_2 = a_1,$ and

$$x_1(t) = 1 + \sum_{r=1}^{\infty}\frac{2^r(-\nu)(2-\nu)\cdots(2r-2-\nu)}{(2r)!}t^{2r},$$

$$x_2(t) = t + \sum_{r=1}^{\infty}\frac{2^r(1-\nu)(3-\nu)\cdots(2r-1-\nu)}{(2r+1)!}t^{2r+1}.$$

It is interesting to note that for ν nonnegative and even, x_1 is a polynomial, whereas for ν positive and odd, x_2 is a polynomial. More specifically,

$$\nu = 0: \qquad x_1(t) = 1$$
$$\nu = 1: \qquad x_2(t) = t$$
$$\nu = 2: \qquad x_1(t) = 1 - 2t^2$$
$$\nu = 3: \qquad x_2(t) = t - \frac{2}{3}t^3$$
$$\nu = 4: \qquad x_1(t) = 1 - 4t^2 + \frac{4}{3}t^4$$
$$\nu = 5: \qquad x_2(t) = t - \frac{4}{3}t^3 + \frac{4}{15}t^5$$

$$\vdots \qquad\qquad\qquad \vdots$$

The multiples of these polynomials for which the coefficient of t^n (where $\nu = n$ is the highest power) is 2^n are called *Hermite polynomials*. They are given by the formula

$$H_k(t) = k! \sum_{i=0}^{N} (-1)^i \frac{2^{k-2i}}{i!(k-2i)!} t^{k-2i},$$

where $N = k/2$ if k is even and $N = (k-1)/2$ if k is odd. The Hermite polynomials offer a convenient way to approximate the solution in several physical applications of Hermite's equation.

PROBLEMS

Determine the radius of convergence of the power series solutions for the equations in Problems 1 to 12.

1. $x'' + 4tx' - \frac{1}{3}x = 0$

2. $5u'' - 6t^2u' + 2tu = 0$

3. $(1 + x)v'' + (3 + x)v' - 7v = 0$

4. $(3 - 2\theta)^2 r'' + 2\theta r' - \frac{2t^2}{7}r = 0$

5. $(t^2 + 4)y'' - (\frac{1}{2}t + t^3)y' + 3ty = 0$

6. $y'' + \frac{1}{(x - 6)^2}y = 0$

7. $(z - 1)(z^2 + \frac{1}{4})w'' + 4zw' + (2z^5 + 1)w = 0$

8. $3tz'' + 5z' + z = 0$

9. $(2u - 1)^3 P'' + \frac{2}{u^2 + 1}P' + 2P = 0$

10. $r(r + 1)\chi'' + r\chi' = \chi$

11. $5f'' + \frac{g}{g^4 - 1}f' + \frac{1}{g^2 - 4}f = 0$

12. $m'' + xm' - \frac{1}{x^2 + 1}m = 0$

Find a polynomial approximation of degree 4 for the power series solutions of the equations in Problems 13 to 18. In each case determine an open interval in which the power series is convergent.

13. $x'' + (2t - 1)x' + x = 0$

14. $3x'' - 3tx' - \frac{3}{2}x = 0$

15. $(t^2 + 4)x'' + tx = 0$

16. $(t^2 - 2)x'' + (t + 1)x = 0$

17. $(t^2 + 9)x'' + tx' + x = 0$

18. $(t^2 - 9)x'' - tx' + tx = 0$

19. In the pole vaulting problem assume that the constants are $k = \frac{1}{2}$ and $L = 6$. What is a degree 8 polynomial approximation of a power series solution in this case?

20. The motion of a pendulum without friction in which the length of the rod varies in time is given by the second-order equation $\theta'' + (\cos t)\theta = 0$, where k is a constant. Find a polynomial approximation of degree 5 for the general power series solution of this equation. What is the physical interpretation of the possible solutions? (*Hint:* Use the fact that $\cos t = \sum_{n=0}^{\infty} (-1)^n t^{2n}/(2n)!$.)

21. In the arms race problem, compare the obtained power series solution for values of t in the interval $(0, \frac{1}{2})$ with the series $\sum_{n=0}^{\infty} 1/2^n$ and give an estimate of the error in taking a seventh-degree polynomial approximation.

22. In the arms race problem, if the expenditures in tens of billions of dollars are given by the initial value problem

$$\begin{cases} x' = \frac{4}{5}y \\ y' = \frac{5}{2}tx - \frac{2}{3}y + 1 \end{cases}, \quad x(0) = y(0) = 1,$$

what is a lower estimate of the seventh-degree polynomial approximation for the expected expenditures x of country A after 9 months?

23. Show that for every nonnegative integer α, the Chebyshev equation

$$(1 - t^2)x'' = tx' - \alpha^2 x$$

has a polynomial solution. (*Hint:* See Problem 36, Section 7.1.)

24. Show that *Gauss's hypergeometric equation,*

$$t(1 - t)x'' + [c - (a + b + 1)t]x' - abx = 0,$$

named after the German mathematician Carl Friedrich Gauss (1777–1855), has a solution of the form

$$x(t) = 1 + \frac{ab}{c}t + \frac{a(a + 1)b(b + 1)}{c(c + 1)}\frac{t^2}{2!}$$

$$+ \frac{a(a + 1)(a + 2)b(b + 1)(b + 2)}{c(c + 1)(c + 2)}\frac{t^3}{3!} + \cdots.$$

What is the radius of convergence of this power series?

7.3 Regular Singular Points

In the previous sections we obtained power series solutions in the neighborhood of regular points. But the method we used there fails near a singular point. In this section we will define regular singular points for second-order equations and present the method of Frobenius, which provides power series solutions near such points. In the end we will apply this method to *RLC* circuits with variable resistance and capacitance, which cannot be solved in general with other methods.

Regular Singular Points   Recall from Section 7.1 that the *singular points* of the equation

$$a(t)x'' + b(t)x' + c(t)x = 0, \tag{1}$$

where a, b, and c are polynomial functions, are the real roots of $a(t)$, i.e., the real solutions of the equation $a(t) = 0$. Such a solution t_0 is called a *regular singular point* if, after reduction to lowest terms, $t - t_0$ appears at the denominator of $b(t)/a(t)$ to the power 1 at most and at the denominator of $c(t)/a(t)$ to the power 2 at most. Notice that for such points the convergence theorem of Section 7.2 does not apply.

EXAMPLE 1   The equation $(t^2 - 1)x'' + tx' - (t + 1)^3x = 0$ can be rewritten as

$$x'' + \frac{t}{(t - 1)(t + 1)}x' - \frac{(t + 1)^2}{t - 1}x = 0,$$

which shows that both 1 and -1 are regular singular points.

EXAMPLE 2   The equation $t^2(t + 3)^2x'' - tx' + 5x = 0$ can be rewritten as

$$x'' - \frac{1}{t(t + 3)^2}x' + \frac{1}{t^2(t + 3)^2}x = 0,$$

which shows that 0 is a regular singular point but -3 is not.

The method providing the solution near regular singular points was published in 1878 by the German mathematician Ferdinand Georg Frobenius (1849–1917) and is based on an idea that goes back to Euler.

Notice first that if $t_0 \neq 0$ is a regular singular point of equation (1), we can always use the change of variable $\tau = t - t_0$ to transform the given equation into one having 0 as a regular singular point.

EXAMPLE 3   The study of the equation $(t + 2)x'' + tx' - 3x = 0$ near the regular singular point -2 can be reduced by the change of variable $\tau = t + 2$ to that of the equation $\tau y'' + (\tau - 2)y' - 3y = 0$ near the regular singular point $\tau = 0$ (where $y(\tau) = x(\tau - 2) = x(t)$). For convenience, we can then rewrite the new equation in terms of x and t as $tx'' + (t - 2) \times x' - 3x = 0$.

The Method of Frobenius

The idea of this method is to check in the equation near $t = 0$ a solution of the form $\sum_{n=0}^{\infty} a_n t^{n+r}$ and then to determine r and $a_1, a_2, \ldots$, which depend on r, such that we obtain two linearly independent solutions.

Step 1. Find the regular singular points of equation (1) and decide near which one you want to obtain a solution. If this point is $t_0 \neq 0$, use the change of variable $\tau = t - t_0$ to make your regular singular point 0.

Step 2. Check a solution of the form $x(t) = \sum_{n=0}^{\infty} a_n t^{n+r}$. This leads to a quadratic equation in r, called an *indicial equation*, and to a recurrence relation involving r and the coefficients $a_1, a_2, \ldots$.

Step 3. Take $a_0 = 1$, solve the indicial equation, and obtain the real solutions $r_1 \geq r_2$ (the case of nonreal solutions is beyond our scope). Then introduce r_1 in the recurrence relation and determine the coefficients $a_1(r_1), a_2(r_1), \ldots$ that give a solution of the form

$$x_1(t) = t^{r_1} + \sum_{n=1}^{\infty} a_n(r_1)t^{n+r_1},$$

convergent in some interval $(0, \rho)$, for $\rho > 0$.

Step 4. If $r_1 - r_2$ is an integer (including 0), go to Step 5. If $r_1 - r_2$ is not an integer, introduce r_2 into the recurrence relation and determine the coefficients $a_1(r_1), a_2(r_2), \ldots$ that give a second (linearly independent) solution

$$x_2(t) = t^{r_2} + \sum_{n=1}^{\infty} a_n(r_2)t^{n+r_2},$$

convergent at least in $(0, \rho)$.

Step 5. If $r_1 - r_2$ is an integer, use the reduction of order of Section 3.2 to obtain a second linearly independent solution.

Remark. If $r_1 - r_2$ is an integer, there is a way of finding a second linearly independent power series solution, but it leads to tedious, sometimes insurmountable, computations. Another way is to use the reduction of order studied in Section 3.2. But this is also difficult to complete unless we seek only an approximate solution. Because these aspects are beyond our introductory scope, we will deal here only with the case where $r_1 - r_2$ is not an integer.

EXAMPLE 4 Let us use the method of Frobenius to obtain a power series solution for

$$3tx'' + x' - x = 0. \tag{2}$$

Step 1. Obviously, $t = 0$ is the only regular singular point.

Step 2. Substitute $x(t) = \sum_{n=0}^{\infty} a_n t^{n+r}$, $x'(t) = \sum_{n=0}^{\infty} (n + r)a_n t^{n+r-1}$, and $x''(t) = \sum_{n=0}^{\infty} (n+r)(n+r-1)a_n t^{n+r-2}$ into equation (2) and obtain

$$\sum_{n=0}^{\infty} 3(n + r)(n + r - 1)a_n t^{n+r-1} + \sum_{n=0}^{\infty} (n + r)a_n t^{n+r-1} - \sum_{n=0}^{\infty} a_n t^{n+r} = 0,$$

which is equivalent to

$$t^r \left[r(3r - 2)a_0 t^{-1} + \sum_{n=1}^{\infty} (n + r)(3n + 3r - 2)a_n t^{n-1} - \sum_{n=0}^{\infty} a_n t^n \right] = 0.$$

Making $n = k + 1$ in the first series and $n = k$ in the second, we obtain

$$t^r \left[r(3r - 2)a_0 t^{-1} + \sum_{k=0}^{\infty} ((k + r + 1)(3k + 3r + 1)a_{k+1} - a_k)t^k \right] = 0.$$

Since the above equation must be satisfied for all t for which the series are convergent, we write the indicial equation $r(3r - 2)a_0 = 0$ and the recurrence relation

$$a_{k+1} = \frac{a_k}{(k + r + 1)(3k + 3r + 1)}, \qquad k = 0, 1, 2, \ldots.$$

Step 3. We now take $a_0 = 1$, so the indicial equation has the solutions $r_1 = \frac{2}{3}$ and $r_2 = 0$. Substituting $r_1 = \frac{2}{3}$ into the recurrence relation, we obtain

$$a_{k+1} = \frac{a_k}{(k + 1)(3k + 5)}, \qquad k = 0, 1, 2, \ldots,$$

which provides the coefficients

$$a_1 = \frac{1}{1 \cdot 5} = \frac{1}{1! \cdot 5}$$

$$a_2 = \frac{a_1}{2 \cdot 8} = \frac{1}{2! \cdot 5 \cdot 8}$$

$$a_3 = \frac{a_2}{3 \cdot 11} = \frac{1}{3! \cdot 5 \cdot 8 \cdot 11}$$

$$\vdots$$

$$a_n = \frac{1}{n! \cdot 5 \cdot 8 \cdot 11 \cdots (3n + 2)},$$

and consequently the solution

$$x_1(t) = t^{2/3} + \sum_{n=1}^{\infty} \frac{1}{n! \cdot 5 \cdot 8 \cdot 11 \cdots (3n + 2)} t^{n+2/3}.$$

This can be written as

$$x_1(t) = t^{2/3}\left(1 + \sum_{n=1}^{\infty} \frac{1}{n! \cdot 5 \cdot 8 \cdot 11 \cdots (3n+2)} t^n\right),$$

a series that, according to the ratio test, is convergent for all $t > 0$.

Step 4. Since $r_1 - r_2 = \frac{2}{3}$ is not an integer, we substitute $r_2 = 0$ into the recurrence relation and obtain

$$a_{k+1} = \frac{a_k}{(k+1)(3k+1)}, \quad k = 0, 1, 2, \ldots,$$

with $a_0 = 1$, which successively leads to the coefficients

$$a_1 = \frac{1}{1 \cdot 1} = \frac{1}{1! \cdot 1}$$

$$a_2 = \frac{a_1}{2 \cdot 4} = \frac{1}{2! \cdot 1 \cdot 4}$$

$$a_3 = \frac{a_2}{3 \cdot 7} = \frac{1}{3! \cdot 1 \cdot 4 \cdot 7}$$

$$\vdots$$

$$a_n = \frac{1}{n! \cdot 1 \cdot 4 \cdot 7 \cdots (3n-2)}.$$

This means that a second linearly independent solution is

$$x_2(t) = 1 + \sum_{n=1}^{\infty} \frac{1}{n! \cdot 1 \cdot 4 \cdot 7 \cdots (3n-2)} t^n,$$

which by the ratio test is convergent for $t > 0$. Thus the general solution of equation (2) has the form $c_1 x_1(t) + c_2 x_2(t)$, where c_1 and c_2 are constants, and is valid for all $t > 0$.

EXAMPLE 5 To find a power series solution of the equation

$$2tx'' + (t+1)x' + x = 0 \tag{3}$$

near the regular singular point $t = 0$, substitute $x(t) = \sum_{n=0}^{\infty} a_n t^{n+r}$ and the corresponding $x'(t)$ and $x''(t)$ into equation (3) and obtain

$$\sum_{n=0}^{\infty} 2(n+r)(n+r-1)a_n t^{n+r-1} + \sum_{n=0}^{\infty}(n+r)a_n t^{n+r}$$

$$+ \sum_{n=0}^{\infty}(n+r)a_n t^{n+r-1} + \sum_{n=0}^{\infty} a_n t^{n+r} = 0,$$

which is equivalent to the equation

$$t^r\left[r(2r-1)a_0 t^{-1} + \sum_{n=1}^{\infty}(n+r)(2n+2r-1)a_n t^{n-1}\right.$$

$$\left. + \sum_{n=1}^{\infty}(n+r+1)a_n t^n\right] = 0.$$

Making $n = k + 1$ in the first series and $n = k$ in the second, we obtain the equation

$$t^r \left[r(2r - 1)a_0 t^{-1} + \sum_{k=0}^{\infty} (k + r + 1) \right.$$

$$\left. \times ((2k + 2r + 1)a_{k+1} + (k + r + 1)a_k)t^k \right] = 0.$$

With the choice $a_0 = 1$, this leads to the indicial equation $r(2r - 1) = 0$ and to the recurrence relation

$$a_{k+1} = -\frac{1}{2k + 2r + 1} a_k, \quad k = 0, 1, 2, \ldots.$$

The indicial equation has two solutions, $r_1 = \frac{1}{2}$ and $r_2 = 0$. For $r_1 = \frac{1}{2}$ the recurrence relation becomes $a_{k+1} = -a_k/[2(k + 1)]$, and since $a_0 = 1$, it leads to the coefficients

$$a_1 = -\frac{1}{2 \cdot 1} = -\frac{1}{2^1 \cdot 1!},$$

$$a_2 = -\frac{a_1}{2 \cdot 2} = \frac{1}{2^2 \cdot 2!},$$

$$a_3 = -\frac{a_2}{2 \cdot 3} = -\frac{1}{2^3 \cdot 3!},$$

$$\vdots$$

$$a_n = \frac{(-1)^n}{2^n n!}.$$

Thus, according to Step 3 of the algorithm, one solution of equation (3) is given by

$$x_1(t) = \sum_{n=0}^{\infty} \frac{(-1)^n}{2^n n!} t^{n+1/2}.$$

This can also be written as

$$x_1(t) = t^{1/2} \sum_{n=0}^{\infty} \frac{(-1)^n}{2^n n!} t^n,$$

which due to the factor $t^{1/2}$ makes no sense for $t < 0$, but by the ratio test is convergent for all $t > 0$.

To find a second linearly independent solution, take $r_2 = 0$ into the recurrence relation, which becomes $a_{k+1} = -a_k/(2k + 1)$ and for $a_0 = 1$ leads to

$$a_1 = -\frac{1}{1},$$

$$a_2 = -\frac{a_1}{3} = \frac{1}{1 \cdot 3},$$

$$a_3 = -\frac{a_2}{5} = -\frac{1}{1 \cdot 3 \cdot 5},$$

$$\vdots$$

$$a_n = \frac{(-1)^n}{1 \cdot 3 \cdot 5 \cdots (2n - 1)}.$$

So a second solution of equation (3) has the form

$$x_2(t) = 1 + \sum_{n=1}^{\infty} \frac{(-1)^n}{1 \cdot 3 \cdot 5 \cdots (2n-1)} t^n$$

and converges for all t. Thus the general solution of equation (3) is $x(t) = c_1 x_1(t) + c_2 x_2(t)$ and converges for all $t > 0$.

Applications

An *RLC* circuit with variable resistance and capacitance

In Section 3.4 we saw that the charge q in the plates of an *RLC* circuit is described by the equation

$$q'' = -\frac{R}{L} q' - \frac{1}{LC} q,$$

where R is the resistance of the resistor, L the inductance of the inductor, and C the capacitance of the capacitor. Assume now that the resistance changes in time: It is very high in the beginning and then decreases like $1/(2t)$. It is known that the capacitance varies linearly with the distance between the plates. Assume therefore that the plates move slowly and uniformly in time such that $C(t) = t$. If the inductance is $L = \frac{1}{3}$ H (henry), the equation describing the charge q of such an *RLC* circuit is $q'' + (1/2t)q' + (3/t)q = 0$, or

$$tq'' + \frac{1}{2} q' + 3q = 0. \tag{4}$$

Obviously, $t = 0$ is a regular singular point. We are interested in determining q for the first hour for various initial conditions, and for this we need to find the general solution, or an approximation of it, in the interval $(0, 1)$. Substituting $x(t) = \sum_{n=0}^{\infty} a_n t^{n+r}$ and the corresponding $x'(t)$ and $x''(t)$ into equation (4), we obtain

$$\sum_{n=0}^{\infty} (n+r)(n+r-1) a_n t^{n+r-1} + \sum_{n=0}^{\infty} \frac{1}{2}(n+r) a_n t^{n+r-1} + \sum_{n=0}^{\infty} 3 a_n t^{n+r} = 0,$$

which is equivalent to

$$t^r \left[r\left(r - \frac{1}{2}\right) a_0 t^{-1} + \sum_{n=1}^{\infty} (n+r)\left(n+r-\frac{1}{2}\right) a_n t^{n-1} + \sum_{n=0}^{\infty} 3 a_n t^n \right] = 0.$$

Making the substitutions $a_0 = 1$, $n = k+1$ in the first series and $n = k$ in the second, the equation becomes

$$t^r \left[r\left(r - \frac{1}{2}\right) t^{-1} + \sum_{k=0}^{\infty} (k+1+r)\left(k+r+\frac{1}{2}\right) a_{k+1} t^k + \sum_{k=0}^{\infty} 3 a_k t^k \right] = 0.$$

From this we obtain the indicial equation $r(r - \frac{1}{2}) = 0$, with solutions $r_1 = \frac{1}{2}$ and $r_2 = 0$, and the recurrence relation

$$a_{k+1} = -\frac{3a_k}{(k+r+1)(k+r+\frac{1}{2})}.$$

For $r = \frac{1}{2}$ this relation becomes

$$a_{k+1} = -\frac{6a_k}{(k+1)(2k+3)}, \qquad k = 0, 1, 2, \ldots,$$

which successively provides

$$a_1 = -\frac{6a_0}{1 \cdot 3} = \frac{(-1)^1 \cdot 6^1}{1! \cdot 3} = -2,$$

$$a_2 = -\frac{6 \cdot (-2)}{2 \cdot 5} = \frac{(-1)^2 \cdot 6^2}{2! \cdot 3 \cdot 5} = \frac{6}{5},$$

$$a_3 = -\frac{6 \cdot \frac{6}{5}}{3 \cdot 7} = \frac{(-1)^3 \cdot 6^3}{3! \cdot 3 \cdot 5 \cdot 7},$$

$$\vdots$$

$$a_n = \frac{(-1)^n 6^n}{n! \cdot 3 \cdot 5 \cdot 7 \cdots (2n+1)}, \qquad n = 1, 2, 3, \ldots.$$

Thus one solution has the form

$$x_1(t) = t^{1/2}\left[1 + \sum_{n=1}^{\infty} \frac{(-1)^n 6^n}{n! \cdot 3 \cdot 5 \cdot 7 \cdots (2n+1)} t^n\right]$$

and according to the ratio test is convergent for all $t > 0$—in particular, for all t in the interval $(0, 1)$.

For $r = 0$ the recurrence relation becomes

$$a_{k+1} = -\frac{6a_k}{(k+1)(2k+1)}, \quad k = 0, 1, 2, \ldots,$$

which successively provides

$$a_1 = -\frac{6a_0}{1 \cdot 1} = \frac{(-1)^1 \cdot 6^1}{1!} = -6,$$

$$a_2 = -\frac{6 \cdot (-6)}{2 \cdot 3} = \frac{(-1)^2 \cdot 6^2}{2! \cdot 3} = 6,$$

$$a_3 = -\frac{6 \cdot 6}{3 \cdot 5} = \frac{(-1)^3 \cdot 6^3}{3! \cdot 3 \cdot 5},$$

$$\vdots$$

$$a_n = \frac{(-1)^n 6^n}{n! \cdot 1 \cdot 3 \cdot 5 \cdots (2n-1)}, \qquad n = 1, 2, 3, \ldots.$$

The second linearly independent solution therefore has the form

$$x_2(t) = 1 + \sum_{n=1}^{\infty} \frac{(-1)^n 6^n}{n! \cdot 1 \cdot 3 \cdot 5 \cdots (2n-1)} t^n.$$

The general solution is $x(t) = c_1 x_1(t) + c_2 x_2(t)$, where c_1, c_2 are constants, and is convergent for all t in the interval of interest, $(0, 1)$.

PROBLEMS

In Problems 1 to 6 use a change of variable to transform the given equation into one that has the singular point at 0.

1. $(1 - t)x'' + 3x' + x = 0$

2. $(2t - 1)u'' + tu' + \frac{1}{3}u = 0$

3. $(x + 5)w'' + 3xw' + xw = 0$

4. $(5z + 1)^2\theta'' + \theta' - 2\theta = 0$

5. $\dfrac{\theta + 1}{\theta - 1}r'' - 2\theta^2 r' + r = 0$

6. $(1 - 4s)v'' + sv = 0$

In Problems 7 to 12 determine whether 0 is a regular singular point.

7. $t^3x'' + 3t^2x' + 5x = 0$

8. $2su'' - \dfrac{1}{(s + 1)^2}u = 0$

9. $y'' + x^{-1}y' + (x + 1)^{-2}y = 0$

10. $z(z^2 + 1)w'' - 2w = 0$

11. $-10\phi v'' + \frac{2}{3}\phi v' - \frac{3}{4}\phi^2 v = 0$

12. $v^2(2v + 1)U'' + vU' + U = 0$

In Problems 13 to 24 use the method of Frobenius to obtain two linearly independent power series solutions near the regular singular point 0. Find in each case an interval of convergence.

13. $2t^2x'' + t(t - 1)x' + x = 0$

14. $2tu'' + \frac{3}{2}u' + u = 0$

15. $4tv'' + 5v' + tv = 0$

16. $3xy'' + (2 - x)y' + 2y = 0$

17. $3z^2w'' - zw' - 4w = 0$

18. $t^2\theta'' + (\frac{2}{9} - t)\theta = 0$

19. $r(r + 1)s'' + \frac{1}{2}s' - 2s = 0$

20. $9v^2x'' + 9v^2x' + 2x = 0$

21. $2sq'' - (3 + 2s)q' + 2q = 0$

22. $v^2y'' - vy' + (2v + \frac{5}{9})y = 0$

23. $2\theta^2r'' - \theta r' + (\theta - 5)r = 0$

24. $x'' + \dfrac{u + 1}{2u}x' + \dfrac{3}{2u}x = 0$

25. Explain why the method of Frobenius cannot be applied to the equation

$$t^4x'' + 2t^3x' - x = 0.$$

Then find a solution by assuming that $x(t) = \sum_{n=0}^{\infty} a_n t^{-n}$.

26. Explain why the method of Frobenius cannot be applied to the equation

$$t^3x'' - tx' + x = 0.$$

Find a and b such that $x_1(t) = at + b$ is a solution of this equation. With the help of this solution, find a second linearly independent solution of the equation. Is this a power series solution?

27. Assume that the charge q of an *RLC* circuit with variable resistance and capacitance is given by the equation

$$tq'' + \frac{1}{5}q' + 2q = 0.$$

Use the method of Frobenius to find the general power series solution of this equation.

28. Assume that the charge q of an *LC* circuit with variable capacitance is given by the equation

$$q'' + \frac{2}{3t}q = 0.$$

Find one power series solution of this equation.

29. The *Bessel equation*,

$$t^2x'' + tx' + (t^2 - \nu^2)x = 0,$$

named after the German astronomer Friedrich Wilhelm Bessel (1784–1846), where ν is a constant, was first studied by the Swiss mathematician Jakob Bernoulli (1654–1705) in connection with the oscillatory behavior of a hanging chain. Bessel arrived at it later on in dealing with planetary motion.

(a) In case $\nu = 0$, show that one solution for $t > 0$ is

$$J_0(t) = 1 + \sum_{n=1}^{\infty}(-1)^n \frac{t^{2n}}{2^{2n}(n!)^2},$$

called the *Bessel function of the first kind of order zero.*

(b) In case $\nu = 1$, show that one solution for $t > 0$ is

$$J_1(t) = \frac{t}{2} \sum_{n=0}^{\infty} (-1)^n \frac{t^{2n}}{n!(n+1)!2^{2n}},$$

called the *Bessel function of the first kind of order 1*.

(c) Find one power series solution of Bessel's equation.

30. The *Laguerre equation*,

$$tx'' + (1-t)x' + \lambda x = 0,$$

named after the French mathematician Edmond Laguerre (1834–1886), where λ

is a constant, came up in connection with *Laguerre's polynomials*, which appear as solutions if λ is a positive integer. Find one power series solution of this equation.

31. Consider the second-order equation

$$t^2 x'' + \lambda t x' + \nu x = 0, \quad \text{for } t > 0,$$

where λ and ν are constants. Using the reduction of order of Section 3.2, show that if the solutions of the corresponding indicial equation are equal, say, to r, then $x_1(t) = t^r$ and $x_2(t) = t^r \ln t$ are solutions of the given equation.

7.4 **Computer Applications**

In this section we will use Maple, *Mathematica*, and MATLAB to obtain approximate power series solutions for differential equations and initial value problems near ordinary and regular singular points. In the latter case we provide programs that apply the method of Frobenius.

Maple

Maple is endowed with a package that deals with power series, which can be called with the command `with(powseries)`. This package makes Maple much better suited than *Mathematica* or MATLAB for solving differential equations with the help of power series. Within this package we can use the command `powsolve` to obtain the first few terms of a power series solution for linear differential equations and initial value problems of any order.

Solutions near ordinary points

Near ordinary points we can use the commands `powsolve` and `tpsform`. The first solves the equation using the power series method; the second provides a truncated power series form of the solution.

> **EXAMPLE 1** An approximate power series solution to a simple linear equation like $x' = 2x$ can be obtained as follows.
>
> ```
> > with(powseries):
> > gensol:=powsolve(diff(x(t),t)=2*x(t)):
> > tpsform(gensol, t);
> ```
>
> and Maple displays its answer:
>
> $$C0 + 2C0t + 2C0t^2 + \frac{4}{3}C0t^3 + \frac{2}{3}C0t^4 + \frac{4}{15}C0t^5 + O(t^6)$$
>
> where $C0$ is a real constant and $O(t^6)$ means terms of order 6 or higher. We recognize here the first terms of the Taylor series expansion about 0 of the function $c_0 e^{2t}$, which is the exact solution of the above equation.

In case of initial value problems, the initial condition must be taken at $t = 0$. Therefore if an initial condition is given at $t = t_0$, the equation must first be transformed with the help of the change of variable $\tau = t - t_0$ into one that has the initial condition at $\tau = 0$. The command `powsolve` is also used for initial value problems.

EXAMPLE 2 We can obtain an approximate power series solution of the initial value problem $x' = tx$, $x(0) = 1$ as follows:

```
> with(powseries):
> sol:=powsolve({diff(x(t),t)=t*x(t),
  x(0)=1}):
> tpsform(sol, t);
```

Maple displays its answer:

$$1 + \frac{1}{2}t^2 + \frac{1}{8}t^4 + O(t^6)$$

where $O(t^6)$ means terms of order 6 or higher. Again, this approximation represents the first few terms of the Taylor series expansion about 0 of the exact solution.

Let us see now how Maple deals with second-order equations and initial value problems. The step to higher-order equations will be obvious. Again, the initial conditions must be set at $t = 0$.

EXAMPLE 3 An approximate power series solution of the second-order equation $x'' + 3t^2x' - 2x = 0$ can be obtained as follows:

```
> with(powseries):
> gensol:=powsolve(diff(x(t),t$2)+3*t^2*diff(x(t),t)
  -2*x(t)=0):
> tpsform(gensol, t);
```

Maple displays the general solution

$$C0 + C1t + C0t^2 + \frac{1}{3}C1t^3 + \left(\frac{1}{6}C0 - \frac{1}{4}C1\right)t^4$$

$$+ \left(\frac{1}{30}C1 - \frac{3}{10}C0\right)t^5 + O(t^6)$$

where $C0$ and $C1$ are real constants and $O(t^6)$ represents terms of order 6 or higher.

EXAMPLE 4 An approximate power series solution of the initial value problem $x'' - 5x' + (t - 1)x = 0$, $x(0) = 1$, $x'(0) = -\frac{1}{2}$ can be obtained as follows:

```
> with(powseries):
> sol:=powsolve({diff(x(t),t$2)-5*diff(x(t),t)
  +(t-1)*x(t)=0, x(0)=1, D(x)(0)=-1/2}):
> tpsform(sol, t);
```

Maple displays the solution

$$1 - \frac{1}{2}t - \frac{3}{4}t^2 - \frac{3}{2}t^3 - \frac{91}{48}t^4 - \frac{29}{15}t^5 + O(t^6)$$

where $O(t^6)$ represents terms of order 6 or higher.

There is a way of avoiding the with(powseries) package—by using the dsolve command with the specification series. Instead of constants, however, this procedure will express the solution in terms of $x(0)$ and $x'(0)$.

EXAMPLE 5 With the help of the dsolve command, the second-order equation in Example 3 can be solved as follows:

```
> eq:=diff(x(t),t$2)+3*t^2*diff(x(t),t)-2*x(t)=0:
> dsolve(eq, x(t), series);
```

Maple displays the answer

$$x(t) = x(0) + D(x)(0)t + x(0)t^2 + \frac{1}{3}D(x)(0)t^3$$
$$+ \left(\frac{1}{6}x(0) - \frac{1}{4}D(x)(0)\right)t^4 + \left(-\frac{3}{10}x(0) + \frac{1}{30}D(x)(0)\right)t^5 + O(t^6)$$

where $O(t^6)$ means terms of order 6 or higher. In case you would like the equation displayed after writing it, replace the colon at the end of the first command with a semicolon.

The method of Frobenius Unfortunately, Maple is unable to directly apply the method of Frobenius for obtaining power series solutions near regular singular points. The attempt to use powsolve for an equation like $tx'' + (t + 1)x' + 2x = 0$ leads to an error answer. So to use the method, we need to write a special program. A possibility is suggested in the following example.

EXAMPLE 6 To find an approximate power series solution for the equation $2tx'' + (t + 1)x' + x = 0$ (already solved in Example 3 of Section 7.3) for, say, terms up to order 5 (i.e., a polynomial of degree 5), we proceed as follows.
First we use the command array(0..5), which creates the vector of the coefficients $(a_0, a_1, a_2, a_3, a_4, a_5)$. Since the above equation has $r_1 = \frac{1}{2}$ and $r_2 = 0$ as solutions of the indicial equation, with the help of this vector we then define x, the degree 5 polynomial corresponding to the solution $r_1 = \frac{1}{2}$, which Maple displays as follows.

```
> a:=array(0..5):
> x:=t^(1/2)*sum(a[n]*t^n,n=0..5);
```

$$x := \sqrt{t}(a_0 + a_1 t + a_2 t^2 + a_3 t^3 + a_4 t^4 + a_5 t^5)$$

We then ask Maple to substitute the polynomial x into the expression defining the equation and simplify this expression, which Maple displays as follows.

```
> eq1:=simplify(2*t*diff(x,t$2)+(t+1)*diff(x,t)+x);
```

$$eq1 := \frac{1}{2}\sqrt{t}(3a_0 + 5a_1 t + 7a_2 t^2 + 9a_3 t^3 + 11a_4 t^4 + 13a_5 t^5$$
$$+ 6a_1 + 20a_2 t + 42a_3 t^2 + 72a_4 t^3 + 110a_5 t^4)$$

We then use the command `collect` to group the coefficients having the same power. Maple displays the new equation, shown below.

```
> eq2:=collect(t^(-1/2)*eq1,t);
```

$$eq2 := \frac{13}{2}a_5 t^5 + \left(\frac{11}{2}a_4 + 55a_5\right)t^4 + \left(36a_4 + \frac{9}{2}a_3\right)t^3$$
$$+ \left(21a_3 + \frac{7}{2}a_2\right)t^2 + \left(\frac{5}{2}a_1 + 10a_2\right)t + \frac{3}{2}a_0 + 3a_1$$

We then form the algebraic system that makes the coefficients of every power of t equal to 0. For this we first redefine the new coefficients of the series as $(b_0, b_1, b_2, b_3, b_4, b_5)$ with the help of the command `do`. Then we form the system with the help of the command `convert` with the specification `set`, which displays the equations in the form of a set. (Why do we go from 0 to 4 and not from 0 to 5?)

```
> b:=array(0..4):
> for n from 0 to 4 do b[n]:=coeff(eq2,t,n)=0: od:
> sys:=convert(b,set);
```

$$sys := \left\{ \frac{3}{2}a_0 + 3a_1 = 0, \frac{5}{2}a_1 + 10a_2 = 0, 21a_3 + \frac{7}{2}a_2 = 0, \right.$$
$$\left. 36a_4 + \frac{9}{2}a_3 = 0, \frac{11}{2}a_4 + 55a_5 = 0 \right\}$$

We further need to tell Maple to consider the set of unknowns a_1, a_2, a_3, a_4, a_5, which we would like to obtain with respect to a_0. For this we again use the command `convert` with the specification `set`. Maple displays this set as follows.

```
> unknowns:=convert([seq(a[n],n=1..5)],set);
```

$$unknowns := \{a_5, a_1, a_2, a_3, a_4\}$$

Using the command `solve`, we then ask Maple to obtain the solution of the system and display it.

```
> coefficients:=solve(sys,unknowns);
```

$$coefficients := \left\{ a_1 = -\frac{1}{2}a_0, a_2 = \frac{1}{8}a_0, a_3 = -\frac{1}{48}a_0, a_4 \right.$$
$$\left. = \frac{1}{384}a_0, a_5 = -\frac{1}{3840}a_0 \right\}$$

Finally, Maple writes the approximate solution if we ask it to substitute with the help of the command subs the values of the coefficients into the initial expression of x.

```
> approxsol:=subs(coefficients,x);
```

$$\text{approxsol} := \sqrt{t}\left(a_0 - \frac{1}{2}a_0t + \frac{1}{8}a_0t^2 - \frac{1}{48}a_0t^3 + \frac{1}{384}a_0t^4 - \frac{1}{3840}a_0t^5\right)$$

This is the approximation of the power series solution for a degree 5 polynomial. Of course, with obvious changes we can improve the approximationto higher-degree polynomials. The second linearly independent solution can be obtained following the same steps for the second solution, $r_2 = 0$, of the indicial equation.

Mathematica Unlike Maple, *Mathematica* lacks a power series package, so we have to obtain the power series solution step by step. We will first write a truncated series, then substitute it into the equation, group the coefficients for the same powers, solve the corresponding system for the unknown coefficients, and finally write the power series solution.

Solutions near ordinary points The computation algorithm presented above works in the case of ordinary points as well as that of regular singular points. Let us present a few examples of the former case.

EXAMPLE 7 An approximate power series solution to a simple linear equation like $x' = 2x$ can be obtained as follows.

```
x=Sum[a[n]t^n,{n,0,5}]+O[t]^6
```

$$a[0] + a[1]t + a[2]t^2 + a[3]t^3 + a[4]t^4 + a[5]t^5 + O[t]^6$$

```
eq=D[x,t]==2x
```

$$a[1] + 2a[2]t + 3a[3]t^2 + 4a[4]t^3 + 5a[4]t^4 + O[t]5$$
$$== 2a[0] + 2a[1]t + 2a[2]t^2 + 2a[3]t^3 + 2a[4]t^4 + 2a[5]t^5 + O[t]^6$$

```
coeffeq=LogicalExpand[eq]
```

$$2a[0] - a[1] == 0 \ \&\& \ 2a[1] - 2a[2] == 0 \ \&\&$$
$$2a[2] - 3a[3] == 0 \ \&\&$$
$$2a[3] - 4a[4] == 0 \ \&\& \ 2a[4] - 5a[5] == 0$$

```
unknowns=Table[a[n],{n,1,5}]
```

$$\{a[1], a[2], a[3], a[4], a[5]\}$$

```
coeff=Solve[coeffeq, unknowns]
```

$$\left\{\left\{a[1] \to 2a[0], a[2] \to 2a[0], a[3] \to \frac{4a[0]}{3}, \right.\right.$$
$$\left.\left. a[4] \to \frac{2a[0]}{3}, a[5] \to \frac{4a[0]}{15}\right\}\right\}$$

```
approxsol=x /. coeff
```

Mathematica displays the approximation of the power series solution,

$$\left\{ a[0] + 2a[0]t + 2a[0]t^2 + \frac{4}{3}a[0]t^3 + \frac{2}{3}a[0]t^4 + \frac{4}{15}a[0]t^5 + O[t]^6 \right\}$$

In the case of initial value problems, the initial condition can always be taken at $t = 0$. Indeed, if an initial condition is given at $t = t_0$, the equation can be transformed with the help of the change of variable $\tau = t - t_0$ into one that has the initial condition at $\tau = 0$. This allows us to conclude that $a_0 = x(0)$ in case of a first-order equation, $a_0 = x(0)$ and $a_1 = x'(0)$ for a second-order equation, etc.

EXAMPLE 8 We can obtain an approximate power series solution of the initial value problem $x' = tx$, $x(0) = 1$ as shown below.

```
x=1+Sum[a[n]t^n,{n,1,5}]+O[t]^6
```

$$1 + a[1]t + a[2]t^2 + a[3]t^3 + a[4]t^4 + a[5]t^5 + O[t]^6$$

```
eq=D[x,t]==t*x
```

$$a[1] + 2a[2]t + 3a[3]t^2 + 4a[4]t^3 + 5a[5]t^4 + O[t]^5$$
$$== t + a[1]t^2 + a[2]t^3 + a[3]t^4 + a[4]t^5 + a[5]t^6 + O[t]^7$$

```
coeffeq=LogicalExpand[eq]
```

$$-a[1] == 0 \;\&\&\; 1 - 2a[2] == 0 \;\&\&\; a[1] - 3a[3] == 0$$
$$a[2] - 4a[4] == 0 \;\&\&\; a[3] - 5a[5] == 0$$

```
unknowns=Table[a[n],{n,1,5}]
```

$$\{a[1], a[2], a[3], a[4], a[5]\}$$

```
coeff=Solve[coeffeq, unknowns]
```

$$\left\{ \left\{ a[1] \to 0, a[2] \to \frac{1}{2}, a[3] \to 0, a[4] \to \frac{1}{8}, a[5] \to 0 \right\} \right\}$$

```
approxsol=x /. coeff
```

Mathematica displays the approximation of the power series solution,

$$\left\{ 1 + \frac{t^2}{2} + \frac{t^4}{8} + O[t]^6 \right\}$$

Let us now see how *Mathematica* can deal with second-order equations and initial value problems. The step to higher-order equations will be obvious. Again, we can take the initial conditions at $t = 0$.

EXAMPLE 9 An approximate power series solution of the second-order equation $x'' + 3t^2x' - 2x = 0$ can be obtained as follows:

```
x=Sum[a[n]t^n,{n,0,5}]+O[t]^6
```

$$a[0] + a[1]t + a[2]t^2 + a[3]t^3 + a[4]t^4 + a[5]t^5 + O[t]^6$$

```
eq=D[x,t,t]+3t^2D[x,t]-2x==0
```

$$(-2a[0] + 2a[2]) + (-2a[1] + 6a[3])t + (3a[1] - 2a[2] + 12a[4])t^2$$
$$+(6a[2] - 2a[3] + 20a[5])t^3 + O[t]^4 = = 0$$

```
coeffeq=LogicalExpand[eq]
```

$$-2a[0] + 2a[2] = = 0 \ \&\& \ -2a[1] + 6a[3] = = 0 \ \&\&$$
$$6a[2] - 2a[3] + 20a[5] = = 0 \ \&\& \ 3a[1] - 2a[2] + 12a[4] = = 0$$

```
unknowns=Table[a[n],{n,2,5}]
```

$$\{a[2], a[3], a[4], a[5]\}$$

```
coeff=Solve[coeffeq, unknowns]
```

$$\left\{\left\{a[2] \to a[0], a[3] \to \frac{a[1]}{3}, a[4] \to \frac{1}{12}(2a[0] - 3a[1]),\right.\right.$$
$$\left.\left. a[5] \to \frac{1}{30}(-9a[0] + a[1])\right\}\right\}$$

```
approxsol=x /. coeff
```

Mathematica displays the approximation to the power series solution,

$$\left\{a[0] + a[1]t + a[0]t^2 + \frac{1}{3}a[1]t^3 + \frac{1}{12}(2a[0] - 3a[1])t^4\right.$$
$$\left. + \frac{1}{30}(-9a[0] + a[1])t^5 + O[t]^6\right\}$$

EXAMPLE 10   An approximate power series solution of the initial value problem $x'' - 5x' + (t - 1)x = 0$, $x(0) = 1$, $x'(0) = -\frac{1}{2}$ can be obtained as follows:

```
x=1-(1/2)t+Sum[a[n]t^n,{n,2,5}]+O[t]^6
```

$$1 - \frac{t}{2} + a[2]t^2 + a[3]t^3 + a[4]t^4 + a[5]t^5 + O[t]^6$$

```
eq=D[x,t,t]-5t^2D[x,t]+(t-1)x==0
```

$$(-1 + 2a[2]) + \left(\frac{3}{2} + 6a[3]\right)t + (2 - a[2] + 12a[4])t^2$$
$$+(-9a[2] - a[3] + 20a[5])t^3 + O[t]^4 = = 0$$

```
coeffeq=LogicalExpand[eq]
```

$$-1 + 2a[2] = = 0 \ \&\& \ \frac{3}{2} + 6a[3] = = 0 \ \&\&$$
$$2 - a[2] + 12a[4] = = 0 \ \&\& \ -9a[2] - a[3] + 20a[5] = = 0$$

```
unknowns=Table[a[n],{n,2,5}]
```

$$\{a[2], a[3], a[4], a[5]\}$$

```
coeff=Solve[coeffeq, unknowns]
```

$$\left\{\left\{a[2] \to \frac{1}{2}, a[3] \to -\frac{1}{4}, a[4] \to -\frac{1}{8}, a[5] \to \frac{17}{80}\right\}\right\}$$

```
approxsol=x /. coeff
```

Mathematica displays the approximation of the power series solution,

$$\left\{1 - \frac{t}{2} + t^2 - \frac{t^3}{4} - \frac{t^4}{8} + \frac{17t^5}{80} + O[t]^6\right\}$$

The method of Frobenius   The idea used for ordinary points also works for finding power series solutions near regular singular points. We first compute by hand the roots of the indicial equation and then proceed as before.

EXAMPLE 11   To find an approximate power series solution for the equation $2tx'' + (t + 1)x' + x = 0$ (already solved in Example 3 of Section 7.3) for, say, terms up to order 5 (i.e., a polynomial of degree 5), we proceed as follows. We compute by hand the roots of the indicial equation and obtain $r_1 = \frac{1}{2}$ and $r_2 = 0$. Let us first find a power series solution for r_1.

```
x=t^(1/2)Sum[a[n]t^n,{n,0,5}]+O[t]^6
```

$$a[0]\sqrt{t} + a[1]t^{3/2} + a[2]t^{5/2} + a[3]t^{7/2} + a[4]t^{9/2} + a[5]t^{11/2} + O[t]^6$$

```
eq=2t*D[x,t,t]+(t+1)D[x,t]+x==0
```

$$\left(\frac{3a[0]}{2} + 3a[1]\right)\sqrt{t} + \left(\frac{5a[1]}{2} + 10a[2]\right)t^{3/2} + \left(\frac{7a[2]}{2} + 21a[3]\right)t^{5/2}$$

$$+ \left(\frac{9a[3]}{2} + 36a[4]\right)t^{7/2} + \left(\frac{11a[4]}{2} + 55a[5]\right)t^{9/2} + O[t]^5 == 0$$

```
coeffeq=LogicalExpand[eq]
```

$$\frac{3a[0]}{2} + 3a[1] == 0 \;\&\&\; \frac{5a[1]}{2} + 10a[2] == 0 \;\&\&$$
$$\frac{7a[2]}{2} + 21a[3] == 0 \;\&\&\; \frac{9a[3]}{2} + 36a[4] == 0 \;\&\&$$
$$\frac{11a[4]}{2} + 55a[5] == 0$$

```
unknowns=Table[a[n],{n,1,5}]
```

$$\{a[1], a[2], a[3], a[4], a[5]\}$$

```
coeff=Solve[coeffeq, unknowns]
```

$$\left\{\left\{a[1] \to -\frac{1}{2}a[0]t^{3/2} + \frac{1}{8}a[0]t^{5/2} - \frac{1}{48}t^{7/2}\right.\right.$$

$$\left.\left. + \frac{1}{384}t^{9/2} - \frac{a[0]t^{11/2}}{3840} + O[t]^6\right\}\right\}$$

```
approxsol=x /. coeff
```

Mathematica displays the approximation to the power series solution,

$$\left\{ a[0]\sqrt{t} - \frac{1}{2}a[0]t^{3/2} + \frac{1}{8}a[0]t^{5/2} - \frac{1}{48}a[0]t^{7/2} \right.$$

$$\left. + \frac{1}{384}a[0]t^{9/2} - \frac{a[0]t^{11/2}}{3840} + O[t]^6 \right\}$$

MATLAB   Unlike Maple, MATLAB lacks a power series package, so we have to obtain the power series solution step by step. We will first write a truncated series, then substitute it into the equation, group the coefficients for the same powers, and solve the corresponding system for the unknown coefficients. Once we have these coefficients, we know what the approximation of the power series solution looks like.

Solutions near ordinary points   The computation algorithm presented above works in the case of ordinary points as well as that of regular singular points. Let us now present a few examples of the former case.

EXAMPLE 12   An approximate power series solution to a simple linear equation like $x' = 2x$ can be obtained as follows. With the help of the command `syms`, we first tell MATLAB to consider the coefficients of the truncated power series and the independent variable t as symbolic objects. To write the equation, we make use of `diff(x,'t',n)`, which means

$$\frac{d^n x}{dt^n}(t)$$

(the nth derivative of x with respect to t). We use the command `collect` to group the coefficients corresponding to the same power of t and then apply the command `solve` to obtain the solution of the linear algebraic system.

```
>> syms a0 a1 a2 a3 a4 a5 t
>> x=a0+a1*t+a2*t^2+a3*t^3+a4*t^4+a5*t^5;
>> dx=diff(x,'t',1);
>> eq=dx-2*x;
>> eq=collect(eq)
eq =
-2*a5*t^5+(5*a5-2*a4)*t^4+(-2*a3+4*a4)*t^3
+(-2*a2+3*a3)*t^2+(-2*a1+2*a2)*t+a1-2*a0
>> eq1='5*a5-2*a4=0';
>> eq2='-2*a3+4*a4=0';
>> eq3='-2*a2+3*a3=0';
>> eq4='-2*a1+2*a2=0';
>> eq5='a1-2*a0=0';
>> solve(eq1,eq2,eq3,eq4,eq5,'a1','a2','a3','a4','a5')
```

MATLAB displays the coefficients

```
a1=2*a0, a2=2*a0, a3=4*a0/3, a4=2*a0/3, a5=4*a0/15
```

In the case of initial value problems, the initial condition can always be taken at $t = 0$. Indeed, if an initial condition is given at $t = t_0$, the equation can be transformed with the help of the change of variable $\tau = t - t_0$ into one that has the initial condition at $\tau = 0$. This allows us to conclude that $a_0 = x(0)$ in case of a first-order equation, $a_0 = x(0)$ and $a_1 = x'(0)$ for a second-order equation, etc.

EXAMPLE 13 We can obtain an approximate power series solution of the initial value problem $x' = tx,\ x(0) = 1$ as shown below.

```
>> syms  a1 a2 a3 a4 a5 t
>> x=1+a1*t+a2*t^2+a3*t^3+a4*t^4+a5*t^5;
>> dx=diff(x,'t',1);
>> eq=dx-t*x;
>> eq=collect(eq)
eq =
a4*t^5+(a3-5*a5)*t^4+(a2-4*a4)*t^3+(a1-3*a3)*t^2
+(1-2*a2)*t-a1
>> eq1='a1=0';
>> eq2='1-2*a2=0';
>> eq3='a1-3*a3=0';
>> eq4='2-4*a4=0';
>> eq5='a3-5*a5=0';
>> solve(eq1,eq2,eq3,eq4,eq5,'a1','a2','a3','a4','a5')
```

MATLAB displays the coefficients

```
a1=0,  a2=1/2,  a3=0,  a4=1/8,  a5=0
```

Let us now see how MATLAB can deal with second-order equations and initial value problems. The step to higher-order equations will be obvious. Again, we can take the initial conditions at $t = 0$.

EXAMPLE 14 An approximate power series solution of the second-order equation $x'' + 3t^2x' - 2x = 0$ can be obtained as follows:

```
>> syms a0 a1 a2 a3 a4 a5 t
>> x=a0+a1*t+a2*t^2+a3*t^3+a4*t^4+a5*t^5;
>> dx=diff(x,'t',1);
>> ddx=diff(x,'t',2);
>> eq=ddx+3*t^2*dx-2*x;
>> eq=collect(eq)
eq=
(6*a2-2*a3+20*a5)*t^3+(3*a1-2*a2+12*a4)*t^2
+(-2*a1+6*a3)*t-2*a0+2*a2
>> eq1='6*a2-2*a3+20*a5=0';
>> eq2='3*a1-2*a2+12*a4=0';
>> eq3='-2*a1+6*a3=0';
>> eq4='-2*a0+2*a2=0';
>> solve(eq1,eq2,eq3,eq4,'a2','a3','a4','a5')
```

MATLAB displays the coefficients

```
a2=a0, a3=(1/3)*a1, a4=(1/6)*a0-(1/4)*a1,
a5=-(3/10)*a0+(1/30)*a1
```

EXAMPLE 15 An approximate power series solution of the initial value problem $x'' - 5x' + (t-1)x = 0$, $x(0) = 1$, $x'(0) = -\frac{1}{2}$ can be obtained as follows:

```
>> syms a0 a1 a2 a3 a4 a5 t
>> x=1-(1/2)*t+a2*t^2+a3*t^3+a4*t^4+a5*t^5;
>> dx=diff(x,'t',1);
>> ddx=diff(x,'t',2);
>> eq=ddx-5*dx+(t-1)*x;
>> eq=collect(eq)
eq=
(-9*a2-a3+20*a5)*t^3+(2-a2+12*a4)*t^2
+(3/2+6*a3)*t-1+2*a2
>> eq1='-9*a2-a3+20*a5=0';
>> eq1='2-a2+12*a4=0';
>> eq1='3/2+6*a3=0';
>> eq1='-1+2*a2=0';
>> solve(eq1,eq2,eq3,eq4,'a2','a3','a4','a5')
```

MATLAB displays the coefficients

```
a2=1/2, a3=-1/4, a4=-1/8, a5=17/80
```

The method of Frobenius The idea used for ordinary points also works for finding power series solutions near regular singular points. We first compute by hand the roots of the indicial equation and then proceed as before.

EXAMPLE 16 To find an approximate power series solution for the equation $2tx'' + (t + 1)x' + x = 0$ (already solved in Example 3 of Section 7.3) for, say, terms up to order 5 (i.e., a polynomial of degree 5), we proceed as follows. We compute by hand the roots of the indicial equation and obtain $r_1 = \frac{1}{2}$ and $r_2 = 0$. Let us first find a power series solution for r_1.

```
>> syms a0 a1 a2 a3 a4 a5 t
>> x=t^(1/2)*(a0+a1*t+a2*t^2+a3*t^3+a4*t^4+a5*t^5);
>> dx=diff(x,'t',1);
>> ddx=diff(x,'t',2);
>> eq=2*t*ddx+(t+1)*dx+x;
>> eq=collect(eq)
eq=
(11*a4/2+55*a5)*t^(9/2)+(9*a3/2+36*a4)*t^(7/2)
+(7*a2/2+21*a3)*t^(5/2)+(5*a1/2+10*a2)*t^(3/2)
+(3*a0/2+3*a1)*t^(1/2)
```

```
>> eq1='11*a4/2+55*a5=0';
>> eq2='9*a3/2+36*a4=0';
>> eq3='7*a2/2+21*a3=0';
>> eq4='5*a1/2+10*a2=0';
>> eq5='3*a0/2+3*a1=0';
>> solve(eq1,eq2,eq3,eq4,eq5,'a1','a2','a3','a4','a5')
```

MATLAB displays the coefficients

```
a1=-(1/2)*a0, a2=(1/8)*a0, a3=-(1/48)*a0, a4=(1/384)*a0,
a5=-(1/3840)*a0
```

PROBLEMS

For the first-order equations in Problems 1 to 8, use Maple, Mathematica, or MATLAB to obtain an approximation of degree 8 for the corresponding power series solution.

1. $x' = 2(3t - 1)^3x + e^{-t}$

2. $u' = 3u + 7tu + 1$

3. $v' = 8v + 25t^3 - 14t^2 - 2$

4. $z' = 9xz - 2x$

5. $x' = 5(t - 2)x + t^2$, $x(0) = 2$

6. $w' = (3t^2 + 1)w + 2$, $w(0) = 1$

7. $r' = 5r - 1$, $r(0) = 0$

8. $u' = (t + 2)u - 5$, $u(0) = 1/2$

For the second-order equations in Problems 9 to 16, use Maple, Mathematica, or MATLAB to obtain an approximation of degree 9 for the corresponding power series solution.

9. $x'' = -3tx' - 3x$

10. $(1 + 4t^2)u'' = 8u$

11. $(1 - t^2)z'' = 4tz' - 6z$

12. $(y^2 + 4)w'' = -2yw' + 12w$

13. $x'' = -t^2x$,
    $x(0) = x'(0) = 1$

14. $v'' = -tv' - 3v + t^2$,
    $v(0) = -1$, $v'(0) = 2$

15. $(t^2 + 4)u'' = -tu' + 9u$, $u(0) = 1$,
    $u'(0) = -\frac{1}{2}$

16. $(1 + 3x^2)q'' = -15xq' - 7q$,
    $q(0) = q'(0) = -1$

Solve the initial value problems in Problems 17 to 20 and obtain exact solutions. Then find approximations of degrees 4, 5, 6, and 7 for the corresponding power series solutions. Compare the exact solutions and their approximations by plotting the graphs. Use Maple, Mathematica, or MATLAB to achieve these goals.

17. $x'' = 4x$, $x(0) = 1$,
    $x'(0) = -1$

18. $u'' = 3u' - 2u$,
    $u(0) = u'(0) = 2$

19. $v'' = -2v$, $v(0) = 2$,
    $v'(0) = -2$

20. $w'' = -6w' - 9w$, $w(0) = 0$,
    $w'(0) = 1$

21. Use Maple, Mathematica, or MATLAB to compute a power series approximation of degree 9 for the equation

$$t(1 - t)x'' = (4t - 1)x' + 2x.$$

Then show that this is an approximation of the general solution

$$x(t) = (1 - t)^{-2}[c_1 + c_2(\ln t - t)].$$

In Problems 22 to 25, use Maple, Mathematica, or MATLAB to obtain a power series approximation of degree 10 for each of the given initial value problems. Plot the graphs of the solutions in each case. What is the domain of each solution?

22. $x'' = (t - 1)x' - x$, $x(1) = 1$,
    $x'(1) = 2$

23. $u'' = (2t - 1)u' + u$,
    $u(1/2) = u'(1/2) = -1$

24. $v'' = (t^2 - 4)v$, $v(2) = 0$,
    $v'(2) = -2$

25. $w'' = (t + 2)w$, $w(-2) = 0$,
    $w'(-2) = -2$

7.5 Modeling Experiments

Before attempting to work on any of the modeling experiments below, read the introductory paragraph of Section 2.8. This will give you an idea about what is expected from you, how to proceed, and what to emphasize.

The Pole Vault

Consider the pole vault equation of Section 7.2,

$$\theta'' = k(L - x)\theta,$$

where k is a constant, L is the length of the pole, x is the distance along the pole measured from the bottom, and $\theta(x)$ is the angular deflection from the vertical at the point of distance x (see Figure 7.2.2). Take L to be 4 meters, $k = 0.01$ and start with the initial conditions $\theta(0) = -\frac{1}{2}$ and $\theta'(0) = \frac{1}{2}$. Obtain an eighth-degree polynomial approximation of the power series solution, and then find a numerical solution of the equation. Compare the two solutions. Is the error between the two solutions reasonably small? Draw the approximate shape of the pole vault. Then proceed in the same way for pole vaults of different lengths but with the same constant $k = 0.01$. What can you tell about the flexibility of the pole? Then change the constant k to other values and repeat the analysis. You can use the computer programs of Section 7.4 to help you with the computations. Can you draw any conclusions? Can you state any results? If so, can you prove them?

Arms Race

Study a more general form of the arms race equations considered in Section 7.2, namely the system

$$\begin{cases} x' = ay \\ y' = btx - cy + 1, \end{cases}$$

where x and y represent the expenditures of two countries and $a, b,$ and c are positive constants. What do these constants represent from the point of view of the arms race? Use, if you wish, one of the computer programs in Section 7.4 to obtain a polynomial approximation of the power series solution for some reasonable values of the constants $a, b,$ and c and for some initial conditions. See how far you can go with the approximation in terms of the polynomial's degree. Fix some initial conditions and estimate the expenditures for the first 6 months after the start of the arms race. Compare your polynomial with a suitable power series and estimate the error. Then compute a numerical solution and compare the result with the one obtained using the polynomial. Is the error between the two solutions within reasonable limits? Keep two of the constants fixed and vary the third. Repeat the whole procedure. What results do you obtain? Can you find a rule that connects the variation of the constant and the outcome? What happens if you keep other two constants fixed and vary the third? Can you state a theorem? Can you prove it? Is this a good model for arms races? Can you improve it?

RLC Circuits with Variable Resistance and Capacitance

As we saw in Section 7.3, the equation describing the charge q in the plates of an *RLC* circuit (in which the resistance changes inversely proportional and the capacitance directly proportional with time) is given by the linear second-order equation

$$tq'' + aq' + bq = 0,$$

where a and b are constants. Fix some values for a and b and take some initial conditions. What is a good value of t at which to impose initial conditions? Then, using if you wish one of the computer programs in Section 7.4, find a polynomial approximation of the solution with the help of Frobenius's method. Can you obtain the entire power series? See for what values of a and b you can. What is the behavior of the solution when $t \rightarrow 0$? What happens if you vary a, then b, and then both constants? How does the variation of the constants influence the solution? Obtain numerical solutions in each case and compare them with the ones you already have. What qualitative properties can you find for your solutions? Can you state a theorem? Can you prove it?

SOLUTIONS TO ODD–NUMBERED PROBLEMS

1 Introduction

1. Nonlinear, order 1

3. Linear, order 3

5. Nonlinear, order 1

7. Nonlinear, order 2

9. (a) $x' = kx$ is a single, first-order, ordinary, linear equation.

(b) $T' = k(T - \tau)$ is a single, first-order, ordinary linear equation.

(c) $u' = k(a - u)(b - u)$ is a single, first-order, ordinary, nonlinear equation.

(d) $p' = \lambda p(\alpha - p)$ is a single, first-order, ordinary, nonlinear equation.

(e) $x'' + bx' + kx = \gamma \sin \omega t$ is a single, second-order, ordinary linear equation.

(f) $x'' + \alpha(x^2 - 1)x' + x = \beta \cos \omega t$ is a single, second-order, ordinary, nonlinear equation.

(g) $x'' + \delta x' - x + x^3 = \gamma \cos \omega t$ is a single, second-order, ordinary, nonlinear equation.

(h) $\begin{cases} x' = \sigma(y - x) \\ y' = \rho x - y - xz \\ z' = -\beta z + xy, \end{cases}$

is a three-dimensional system of first-order, ordinary, nonlinear equations.

(i) $\begin{cases} q_{1i}'' = Gm_2 \dfrac{q_{2i} - q_{1i}}{r_{21}^3} + Gm_3 \dfrac{q_{3i} - q_{1i}}{r_{31}^3} \\[2mm] q_{2i}'' = Gm_1 \dfrac{q_{1i} - q_{2i}}{r_{12}^3} + Gm_3 \dfrac{q_{3i} - q_{2i}}{r_{32}^3} \\[2mm] q_{3i}'' = Gm_1 \dfrac{q_{1i} - q_{3i}}{r_{13}^3} + Gm_2 \dfrac{q_{2i} - q_{3i}}{r_{23}^3} \end{cases} \quad (i = 1, 2, 3)$

is a three-dimensional system of second-order, ordinary, nonlinear equations.

(j) $\begin{cases} x' = au(t)y \\ y' = [kv(t) - u(t)]y \end{cases}$

is a two-dimensional system of first-order, ordinary linear equations.

15. The equations are
$$\begin{cases} q_{1i}'' = G\dfrac{m_2}{r^3}[1 + \dfrac{3G}{c^2 r}(m_1 + m_2)](q_{2i} - q_{1i}) \\[2mm] q_{2i}'' = G\dfrac{m_1}{r^3}[1 + \dfrac{3G}{c^2 r}(m_1 + m_2)](q_{1i} - q_{2i}), \end{cases}$$

$(i = 1, 2, 3)$ with $r = \sqrt{(q_{11} - q_{21})^2 + (q_{12} - q_{22})^2 + (q_{13} - q_{23})^2}$.

17. For $n = 2$ particles:
$$\begin{cases} q_{1i}'' = \dfrac{\alpha}{r^{a+1}}(q_{2i} - q_{1i}) + \dfrac{\beta}{r^{b+1}}(q_{2i} - q_{1i}) \\[2mm] q_{2i}'' = \dfrac{\alpha}{r^{a+1}}(q_{1i} - q_{2i}) + \dfrac{\beta}{r^{b+1}}(q_{1i} - q_{2i}), \end{cases}$$

$(i = 1, 2, 3)$ with $r = \sqrt{(q_{11} - q_{21})^2 + (q_{12} - q_{22})^2 + (q_{13} - q_{23})^2}$.
For $n = 3$ particles:

$$\begin{cases} q_{1i}'' = \left[\dfrac{\alpha}{r_{12}^{a+1}} + \dfrac{\beta}{r_{12}^{b+1}}\right](q_{2i} - q_{1i}) + \left[\dfrac{\alpha}{r_{13}^{a+1}} + \dfrac{\beta}{r_{13}^{b+1}}\right](q_{3i} - q_{1i}) \\[3mm] q_{2i}'' = \left[\dfrac{\alpha}{r_{21}^{a+1}} + \dfrac{\beta}{r_{21}^{b+1}}\right](q_{1i} - q_{2i}) + \left[\dfrac{\alpha}{r_{32}^{a+1}} + \dfrac{\beta}{r_{32}^{b+1}}\right](q_{3i} - q_{2i}) \\[3mm] q_{3i}'' = \left[\dfrac{\alpha}{r_{32}^{a+1}} + \dfrac{\beta}{r_{32}^{b+1}}\right](q_{2i} - q_{3i}) + \left[\dfrac{\alpha}{r_{13}^{a+1}} + \dfrac{\beta}{r_{13}^{b+1}}\right](q_{1i} - q_{3i}) \end{cases}$$

with $r_{jk} = \sqrt{(q_{j1} - q_{k1})^2 + (q_{j2} - q_{k2})^2 + (q_{j3} - q_{k3})^2}$ ($j, k = 1, 2, 3$).

Section 2.1 General Aspects

1. An example is $u' + \sin u' - \ln uu' = 0$.

3. The general form is $2x' - x + t - \sqrt{t} = 0$. F is not unique. The normal form is $x' = (x - t + \sqrt{t})/2$. the resulting vector field is

$$f(x, t) = \frac{x - t + \sqrt{t}}{2}.$$

5. The general form is $[(\sqrt{3} - 1)/2]\theta'\theta - t^{3/2} = 0$. F is not unique. The normal form is

$$\theta' = \frac{2t^{3/2}}{(\sqrt{3} - 1)\theta}.$$

7. The general form is $6w' + \frac{3}{2}w'xw^2(2x - 3) - (\sqrt{7} - 4)/(\sqrt{7} + 4) = 0$. F is not unique. The normal form is

$$w' = \frac{\sqrt{7} - 4}{\sqrt{7} + 4} \cdot \frac{1}{6 + \frac{3}{2}xw^2(2x - 3)}.$$

9. The general form is $(v')^3 - 1 = 0$. F is not unique. The normal form is

$$v' = 1.$$

11. $x(t)$ is a solution.

13. Neither $v_1(x)$ nor $v_2(x)$ is a solution.

15. Neither $y_1(t)$ nor $y_2(t)$ is a solution.

17. $x(t) = ct^2$ fills the plane. The singular solution $x(t) = 0$ is contained in the formula for $c = 0$.

19. $u(x) = -ce^{-x}$ fills the plane. The singular solution $u(t) = 0$ is contained in the formula for $c = 0$.

21. $w(y) = \sqrt{2\ln y + c}$ does not fill the plane, and there are no singular solutions.

23. $x' - x/t = 0$

25. $u' - 1/u = 0$

27. $v' - v = 0$

29. $z' + z\tan t = 0$

33. $x(t)$ is a solution.

35. $w(z)$ is a solution.

37. $x(t)$ is not a solution.

Section 2.2 Separable Equations

1. $x(t) = -\dfrac{1}{2t + c}$

3. $\csc v - \cot v = ke^{t/\sqrt{2}}$ (implicit form)

5. $\theta(r) = kr^{(7/3)r}e^{-(7/3)r}$

7. $x(t) = e^{t^2/2} - 1$

9. $u(x) = \dfrac{\pi\sqrt{2}}{4}\sin x$

11. $w(x) = 2\sqrt{3}(\sin x) - 2$

13. General solution:

$$x(t) = \frac{2(1 + ke^{4t})}{1 - ke^{4t}}$$

Singular solutions: $x_1(t) = 2$ and $x_2(t) = -2$. The solution x_1 can be obtained from the general formula for $k = 0$.

15. General solution:

$$w(t) = \frac{3 - 2ke^t}{1 - ke^t}$$

Singular solutions: $w_1(t) = 2$ and $w_2(t) = 3$. The solution w_2 can be obtained from the general formula for $k = 0$.

17. General solution:

$$v(t) = \frac{t - 1 + k}{2(t + k)}$$

Singular solution: $v(t) = \frac{1}{2}$ cannot be obtained from the general formula.

19. $x' = 4t$

21. $\varphi' = \dfrac{1}{\theta - 2} + 3\sqrt{2}\theta^2$

23. $\varphi' = \cos r - \sin r$

25. Separable equation: $x' = 0.055x$. Solution: $x(t) = ke^{0.055t}$. After 3 years the amount is \$5896.96. The doubling time is $T = (\ln 2)/k \simeq 12.6$ years.

27. 96.05%

29. 91.5 days

31. 75.97°F

33. $d' = hd$, where H is Hubble's constant

35. $u(t) = \pm\sqrt{2(\ln t + c)}$; $x(t) = \pm t\sqrt{2(\ln t + c)}$

37. $u = \sin^{-1}(ke^t)$; $x(t) = t\sin^{-1}(ke^t)$

39. $u = \ln\dfrac{1}{|\ln 1/|t| - c|}$ and $x = -t\ln|\ln|t| + c|$

Section 2.3 Linear Equations

1. $x(t) = \dfrac{3}{2}t + \dfrac{3}{4} - \dfrac{e^{-t}}{3} + ce^{+2t}$

3. $p(t) = -r^2 - 3r - 6 - \dfrac{6}{r} + \dfrac{ce^r}{r}$

5. $v(x) = x\sin x + c\sin x$

7. $w(t) = -\dfrac{1}{2t} + ct$

9. $x(t) = 4\exp[\frac{1}{2}(t^2 - 1)] - 2$

11. $w(\theta) = \dfrac{\sin\theta - 1}{\cos\theta}$

13. $x(t) = -\cos t + (10\sqrt{2} + 1)\sin t$

15. $x(t) = e^t(e^t + 1)$

17. Linear nonhomogeneous equations can have singular solutions if they are of the form $x'(t) = xf(t) + kf(t)$, $k \in R$.

19. $v(t) = \dfrac{\sin t - RC\cos t - RCe^{-t/RC}}{1 + R^2C^2}$

21. $65.36m$

23. $x(t) = \pm\dfrac{\sqrt{5}\sqrt{(2 + 5t^5c)t}}{2 + 5t^5c}$

25. $x(t) = \dfrac{1}{\frac{3}{2} + c_1e^{-2t}}$

27. $x(t) = \sqrt{\dfrac{a}{b + ke^{-2t}}}$

29. $x(t) = \dfrac{e^{-t^2}}{\left(t - \dfrac{t^3}{3} + \dfrac{t^5}{10} - \dfrac{t^7}{42} + \cdots\right) + c} + t$

31. $x(t) = \dfrac{1}{c \cos t - \sin t} + \sin t$

33. $x(t) = \dfrac{4}{e^{-4t-c} + 1} + t - 2$

35. $3tx + \dfrac{x^2}{2} + c - t^2 = 0$

37. $tx + \dfrac{t^3}{3} + t \ln t - t + \dfrac{t^2}{2} + c = 0$

39. $x(t) = -\dfrac{3t^2 + c}{\ln 2 - 2}$

Section 2.4 Qualitative Methods

1.

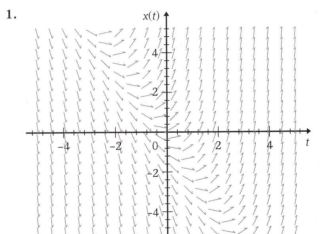

3.

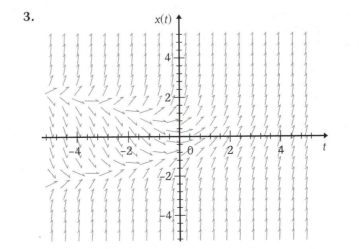

5.

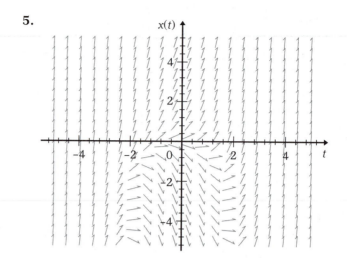

7.

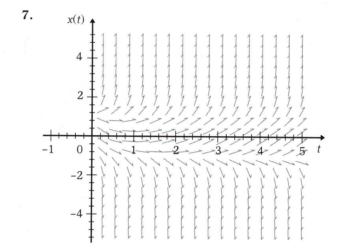

9.

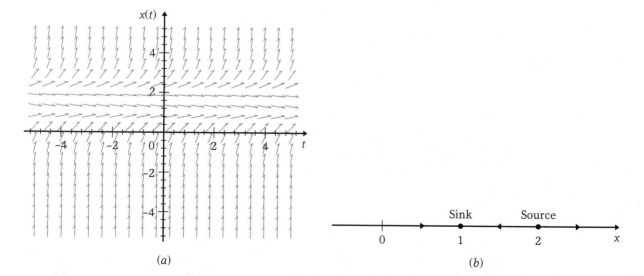

(a)

(b)

11.

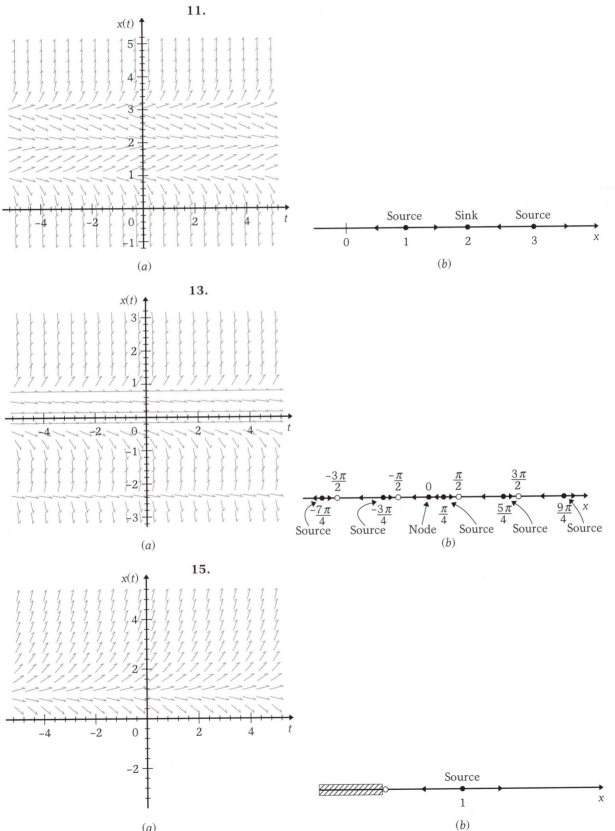

(a)

(b)

13.

(a)

(b)

15.

(a)

(b)

17. $x = -2$ is a sink and $x = 8$ is a source.

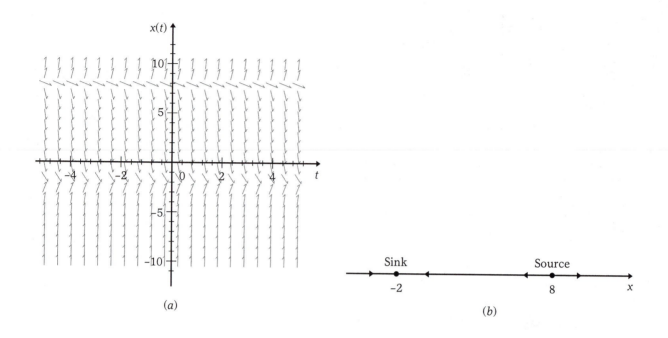

(a)

(b)

19. $x = -\pi, -2\pi, -3\pi, \ldots$ are sinks. $x = \pi, 2\pi, 3\pi, \ldots$ are sources, and $x = 0$ is a node.

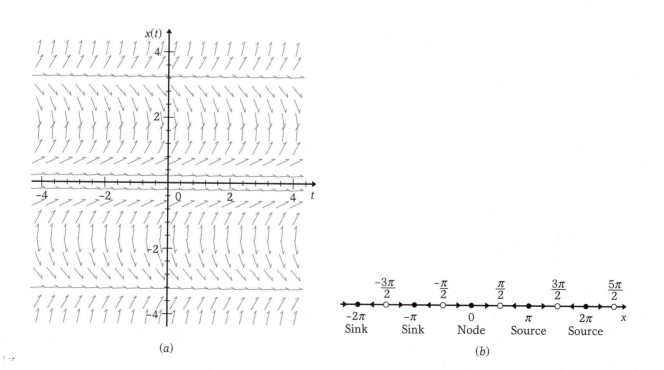

(a)

(b)

21. $x = \pm 2\pi, \pm 4\pi, \pm 6\pi, \ldots$ are sources; $x = \pm \pi, \pm 3\pi, \pm 5\pi, \ldots$ are sinks; $x = 0$ is a node.

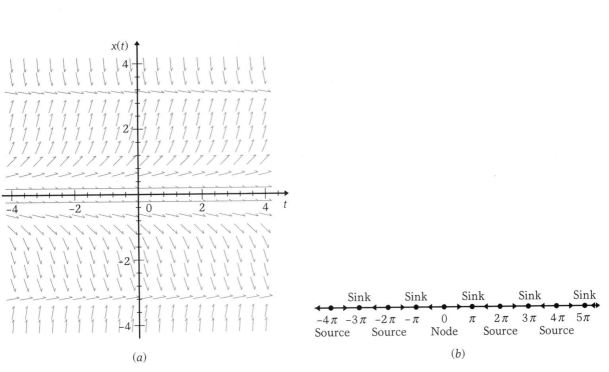

(a)

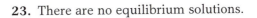

(b)

23. There are no equilibrium solutions.

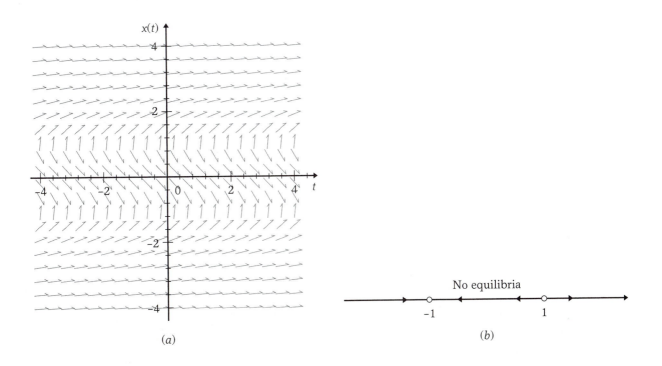

(a)   (b)

25. We can choose $g(x)$ of the form $g(x) = -(x^2 - 1)^2$. The choice is not unique.

27. $x' = |x|, x_0 = 0$

29. $x = -3$ is a sink and $x = 0$ is a source.

31. $x = 0, \pm 2\pi, \pm 4\pi, \ldots, 2k\pi$ are sources and $x = \pm \pi, \pm 3\pi, \pm 5\pi, \ldots,$ $(2k + 1)\pi$ are sinks, where k is an integer.

35. For mass (m), propulsion force (F), weight (mg) constants, the equation of motion is $v' = F/m - g - (kv^2/m)(t - T)$.

37. $\alpha \geq 1$

Section 2.5 Existence and Uniqueness

1. The theorem of Peano and the theorem of Cauchy apply.

3. The theorem of Peano and the theorem of Cauchy do not apply.

5. The theorem of Peano and the theorem of Cauchy apply.

7. The theorem of Peano and the theorem of Cauchy do not apply.

9. The theorem of Cauchy does not apply.

$$\begin{cases} \varphi_2(t) = \left(\dfrac{2}{3}t\right)^{3/2}, & t \geq 0 \\[3mm] \varphi_3(t) = -\left(\dfrac{2}{3}t\right)^{3/2}, & t \geq 0 \end{cases}$$

11. The theorem of Cauchy does not apply; $\varphi_2(t) = \left(\dfrac{5\pi}{7}t\right)^{7/5}$; t is real.

13. The theorem of Cauchy does not apply.

$$\begin{cases} \varphi_2(t) = \left(\dfrac{8}{21}t\right)^{7/4}, & t \geq 0 \\[3mm] \varphi_3(t) = -\left(\dfrac{8}{21}t\right)^{7/4}, & t \geq 0 \end{cases}$$

15. $\begin{cases} x(t) = 0 & t \leq 0 \\ x(t) = ke^{-1/(4t)} & t > 0 \end{cases}$

17. $\varphi_n = 4\left(\dfrac{t}{2}\right)^{n+1} \dfrac{1}{(n+1)!} + 4\left(\dfrac{t}{2}\right)^n \dfrac{1}{n!} + \cdots + 4\left(\dfrac{t}{2}\right)^5 \dfrac{1}{5!} + 4\left(\dfrac{t}{2}\right)^3 \dfrac{1}{3!} + 4\left(\dfrac{t}{2}\right)^2 \dfrac{1}{1!}$

19. $\varphi_n(t) = -\dfrac{t^{3n-1}}{5 \cdot 8 \cdot 11 \cdot 14 \cdots (3n-1)} - \dfrac{t^{3n-4}}{5 \cdot 8 \cdots (3n-4)} - \cdots \dfrac{t^8}{5 \cdot 8} - \dfrac{t^5}{5} - t^2$

21. $\varphi_1(t) = \cos t + 1 - t$

$\varphi_2(t) = \cos t + 1 - t$

$\varphi_3(t) = \cos t - t + 1$

23. $\varphi_1(t) = \dfrac{t^3}{3}$

$\varphi_2(t) = \dfrac{t^7}{7 \cdot 9} + \dfrac{t^3}{3}$

$\varphi_3(t) = \dfrac{t^{15}}{7^2 \cdot 9^2 \cdot 15} + \dfrac{t^7}{3^2 \cdot 7} + \dfrac{2 \cdot t^{11}}{3 \cdot 7 \cdot 9 \cdot 11 \cdot} + \dfrac{t^3}{3}$

25. $\varphi_1(t) = \varphi_2(t) = \varphi_3(t) = 0$

27. $\varphi_1(t) = t$

$\varphi_2(t) = \dfrac{t^3}{3} + \dfrac{t^2}{2} + t$

$\varphi_3(t) = \dfrac{1}{63}t^7 + \dfrac{1}{18}t^6 + \dfrac{11}{60}t^5 + \dfrac{1}{3}t^4 + \dfrac{1}{2}t^3 + \dfrac{1}{2}t^2 + t$

Section 2.6 Numerical Methods

1. (a) $x_{10} = 2.334633363$
  (b) $x_{20} = 2.510662314$
  (c) $x_{10} = 2.709057013$
  (d) $x_{20} = 2.715989839$
3. (a) $x_{10} = 0.4736842105$
  (b) $x_{20} = 0.4871794872$
  (c) $x_{10} = 0.5000000000$
  (d) $x_{20} = 0.5000000002$
5. (a) $x_{10} = 0.7112128253$
  (b) $x_{20} = 0.7019909764$
  (c) $x_{10} = 0.6929398593$
  (d) $x_{20} = 0.6930951588$
7. (a) $x_{10} = 0.7359335000$
  (b) $x_{20} = 0.7897095333$
  (c) $x_{10} = 0.8276980489$
  (d) $x_{20} = 0.8376342081$
9. (a) $x_n = 3$ for all n. The initial value problem has no solution.
  (b) $x_{50} = 2.999999999$. The exact solution is
$$x(t) = \frac{3e^{4t + \ln(-1/3)}}{-1 + e^{4t + \ln(-1/3)}}.$$
11. (a) -870
  (b) -866
  (c) -866.0000000
13. (a) -94
  (b) -94.0
  (c) -93.984768
15. (a) $x_5 = 3.23248$
  (b) $x_{10} = 3.390613691$
  (c) $x_5 = 3.554062245$
  (d) $x_{10} = 3.571121273$

17. (a) $x_5 = 0.042075392$
   (b) $x_{10} = 0.04418944305$
   (c) $x_5 = 0.0438098174$
   (d) $x_{10} = 0.04507589990$

19. (a) $x_5 = 1.828061748$
   (b) $x_{10} = 1.882180558$
   (c) $x_5 = 1.936450260$
   (d) $x_{10} = 1.939402555$

21. $x_n = 0$ for all n

23. For Euler's method, make $h^3 < 0.5$ and for the Runge-Kutta method, make $h^4 < 0.5$.

Section 3.1 Homogeneous Equations

1. $x'' = \dfrac{-2x' + x^2 - \frac{2}{5}x + 4t}{5(t + 1)}$

3. $v'' = \dfrac{2v' - e^{-x}v}{(1 + 3x^2)}$

5. $y'' = \dfrac{3y' - 5y + 4t^5 - 6}{7ty - 1}$

7. $\varphi(t)$ is a solution.

9. $r(z)$ is a solution.

11. $s(p)$ is a solution.

13. $\varphi(t)$ is not a solution.

15. $r(\theta)$ is a solution.

17. $\nu(t)$ is a solution.

19. $I = (-\infty, +\infty)$

21. $I = (0, \infty)$

23. $I = (-\infty, 1)$

25. $f(t)$ and $g(t)$ are linearly dependent on all given intervals.

27. $\tan t$ and $\cot t$ are linearly independent on all given intervals.

29. $f(t)$ and $g(t)$ are linearly dependent on $(-\infty, 0)$ and $(0, +\infty)$. $f(t)$ and $g(t)$ are linearly independent on $(-\infty, +\infty)$.

31. $f(t)$ and $g(t)$ are linearly dependent on $(-\infty, 0)$ and $(0, +\infty)$. $f(t)$ and $g(t)$ are linearly independent on $(-\infty, +\infty)$.

33. $f(t)$ and $g(t)$ are linearly independent on all given intervals.

35. $f(t)$ and $g(t)$ are linearly independent on all given intervals.

37. $fg' - f'g = 0$. the converse of Theorem 3.11 is not true.

39. $fg' - f'g = -2t$.

43. $y_1(t), y_2(t)$ form a fundamental set of solutions.

45. $z_1(t), z_2(t)$ form a fundamental set of solutions.

Section 3.2 Integrable Cases

1. $x(t) = c_1 e^{4t} + c_2 e^{3t}$

3. $x(t) = e^{3t}(c_1 + c_2 t)$

5. $x(t) = e^{-3t/2}(c_1 \cos \sqrt{11}t + c_2 \sin \sqrt{11}t)$

7. $x(t) = e^{2t}$

9. $x(t) = 5e^{5t-4} - 4te^{5t-4}$

11. $x(t) = \left(\dfrac{-5}{42}\sqrt{21} + \dfrac{1}{2}\right)e^{[(-7-\sqrt{21})/2]t} + \dfrac{\sqrt{21}}{42}(5 + \sqrt{21})e^{[(-7+\sqrt{21})/2]t}$

13. $x(t) = \dfrac{k}{3}t^3 + c$

15. $x(t) = \dfrac{5k}{2}t^2 - k \cos t + c$

17. $x(t) = \ln\left|\dfrac{1}{\cos t} + \tan t\right| + 1$

19. $x(t) = \tan^{-1}\left(\dfrac{t}{2}\right) + 1$

21. $x(t) = c_1 e^t t + c_2 t$

23. $x(t) = c_1 t + c_2 \dfrac{1}{t}$

25. $x(t) = c_1 \sin(t^2) + c_2 \cos(t^2)$

27. $x(t) = c_1 t^3 + c_2 t^2$

31. $x(t) = c_1 t^{1/2} + c_2 t^{3/2}$

33. $x(t) = t^2(c_1 + c_2 \ln t)$

35. $x(t) = t^{-1/4}\left[c_1 \cos\left(\dfrac{\sqrt{39}}{4}\ln t\right) + c_2 \sin\left(\dfrac{\sqrt{39}}{4}\ln t\right)\right]$

41. $\phi(t) = \dfrac{\pi}{90}\cos 3t$

Section 3.3 Nonhomogeneous Equations

1. $x(t) = -\cos t \ln \dfrac{1 + \sin t}{\cos t} + c_1 \cos t + c_2 \sin t$

3. $x(t) = \dfrac{e^t}{6} + c_1 e^{-2t} + c_2 e^{-t}$

5. $x(t) = -(\frac{1}{2}t^2 \arctan t - \frac{1}{2}t + \frac{1}{2}\arctan t)e^t + (t \arctan t - \frac{1}{2}\ln(t^2 + 1))te^t$
$+ c_1 e^t + c_2 te^t$

7. $x(t) = e^t \ln|t| + c_1 e^t + c_2 e^{-t}$

9. $x(t) = \dfrac{1}{2} - \dfrac{1}{3}t^2 + t^2 \ln t + \dfrac{c_1}{t} + c_2 t^2$

11. $x(t) = t^2 e^{-t} + c_1 e^t + c_2 t$

13. $x(t) = \dfrac{t^4}{12} + \dfrac{1}{24}t^{-2} + \dfrac{7}{8}t^2$

15. $x(t) = \dfrac{-1}{36} - \dfrac{1}{12}t + \dfrac{358}{441}e^{3t-6} + \dfrac{5}{28}te^{-4t+8} + \dfrac{5}{196}e^{-4t+8}$

17. $x(t) = c_1 e^{3t} + c_2 e^{-3t} - \dfrac{7}{9}$

19. $x(t) = c_1 e^{-t/2} \cos\left(\dfrac{\sqrt{3}}{2}t\right) + c_2 e^{-t/2} \sin\left(\dfrac{\sqrt{3}}{2}t\right) + \dfrac{2}{13}e^{3t}$

21. $x(t) = c_1 e^t + c_2 t e^t + \cos t - 1$

23. $x(t) = c_1 \cos 3t + c_2 \sin 3t + \dfrac{13}{6}t \sin 3t$

25. $x(t) = \dfrac{5}{8}e^{-4t} + \dfrac{5}{8}e^{4t} - \dfrac{1}{4}$

27. $x(t) = \dfrac{79}{243}e^{9t} - \dfrac{79}{243} - \dfrac{1}{6}t^2 + \dfrac{2}{27}t$

29. $x(t) = -\dfrac{1}{2}t \cos t + \left(1 + \dfrac{\pi}{4}\right)\cos t$

31. $x(t) = \dfrac{-1}{4}e^t t^2 + \dfrac{1}{4}te^t + \dfrac{3}{8}e^t - \dfrac{3}{8}e^{-t}$

33. Yes, the answer is the same as in the text.

35. The general solution is $\phi(t) = \dfrac{4\pi^2}{4\pi^2 - 1} \cos t - \dfrac{1}{4\pi^2 - 1} \cos 2\pi t.$

Section 3.4 Harmonic Oscillators

1. $k \simeq 15.31\dfrac{N}{m}$

3. $x(t) = 0.25 \cos \sqrt{19.6}t$

5. $\omega_0 = \sqrt{\dfrac{g}{L}}$ and $t_0 = 2\pi\sqrt{\dfrac{L}{g}}$

7. $v = 2.8\dfrac{kg}{s}$

9. $x(t) = \dfrac{at}{2\omega_0} \sin \omega_0 t$

11. $x(t) = A(t)\sin\dfrac{(\omega - \omega_0)t}{2}$ with $A(t) = \dfrac{2a}{(\omega^2 - \omega_0^2)} \cos(\omega + \omega_0)\dfrac{t}{2}$

13. $\sin 3t + 2\cos 3t = \sqrt{5}\cos(3t - \cot^{-1} 2)$

15. $5\sin\dfrac{t}{2} - \cos\dfrac{t}{2} = 5.09\cos\left(\dfrac{t}{2} - 1.7\right)$

17. $19.85\cos(6.28t - \cot^{-1}\dfrac{-15}{13})$

19. $5(\cos 2t - \cos 3t) = A(t)\sin\dfrac{5t}{2}$, where $A(t) = 10\sin\dfrac{t}{2}$

21. $\dfrac{\pi}{2}(\cos t - \cos 5\pi t) = A(t)\sin\dfrac{(5\pi + 1)t}{2}$,

where $A(t) = \pi \sin\dfrac{(5\pi - 1)t}{2}$

23. $\cos 6t - \cos\dfrac{17}{2}t = A(t)\sin\dfrac{29t}{4}$, where $A(t) = 2\sin\dfrac{5t}{4}$

27. $\begin{cases} t + 5\cos t - \sin t & t \in [0,\,\pi/2) \\ \pi - t + 3\cos t - \sin t & t \in [\pi/2,\,\pi) \\ 3\cos t & t \in [\pi,\,\infty) \end{cases}$

The solution is continuous.

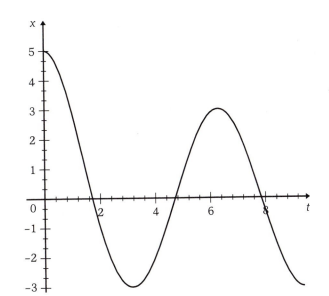

Section 3.5 Qualitative Methods

1.

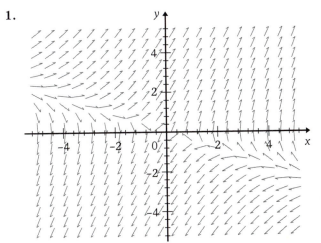

3.

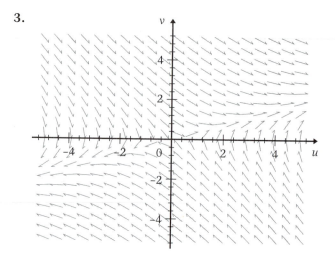

5.

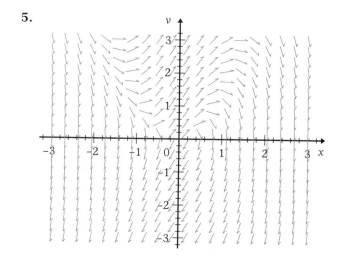

7.

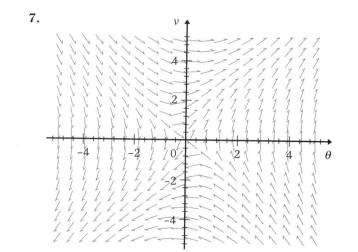

9. $x_1(t) = 0$, $x_2(t) = 1$, $x_3(t) = -1$ are the equilibrium solutions.

11. $z = (4k + 1)\dfrac{\pi}{2}$ are the equilibrium solutions, where k is an integer.

13. $y = \dfrac{-1}{2}$ is the equilibrium solution.

15. If $b \leq 1$, $v = \pm \cos^{-1} b \pm 2k\pi$. If $b > 1$, there is no equilibrium solution.

17. The equilibrium is at $x = \frac{5}{3}$. The new homogeneous equation is $X'' = X' - 3X$, where $X = x + \frac{5}{3}$. The equilibrium $X = 0$ is a spiral source.

19. The equilibrium is at $x = (1 - \sqrt{13})/4$. The new homogeneous equation is $X'' = -X' + 2X$, where $X = x + (1 - \sqrt{13})/4$. The equilibrium $X = 0$ is a saddle.

21. The equilibrium $x = 0$ is a saddle.

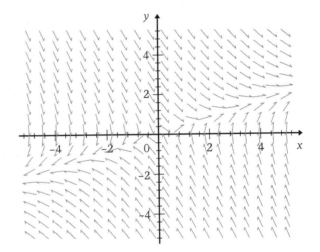

23. The equilibrium $z = 0$ is a source.

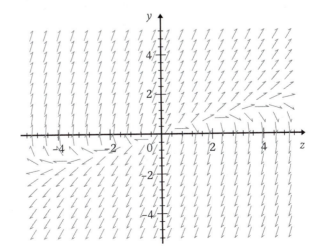

25. The equilibrium $u = 0$ is a saddle.

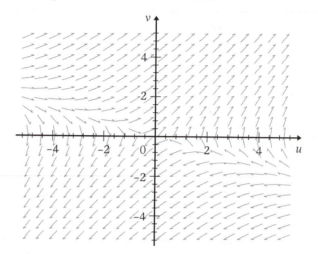

27. The equilibrium $\rho = 0$ is a center.

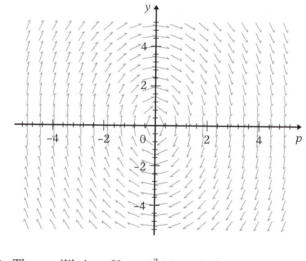

29. The equilibrium $Y = -\frac{2}{3}$ is a spiral source.

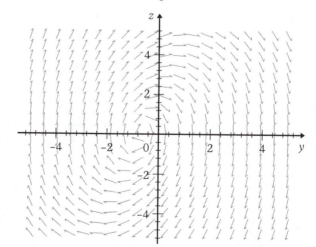

31. The general solution is $x(t) = c_1 + c_2 e^{-bt}$.
   (a) If $b < 0$, then any x is a source.

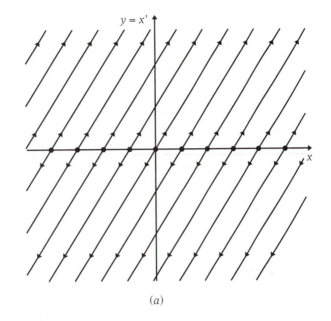

(a)

(b) If $b > 0$ then any x is a sink.

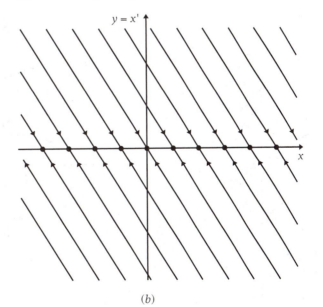

(b)

33. The equations have the same form as those for the *RLC* circuit.

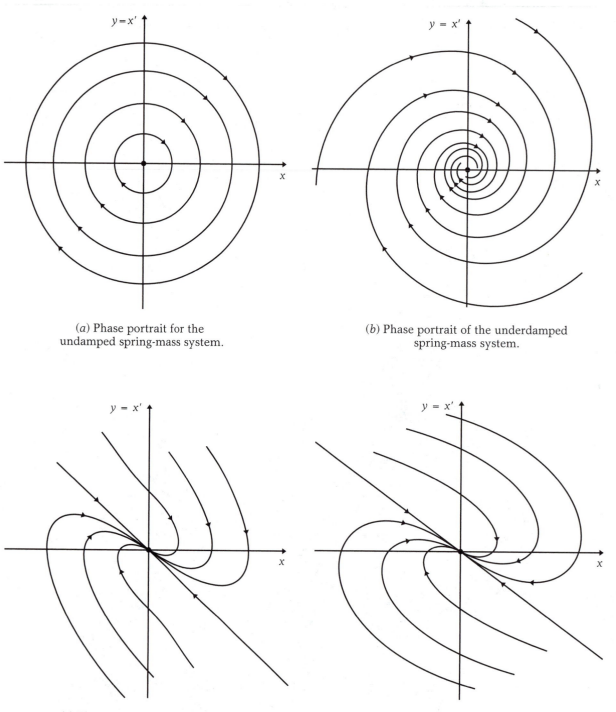

(*a*) Phase portrait for the
undamped spring-mass system.

(*b*) Phase portrait of the underdamped
spring-mass system.

(*c*) Phase portrait of the critically
damped spring-mass system.

(*d*) Phase portrait of the
overdamped spring-mass system.

35. The equations are of the same type as those for the spring-mass
system (see Problem 33).

37. $x'' = -\omega_0^2 x$, where $\omega_0^2 = \rho_{\text{water}}g/(\rho_{\text{cube}}l)$. The equilibrium is $x = 0$ and is a center. The model is structurally unstable.

39. The direction vectors are perpendicular at the x-axis.

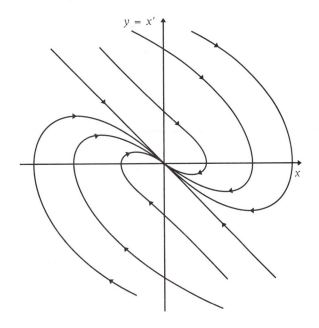

Section 3.6 Numerical Methods

1. (a) $x_5 = 2.48832$, $y_5 = 4.97664$
  (b) $x_5 = 1.610510000$, $y_5 = 3.221020000$
  (c) $x_5 = 2.702708163$, $y_5 = 5.405416327$
  (d) $x_5 = 1.647446766$, $y_5 = 3.294893532$

3. (a) $x_5 = -1.592270000$, $y_5 = 1.695870000$
  (b) $x_5 = -2.013258438$, $y_5 = 1.492574063$
  (c) $x_5 = -1.600824058$, $y_5 = 1.462464036$
  (d) $x_5 = -1.977487820$, $y_5 = 1.194297065$

5. (a) $x_5 = 1.793339781$, $y_5 = 1.188473028$
  (b) $x_5 = 1.441983277$, $y_5 = 1.474448943$
  (c) $x_5 = 1.763751884$, $y_5 = 1.223889880$
  (d) $x_5 = 1.430873319$, $y_5 = 1.491911639$

7. (a) $x_5 = 1.706010786$, $y_5 = 2.045255073$
  (b) $x_5 = 1.301528059$, $y_5 = 1.511628835$
  (c) $x_5 = 1.758134525$, $y_5 = 2.063943847$
  (d) $x_5 = 1.314041930$, $y_5 = 1.510228788$

The values given for Problems 11–17 were computed using Euler's method.

11. $x_{200} = 0.1344922091 \times 10^{30}$

13. $x_{200} = 0.3323342381 \times 10^{9}$

15. $x_{360} = 0.8803441790 \times 10^{18}$

17. $x_{425} = 0.1758761046 \times 10^{11}$

19. **(a)** $x_{20} = 1.636002126$, $y_{20} = 0.2178931513$
   (b) $x_{20} = 1.691930906$, $y_{20} = 0.0036134350$
   (c) $x_{20} = 1.536844525$, $y_{20} = 0.6575431369$
   (d) $x_{20} = 1.444770606$, $y_{20} = 0.4445615959$

4.1 Linear Algebra Review

1. The vector form: $\mathbf{a}_1 = \begin{pmatrix} 5 \\ 3 \end{pmatrix}$, $\mathbf{a}_2 = \begin{pmatrix} -8 \\ \pi \end{pmatrix}$, $\mathbf{b} = \begin{pmatrix} 4 \\ 2\pi \end{pmatrix}$

The matrix form: $\mathbf{Av} = \mathbf{b}$, where $\mathbf{A} = \begin{pmatrix} 5 & -8 \\ 3 & \pi \end{pmatrix}$, $\mathbf{v} = \begin{pmatrix} x \\ y \end{pmatrix}$

3. The vector form: $\mathbf{a}_1 = \begin{pmatrix} 3 \\ 5 \\ 2 \end{pmatrix}$, $\mathbf{a}_2 = \begin{pmatrix} -2 \\ \frac{1}{2} \\ -4 \end{pmatrix}$, $\mathbf{a}_3 = \begin{pmatrix} 4 \\ 2 \\ 2 \end{pmatrix}$, $\mathbf{b} = \begin{pmatrix} -1 \\ -2 \\ -3 \end{pmatrix}$

The matrix form: $\mathbf{Ax} = \mathbf{b}$, where $\mathbf{A} = \begin{pmatrix} 3 & -2 & 4 \\ 5 & \frac{1}{2} & 2 \\ 2 & -4 & 2 \end{pmatrix}$, $\mathbf{x} = \begin{pmatrix} x \\ y \\ z \end{pmatrix}$

5. The vector form: $\mathbf{a}_1 = \begin{pmatrix} 5 \\ 4 \\ 2 \\ 3 \end{pmatrix}$, $\mathbf{a}_2 = \begin{pmatrix} -2 \\ -3 \\ 3 \\ -5 \end{pmatrix}$, $\mathbf{a}_3 = \begin{pmatrix} 7 \\ -2 \\ -4 \\ 2 \end{pmatrix}$,

$\mathbf{a}_4 = \begin{pmatrix} -2 \\ 3 \\ 5 \\ -4 \end{pmatrix}$, $\mathbf{b} = \begin{pmatrix} 2 \\ 1 \\ 4 \\ 8 \end{pmatrix}$

The matrix form: $\mathbf{Ax} = \mathbf{b}$, where $\mathbf{A} = \begin{pmatrix} 5 & -2 & 7 & -2 \\ 4 & -3 & -2 & 3 \\ 2 & 3 & -4 & 5 \\ 3 & -5 & 2 & -4 \end{pmatrix}$,

$\mathbf{x} = \begin{pmatrix} x \\ y \\ z \\ w \end{pmatrix}$

7. $\det\mathbf{A} = -8$, $\det\mathbf{B} = -1$, $\det\mathbf{C} = -26 + 2i$, $\det\mathbf{D} = 11$, $\det\mathbf{E} = -10$, $\det\mathbf{F} = 0$

11. The unique solution is $(\frac{2}{5}, \frac{-3}{5})$, which is the intersection of two lines.

13. The unique solution is $(-\frac{11}{26}, -\frac{1}{13}, \frac{29}{13})$, which is the intersection of three planes.

15. There is no unique solution as the first two equations form two parallel planes.

17. x_1, x_2, x_3 are linearly independent.

19. x_1, x_2, x_3 are linearly dependent.

21. x_1, x_2, x_3 are linearly independent.

23. $\lambda_1 = 1$, $\lambda_2 = -2$. The corresponding eigenvectors are $\mathbf{x}_1 = \begin{pmatrix} 1 \\ 0 \end{pmatrix}$, $\mathbf{x}_2 = \begin{pmatrix} 0 \\ 1 \end{pmatrix}$.

25. $\lambda_1 = 1$, $\lambda_2 = 0$, $\lambda_3 = 2$. The corresponding eigenvectors are

$$\mathbf{x}_1 = \begin{pmatrix} 1 \\ 1 \\ 0 \end{pmatrix}, \mathbf{x}_2 = \begin{pmatrix} 1 \\ 0 \\ -1 \end{pmatrix}, \mathbf{x}_3 = \begin{pmatrix} 1 \\ 0 \\ 1 \end{pmatrix}.$$

27. $\lambda_1 = 0$, $\lambda_2 = 1$, $\lambda_3 = -2$. The corresponding eigenvectors are

$$\mathbf{x}_1 = \begin{pmatrix} 1 \\ 1 \\ 1 \end{pmatrix}, \mathbf{x}_2 = \begin{pmatrix} 0 \\ 0 \\ 1 \end{pmatrix}, \mathbf{x}_3 = \begin{pmatrix} -3 \\ 3 \\ 1 \end{pmatrix}.$$

Section 4.2 Fundamental Results

1. The general solution is defined for all real t.

3. The general solution is defined for all $t > 0$ except $(2k + 1)\pi/2$, $k = 0, 1, 2, \ldots$.

5. The general solution is defined for all real t.

7. The general solution is defined for all $t > 0$ except $t = 1, \pi/2, \pi, 3\pi/2, \ldots$.

9. $\mathbf{v}_1$, $\mathbf{v}_2$ are linearly independent on the given interval.

11. $\mathbf{v}_1$, $\mathbf{v}_2$ are linearly independent on the given interval.

13. $\mathbf{v}_1$, $\mathbf{v}_2$, $\mathbf{v}_3$ are linearly independent on the given interval.

15. $\mathbf{v}_1$, $\mathbf{v}_2$, $\mathbf{v}_3$ are linearly dependent on the given interval.

17. Yes

25. A particular solution is $x_p(t) = \begin{pmatrix} \frac{2}{5}\sin t - \frac{16}{5}\cos t \\ \frac{-1}{5}\sin t - \frac{22}{5}\cos t \end{pmatrix}$

27. A particular solution is $x_p(t) = \begin{pmatrix} -\frac{85}{32} + \frac{3}{8}t - \frac{3}{4}t^2 \\ \frac{75}{32} + \frac{19}{8}t + \frac{1}{4}t^2 \\ \frac{11}{32} - \frac{21}{8}t + \frac{1}{4}t^2 \end{pmatrix}$

29. $x_p(t) = \begin{pmatrix} 6 + 6t + 3t^2 - 2e^t + 3te^t \\ -2 - 2t - \frac{3}{2}t^2 + e^t - te^t \end{pmatrix}$

Section 4.3 Equations with Constant Coefficients

1. $x(t) = c_1 \begin{pmatrix} 1 \\ -8 \end{pmatrix} e^{-6t} + c_2 \begin{pmatrix} 1 \\ 1 \end{pmatrix} e^{3t}$

3. $x(t) = c_1 \begin{pmatrix} 0 \\ 0 \\ 1 \end{pmatrix} e^{t} + c_2 \begin{pmatrix} 3 \\ 1 \\ \frac{3}{2} \end{pmatrix} e^{3t} + c_3 \begin{pmatrix} -2 \\ 1 \\ -1 \end{pmatrix} e^{-2t}$

5. $x(t) = \begin{pmatrix} c_1 \\ -c_1 + c_2 \end{pmatrix} e^{3t} + c_2 \begin{pmatrix} 1 \\ -1 \end{pmatrix} te^{3t}$

7. $x(t) = c_1 \begin{pmatrix} 11 \\ -2 \\ 3 \end{pmatrix} e^{4t} + c_2 \begin{pmatrix} 1 \\ 0 \\ 0 \end{pmatrix} e^{7t} + c_3 \begin{pmatrix} -9 \\ 0 \\ 1 \end{pmatrix} e^{8t}$

9. $x(t) = c_1 \begin{pmatrix} 1 \\ -3 \end{pmatrix} e^{t} + c_2 \begin{pmatrix} 1 \\ 1 \end{pmatrix} e^{5t}$

11. $x(t) = c_1 \begin{pmatrix} \frac{1}{3}\cos(\sqrt{2}t) + \frac{\sqrt{2}}{3}\sin(\sqrt{2}t) \\ 0 \\ \cos(\sqrt{2}t) \end{pmatrix} e^{-t}$

$+ c_2 \begin{pmatrix} \frac{1}{3}\sin(\sqrt{2}t) - \frac{\sqrt{2}}{3}\cos(\sqrt{2}t) \\ 0 \\ \sin(\sqrt{2}t) \end{pmatrix} e^{-t} + c_3 \begin{pmatrix} 1 \\ 3 \\ 5 \end{pmatrix}$

13. The eigenvalues are $\lambda_{1,2} = (b \pm \sqrt{b^2 + 4c}/2)$.

$$\lambda_1 \neq \lambda_2 \rightarrow \begin{cases} x(t) = c_1 e^{\lambda_1 t} + c_2 e^{\lambda_2 t} \\ y(t) = c_1 \lambda_1 e^{\lambda_1 t} + c_2 \lambda_2 e^{\lambda_2 t} \end{cases}$$

$$\lambda_1 = \lambda_2 = \lambda \rightarrow \begin{cases} x(t) = c_1 e^{\lambda t} + c_2 t e^{\lambda t} \\ y(t) = c_1 \lambda e^{\lambda t} + c_2 (\lambda t + 1) e^{\lambda t} \end{cases}$$

$$\lambda_{1,2} = \alpha \pm i\beta \rightarrow \begin{cases} x(t) = c_1 e^{\alpha t} \cos \beta t + c_2 e^{\alpha t} \sin \beta t \\ y(t) = c_1 e^{\alpha t}(\alpha \cos \beta t - \sin \beta t) \\ \quad + c_2 e^{\alpha t}(\alpha \sin \beta t + \cos \beta t) \end{cases}$$

19. $\begin{pmatrix} u(t) \\ v(t) \end{pmatrix} = \begin{pmatrix} 455.55 \\ 205.55 \end{pmatrix} - 490.979 \begin{pmatrix} 1 \\ 0.121 \end{pmatrix} e^{-0.089t}$

$+ 35.423 \begin{pmatrix} 1 \\ -4.121 \end{pmatrix} e^{-0.0512t}$,

where $u(t)$ is the amount of salt in tank A and $v(t)$ is the amount of salt in tank B.

21. The eigenvalues are $\frac{-5}{2} \pm \frac{1}{2}\sqrt{1 + 4\alpha\beta}$. The origin is not a center; therefore, periodic solutions do not exist.

23. $\begin{cases} x(t) = e^{-at}[c_1 \cos t + c_2 \sin t] \\ y(t) = e^{-at}[-c_1 \sin t + c_2 \cos t] \end{cases}$

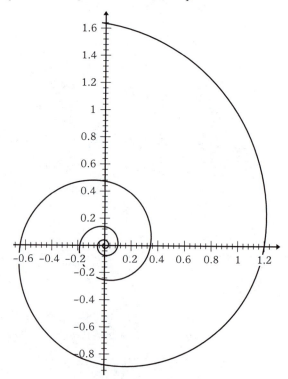

Section 4.4 Qualitative Methods

1.

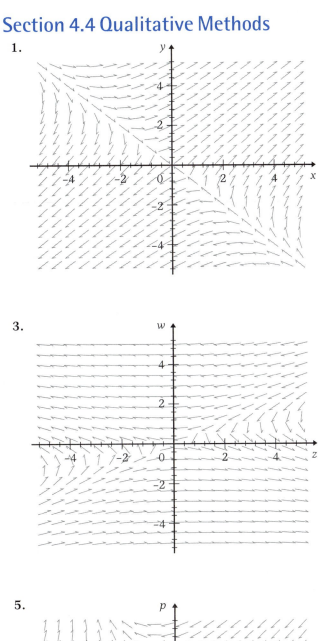

3.

5.

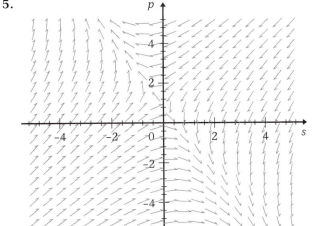

7.

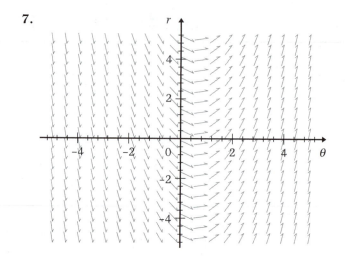

9. Continuous

11. Not continuous

13. Continuous

15. Not continuous

17. The only equilibrium point is the origin.

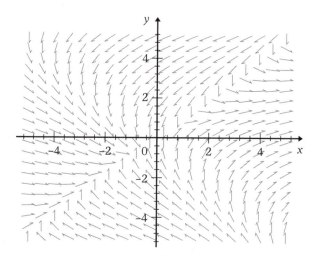

19. The only equilibrium point is $(-2, 1)$.

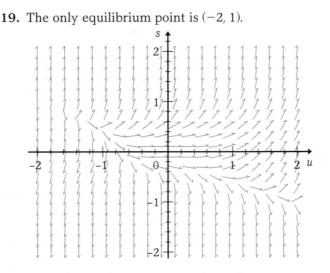

21. The only equilibrium point is $(0, 0, 0)$.

23. The equilibria are at $((2k + 1)\pi/2, (2k + 1)\pi/2, -(2k + 1)\pi)$.

25. $(0, 0)$ is a sink.

27. $(0, 0)$ is a source.

29. $(-1, -1, -1)$ is a saddle.

31. $(0, 0, 0)$ is a saddle.

33. $(0, 0, 0)$ is a saddle.

35. $(0, 0, 0)$ is a saddle.

37. $$\begin{cases} x(t) = -c_1\dfrac{\alpha + \beta - \gamma}{\beta}e^{-(\alpha+\beta)t} \\ y(t) = c_1 e^{-(\alpha+\beta)t} + c_2 e^{-\gamma t} \end{cases}$$
The mice die if $\alpha, \beta, \gamma > 0$.

39. No

Section 4.5 Numerical Methods

1. (a) $x_{10} = 155.7641036, y_{10} = 155.7642084$
  (b) $x_{20} = 269.2366547, y_{20} = 269.2374526$
  (c) $x_{10} = 452.3410223, y_{10} = 452.3453303$
  (d) $x_{20} = 531.7560243, y_{20} = 531.7587984$

3. (a) $x_{10} = 106.5141400, y_{10} = 16.50916531$
  (b) $x_{20} = 138.5194273, y_{20} = 21.36581530$
  (c) $x_{10} = 180.6434678, y_{10} = 27.53350129$
  (d) $x_{20} = 188.8193024, y_{20} = 28.93874122$

5. (a) $x_{10} = -10.88577255, y_{10} = 44.77940540, z_{10} = -6.938794975$
  (b) $x_{20} = -18.24582759, y_{20} = 66.72071441, z_{20} = -9.352323448$
  (c) $x_{10} = -28.91468349, y_{10} = 98.43530115, z_{10} = -12.82895778$
  (d) $x_{20} = -32.53192949, y_{20} = 109.1260551, z_{20} = -13.99490129$

7. (a) $s_{10} = -71.28041646 + 0.1848869191$
    $\cos 1 \ r_{10} = -4.133698206 + 0.1361394742 \cos 1$
    $u_{10} = 34.57698825 + 1.668578607 \cos 1$

(b) $s_{20} = -109.1239976 + 0.1300197468 \cos 1$
$r_{20} = -6.695716070 + 0.1015916493 \cos 1$
$u_{20} = 49.54869152 + 1.230043667 \cos 1$

(c) $s_{10} = -13100.07576 - 406.5970390 \cos 3$
$r_{10} = -423.2025061 - 4.068458039 \cos 3$
$u_{10} = 15656.77495 + 487.1558469 \cos 3$

(d) $s_{20} = -19642.42311 - 348.5234944 \cos 3$
$r_{20} = -268.2835743 - 4.659384904 \cos 3$
$u_{20} = 25217.99414 + 444.8544572 \cos 3$

9. $\text{infected}_5 = 7.774759738$, $\text{quarantined}_5 = 4.106825370$, $\text{infected}_5 = 489.9379204$

11. $\text{infected}_5 = 7.710733049$, $\text{quarantined}_5 = 12.14839754$, $\text{infected}_5 = 571.1414542$

13. $x_{500} = 407.4964584$, $y_{500} = 1250.790766$

15. $x_{500} = -739.6367844$, $y_{500} = 314.2972442$

17. (a) $x_1 = 0.9$, $y_1 = 1.4$
(b) $x_{10} = 0.8201083676$, $y_{10} = 1.575644364$
(c) $x_{100} = 0.8063971091$, $y_{100} = 1.605445960$
(d) $x_1 = 0.8300000000$, $y_1 = 1.555000000$
(e) $x_{10} = 0.8051994389$, $y_{10} = 1.608060717$
(f) $x_{100} = 0.8047649198$, $y_{100} = 1.608989587$

19. (a) $x_{20} = -0.05939808397$, $\rho_{20} = -0.7401729140$
(b) $x_{200} = -0.06369774365$, $\rho_{200} = -0.7425067065$
(c) $x_{2000} = -0.06414667864$, $\rho_{2000} = -0.7427032075$
(d) $x_{20} = -0.06400551007$, $\rho_{20} = -0.7429668829$
(e) $x_{200} = -0.06419491970$, $\rho_{200} = -0.7427270329$
(f) $x_{2000} = -0.06419673717$, $\rho_{2000} = -0.7427246696$

21. (a) $x_8 = -38.52635646$, $y_8 = 102.5487523$, $z_8 = -10.91534118$
(b) $x_{80} = -251.7072839$, $y_{80} = 535.3683065$, $z_{80} = -201.7557547$
(c) $x_{800} = -332.4314736$, $y_{800} = 698.0577989$, $z_{800} = -276.7880618$
(d) $x_8 = -202.4074568$, $y_8 = 436.5594972$, $z_8 = -154.6862224$
(e) $x_{80} = -340.3203774$, $y_{80} = 713.9611717$, $z_{80} = -284.1110083$
(f) $x_{800} = -343.4262103$, $y_{800} = 720.2012311$, $z_{800} = -287.0426332$

23. (a) $\alpha_{10} = 4.187484920$, $\beta_{10} = 4.187484920$, $\theta_{10} = 4.187484920$
(b) $\alpha_{100} = 4.409627655$, $\beta_{100} = 4.409627655$, $\theta_{100} = 4.409627655$
(c) $\alpha_{1000} = 4.433847872$, $\beta_{1000} = 4.433847872$, $\theta_{1000} = 4.433847872$
(d) $\alpha_{10} = 4.428161693$, $\beta_{10} = 4.428161693$, $\theta_{10} = 4.428161693$
(e) $\alpha_{100} = 4.436473727$, $\beta_{100} = 4.436473727$, $\theta_{100} = 4.436473727$
(f) $\alpha_{1000} = 4.436562757$, $\beta_{1000} = 4.436562757$, $\theta_{1000} = 4.436562757$

Section 5.1 Linearization

1. There are three equilibrium points, $(0, 0)$, $(1/16, 1/64)$, $(-1/16, 1/64)$, all of them isolated.

3. The equilibria are all the points on the line $z = 6w$. There are no isolated equilibrium points.

5. There are two isolated equilibrium points at $(0, 0, 0)$ and $(-2, -2, -2)$.

7. There is one isolated equilibrium at $(-1/2, -1/2, 0)$.

9. The linearized system around $(2, 1)$ is $\begin{cases} x' = 2y \\ y' = 4x \end{cases}$. The linearized system around $(-2, -1)$ is $\begin{cases} x' = -2y \\ y' = -4x \end{cases}$. Both equilibria are saddles.

11. The linearized system around $(0, 0)$ is $\begin{cases} x' = x \\ y' = -y \end{cases}$. The linearized system around $(\pi/2, \pi/2)$ is $\begin{cases} x' = -y \\ y' = x \end{cases}$. $(0, 0)$ is a saddle and $(\pi/2, \pi/2)$ is nonhyperbolic.

13. The linearized system around $(0, 0, 0)$ is
$$\begin{cases} x' = -10x + 10y \\ y' = 28x - y \\ z' = -\frac{8}{3}z \end{cases}.$$
The linearized system around $(-6\sqrt{2}, -6\sqrt{2}, 27)$ is
$$\begin{cases} u' = -10u + 10v \\ v' = u - v + 6\sqrt{2}w \\ w' = -6\sqrt{2}u - 6\sqrt{2}v - \frac{8}{3}w \end{cases}.$$
The linearized system around $(6\sqrt{2}, 6\sqrt{2}, 27)$ is
$$\begin{cases} u' = -10u + 10v \\ v' = u - v - 6\sqrt{2}w \\ w' = 6\sqrt{2}u + 6\sqrt{2}v - \frac{8}{3}w. \end{cases}$$
All three equilibrium points are saddles.

15. The linearized system around $(0, 0, 0)$ is $\begin{cases} x' = \frac{1}{4}y \\ y' = \frac{1}{4}z \\ z' = \frac{1}{4}x \end{cases}$.

The linearized system around $(\frac{1}{2}, \frac{1}{2}, \frac{1}{2})$ is $\begin{cases} x' = -\frac{1}{2}y \\ y' = -\frac{1}{2}z \\ z' = -\frac{1}{2}x \end{cases}$.

$(0, 0, 0)$ is a saddle and $(\frac{1}{2}, \frac{1}{2}, \frac{1}{2})$ is nonhyperbolic.

17. The equilibria of the first two systems are saddles at the origin and the third is a spiral source.

19. If $\alpha > 0$, then the isolated equilibrium point $(\sqrt{\alpha}/2, 0)$ is a saddle. If $\alpha = 0$, then the origin is a nonhyperbolic equilibrium. If $\alpha < 0$, there are no equilibria.

23. The two species can survive when $a > b$.

Section 5.2 Periodic Solutions

1. $\begin{cases} r' = (r^2 - 1)(\cos\theta - \sin\theta) + 3r\cos\theta\sin\theta \\ \theta' = \dfrac{\cos\theta(3r\cos\theta - r^2 + 1) - (r^2 - 1)\sin\theta}{r} \end{cases}$

3. $\begin{cases} r' = r\cos r\cos^2\theta - r\cos r + r\sin r\cos^2\theta \\ \theta' = -\sin\theta\cos\theta(\sin r + \cos r) \end{cases}$

7. $T = \pi$

9. One limit cycle at $r = 1$

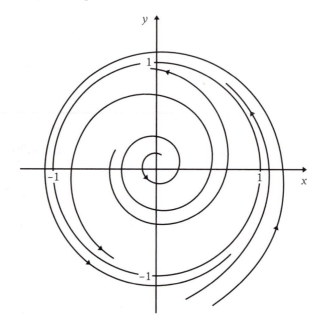

11. An example is $\begin{cases} x' = x\cos\sqrt{x^2+y^2} - y \\ y' = x + y\cos\sqrt{x^2+y^2} \end{cases}$.

The system can't be a polynomial by Dulac's theorem.

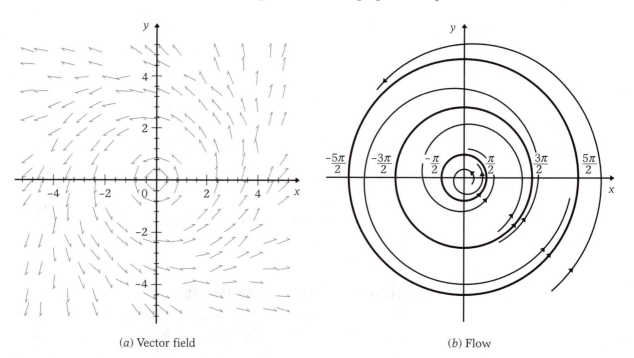

(a) Vector field   (b) Flow

Section 5.3 Gradient and Hamiltonian Systems

1. The system is a gradient system and $G(x, y) = 2 \sin x \sin y + c$.

3. The system is not a gradient system.

5. The system is a gradient system and $G(P, Q) = -2P^4 - 2Q^4 + c$.

7. The system is a Hamiltonian system and $H(z, w) = -w^2/2 + z^2/2 + c$.

9. The system is a Hamiltonian system and $H(x, y) = 14xy + y^2/2 + x^2y - xy^2 + y^3 - x^2/2 - x^3/3 + c$.

11. The system is not a Hamiltonian system.

13. The equilibrium is a spiral source, so the system is not gradient.

15. The equilibrium is a spiral sink, so the system is not gradient.

17. One of the equilibria is a spiral sink, so the system is not Hamiltonian.

19. The equilibrium is a spiral source, so the system is not Hamiltonian.

21. These are systems of the form: $\begin{cases} x' = f(x, y) \\ y' = g(x, y) \end{cases}$ with $f(x, y), g(x, y)$ such that $\dfrac{\partial^2 f}{\partial x^2} + \dfrac{\partial^2 f}{\partial y^2} = 0$ and $\dfrac{\partial^2 g}{\partial x^2} + \dfrac{\partial^2 g}{\partial y^2} = 0$.

23. The Hamiltonian system is $\begin{cases} u' = -2v \\ v' = -2u \end{cases}$.

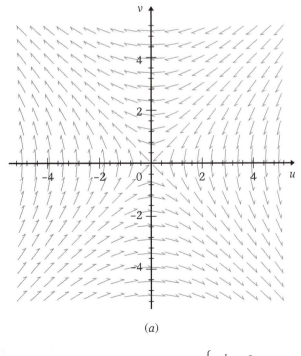

(a)

The reciprocal gradient system is $\begin{cases} u' = 2u \\ v' = -2v \end{cases}$.

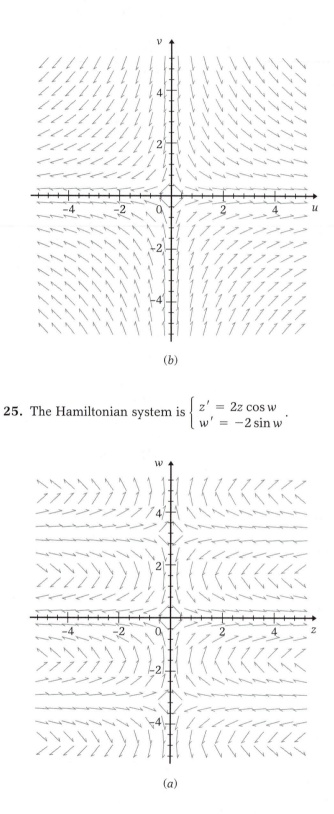

(b)

25. The Hamiltonian system is $\begin{cases} z' = 2z\cos w \\ w' = -2\sin w \end{cases}$.

(a)

The reciprocal gradient system is $\begin{cases} z' = 2\sin w \\ w' = 2z\cos w \end{cases}$.

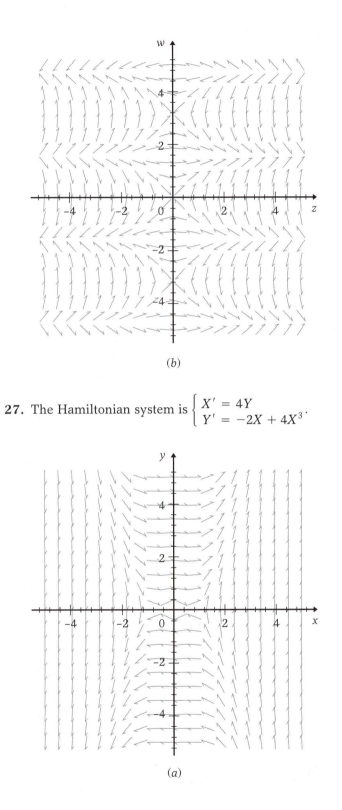

(b)

27. The Hamiltonian system is $\begin{cases} X' = 4Y \\ Y' = -2X + 4X^3 \end{cases}$.

(a)

The reciprocal gradient system is $\begin{cases} X' = 2X - 4X^3 \\ Y' = 4Y \end{cases}$.

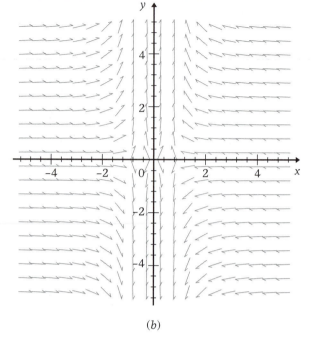

(b)

29. The converse is true.

31. The Hamiltonian function is $H = \dfrac{u^2}{2} + \dfrac{v^2}{2} - \dfrac{Gm}{(x^2 + y^2)^{1/2}} + c$, with $u = x', v = y'$.

Section 5.4 Stability

1. $\dot{V}(x, y) = 0$ for all (x, y) in the plane

3. $\dot{V}(x, y) = 2(x^2 + y^2)^2 > 0$ for all (x, y) in the plane

5. $V(x, y) = -2(ax^4 - 2ax^2y^2 + by^4) < 0$ if $0 < a < b$

9. $V(x, y, z) = au^2 + bv^2 + cw^2$, where $a > 0$, $b = 2a$, $c = a$; it is not a sink because the equilibrium of the linearized system has a zero eigenvalue

11. $V(x, y) = ax^2 + by^2$, where $a = b > 0$

13. Spiral sinks at $\phi = 2k\pi, \varphi = 0$
Saddles at $\phi = (2k + 1)\pi, \varphi = 0$, k is an integer

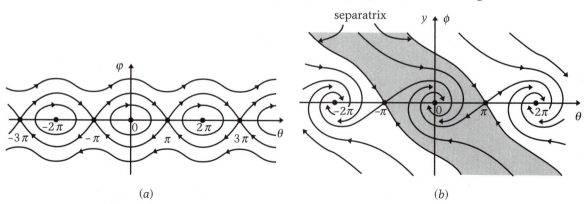

(a)   (b)

15. *Hint:* see Theorem 5.3.1.

17. $V(x, y) = ax^2 + by^2$ with $b > 0$; $a = 4b$ is a Liapunov function for the given system.

Section 5.5 Chaos

1. (a) Initial conditions $(5, 5, 5)$ with 2000 steps.

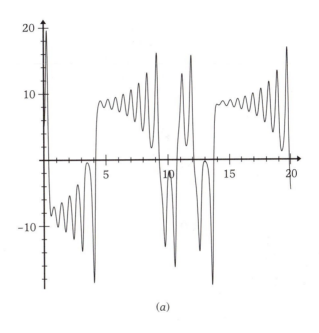

(*a*)

(b) Inital conditions $(5.01, 5, 5)$ with 2000 steps.

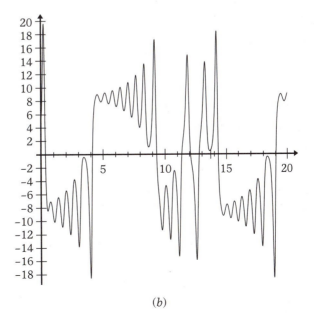

(*b*)

(c) Inital conditions (5, 5, 5) with 1000 steps.

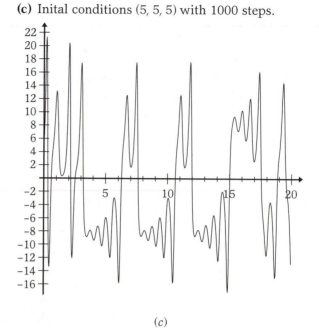

(c)

(d) Inital conditions (5.01, 5, 5) with 1000 steps.

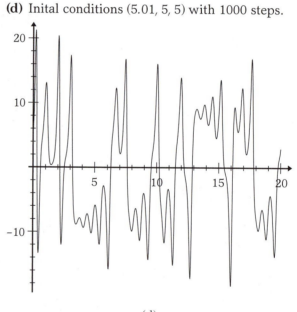

(d)

3. Using Euler's method with step size 0.1:
 (a) $x_{300} = 6.549677718$

 $y_{300} = -0.0247909356$

 $z_{300} = 18.29275056$
 (b) $x_{300} = 6.366010905$

 $y_{300} = -6.185881697$

 $z_{300} = 0.2370566813$

7. $s_i = s_i + 3, i = 0, \pm 1, \pm 2, \pm 3, \ldots$

Section 6.1 Fundamental Properties

1. $f(t)$ is of exponential order.

3. $f(t)$ is of exponential order.

5. $f(t)$ is not of exponential order.

7. $f(t)$ is of exponential order.

9. $f(t)$ is of exponential order.

11. $f(t)$ is of exponential order.

13. If two functions are of exponential order, then so are their sum and product.

15. $\mathcal{L}[t^2] = \dfrac{2!}{s^3}$

17. $\mathcal{L}[t^n] = \dfrac{n!}{s^{n+1}}$

19. $\mathcal{L}\left[\dfrac{3}{3^t}\right] = \dfrac{3}{s + \ln 3}$

21. $\mathcal{L}[t \sin at] = \dfrac{2as}{(s^2 + a^2)^2}$, defined for $s > 0$

23. $\mathcal{L}\left[\dfrac{1}{a} \sin at - t \cos at\right] = \dfrac{2a^2}{(s^2 + a^2)^2}$, where $s > 0$

25. $\mathcal{L}[\cosh t] = \dfrac{s}{s^2 - 1}$

27. The Laplace transform of a product of functions is not equal to the product of the Laplace transform of the respective functions.

29. $f(t) = \frac{2}{5} - \frac{2}{5}e^{-5t}$

31. $f(t) = \frac{3}{2}e^{-t} - 2e^{-2t} + \frac{1}{2}e^{-3t}$

33. $f(t) = 5\cos 2t - 2\sin 2t + 3$

35. $f(t) = -2e^{-2t}\cos t + 5e^{-2t}\sin t$

Section 6.2 Step Functions

1.

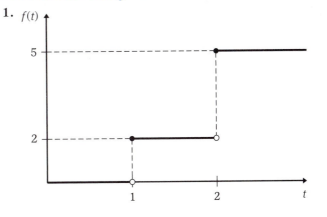

3.

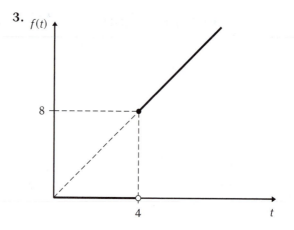

5.

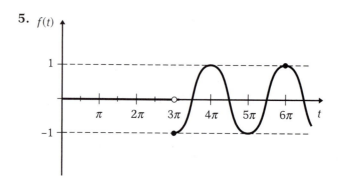

7. $f(t) = 1 + u_1(t) - u_3(t)$

9. $f(t) = 1 + u_1(t) + u_2(t) + u_3(t) + u_4(t) + \cdots$

11. $f(t) = 1 + \sum_{i=1}^{n} (-1)^{i+1}(i+1)u_n(t)$

13. $f(t) = 1 + u_3(t)$

15. $f(t) = \pi + (2 - \pi)u_1(t) - (\frac{1}{3})u_{1.8}(t)$

17. $\mathscr{L}[f(t)] = e^{-3s}\dfrac{2!}{s^3}$

19. $\mathscr{L}[f(t)] = e^{-\pi s}\dfrac{1}{s^2} - e^{-2\pi s}\left(\dfrac{1}{s^2} + \dfrac{\pi}{s}\right)$

21. $\mathscr{L}[f(t)] = 2\dfrac{e^{-s}}{s} - 3\dfrac{u^{-2s}}{s} + 7\dfrac{e^{-7s}}{s}$

23. $\mathscr{L}[f(t)] = \left(\dfrac{1}{s^2} + \dfrac{1}{s}\right)(e^{-3s} - e^{-2s})$

25. $\mathscr{L}^{-1}[F(s)] = \dfrac{1}{6}e^{2t}t^3$

27. $\mathscr{L}^{-1}[F(s)] = u_2(t)e^{-(t-2)}\cos(t-2)$

29. $\mathscr{L}^{-1}[F(s)] = u_1(t) + u_2(t)$

Section 6.3 Initial Value Problems

1. $x(t) = -\frac{6}{5} + \frac{11}{5}e^{5t}$

3. $v(t) = e^{t/2}$

5. $x(t) = 2\sinh 2t + e^{-t}(\cos t - 2\sin t)$

7. $z(t) = \cos t + \dfrac{2}{\sqrt{3}}\sin\left(\dfrac{\sqrt{3}}{2}t\right)e^{-t/2}$

9. $\begin{cases} x(t) = e^{-t}\,[\frac{2}{3}\,\sqrt{6}\sinh\,(\sqrt{6}t) + \cosh\,(\sqrt{6}t)] \\ y(t) = e^{-t}\cosh\,(\sqrt{6}t) - e^{-t}/\sqrt{6}\,\sinh\,(\sqrt{6}t) \end{cases}$

11. $\begin{cases} r(t) = e^{-t} + 4te^{-t} + 2te^{-2t} \\ \Theta(t) = -e^{-2t} + 4te^{-2t} + 6te^{-t} \end{cases}$

13. $x(t) = \sin t - 1 + \cos t + u_\pi(t)[1 - \cos\,(t - \pi)].$

15. $v(t) = -\dfrac{3}{4}\,\sqrt{3}\sinh\left(\dfrac{\sqrt{3}}{3}t\right) + \dfrac{3}{4}\,\sin t - \dfrac{3}{4}\,\sin t\,u_{\pi/2}(t)$

$\qquad + \dfrac{3}{4}\,\cosh\left(\dfrac{\sqrt{3}}{3}\left(t - \dfrac{\pi}{2}\right)\right)u_{\pi/2}(t)$

17. $z(t) = -\frac{1}{2}e^{-t} + \frac{1}{2}\cos t - \frac{1}{2}\sin t - \frac{1}{2}u_\pi e^{-t+\pi}$
$\qquad -\frac{1}{2}\cos tu_\pi + \frac{1}{2}\sin tu_\pi$

19. $q(t) = 2\,\cos(\sqrt{2}t) + 2\,\sqrt{2}\,\sin(\sqrt{2}t) + u_{\pi/4}\,\cos[\sqrt{2}(t - \pi/4)]$
$\qquad -u_{\pi/4}\,\cos(t - \pi/4).$

21. (i)

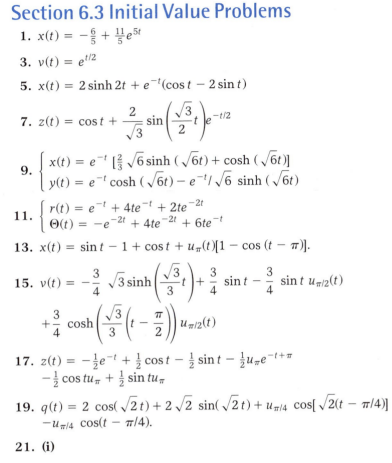

(ii) The solution of the initial value problem is

$$x(t) = 10\sin t - \alpha u_{2.5}(t) + \alpha u_{2.5}(t)\cos(t - 2.5)$$
$$+ \alpha u_{3.5}(t) - \alpha u_{3.5}\cos(t - 3.5).$$

(iii)

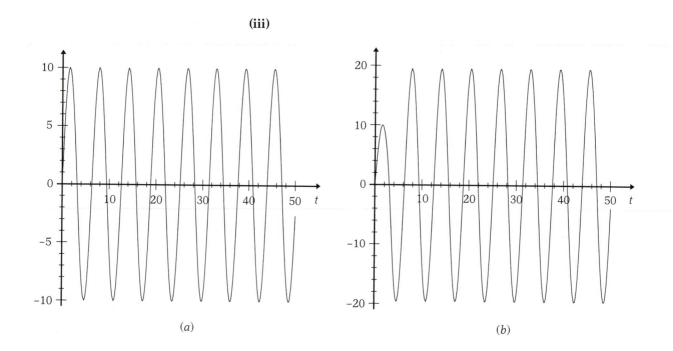

(a)

(b)

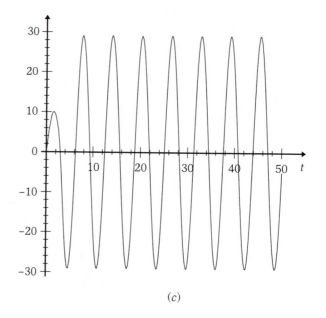

(c)

23. $x(t) = \frac{1}{2}\sin 2t + \frac{1}{2}u_\pi \sin 2t$

25. $x(t) = \frac{1}{2}\sin 2t(1 + u_\pi + u_{2\pi} + u_{3\pi})$

27. $x(t) = u_{2\pi} + 2e^{-t+2\pi}u_{2\pi}\pi - e^{-t+2\pi}u_{2\pi} + 2te^{-t} - e^{-t+2\pi}tu_{2\pi}$

29. $x(t) = -u_{\pi/2}e^{-t+\pi/2}\cos t - \frac{1}{5}e^{-t}\cos t - \frac{3}{5}e^{-t}\sin t + \frac{1}{5}\cos t + \frac{2}{5}\sin t$

Section 7.1 The Power Series Method

1. $1 + \sum_{n=1}^{\infty} \left(\dfrac{n! + (2n)!}{(2n)!n!} \right) t^n$

3. $\sum_{n=1}^{\infty} \left(\dfrac{t^n + 2}{2n} \right) t^n$

5. $\sum_{n=0}^{\infty} \left(\dfrac{2^n + (2n)!t}{(2n)!2^n} \right) t^n$

7. $f(t)g(t) = \sum_{n=0}^{\infty} \left(\dfrac{1}{n!} \right) t^{n+1}$

9. $f(t)g(t) = \sum_{n=0}^{\infty} \left(\dfrac{2^n}{(n+1)(n+2)} \right) t^{n+2}$

11. $f(t)g(t) = 1 + t + \dfrac{t^2}{2} - \dfrac{t^3}{2} - \dfrac{11t^4}{24} + \dfrac{t^5}{24} + \dfrac{29t^6}{720} - \dfrac{t^7}{720} + \cdots$

13. $a_n = (-1)^n n! a_0$

15. The series converges absolutely for $|t - 1| < 1$ or $0 < t < 2$.

17. The series converges on $(-3, 3)$.

19. The series converges absolutely for $-e < t < e$.

21. $x(t) = a_0 \sum_{k=0}^{\infty} \dfrac{1}{(2k)!} t^{2k} + a_1 \sum_{k=0}^{\infty} \dfrac{1}{(2k+1)!} t^{2k+1}$

23. $x(t) = a_0 + a_1 t + \dfrac{a_0}{2} t^2 - \dfrac{a_0}{24} t^4 + \dfrac{a_0}{240} t^6 + \cdots$

25. $x(t) = a_0 + a_1 t - a_0 t^2 - \dfrac{a_1}{2} t^3 + \dfrac{a_0}{3} t^4 + \dfrac{a_1}{8} t^5 + \cdots$

27. $x(t) = \sum_{n=0}^{\infty} \dfrac{(2t)^n}{n!} = e^{2t}$ or $x(t) = e^{2t}$

29. $x(t) = 1 - \tfrac{1}{2} t + \tfrac{3}{2} t^2 + \tfrac{1}{6} t^3 + \tfrac{1}{8} t^4 - \tfrac{1}{240} t^6 + \tfrac{1}{4480} t^8 + \cdots$

31. $x(t) = 2 - \tfrac{1}{3} t + \tfrac{1}{2} t^2 + \tfrac{1}{18} t^3 + \tfrac{5}{144} t^4 + \tfrac{17}{1440} t^5 + \cdots$

35. $x(t) = a_0 \left(1 + \dfrac{t^3}{6} + \dfrac{t^6}{180} + \cdots \right) + a_1 \left(t + \dfrac{t^4}{12} + \dfrac{t^7}{504} + \cdots \right)$; the radius of convergence is infinite.

37. $x_1(t) = 1 - \dfrac{\lambda(\lambda + 1)}{2!} t^2 + \dfrac{\lambda(\lambda + 1)(\lambda - 2)(\lambda + 3)}{4!} t^4 + \cdots$

$x_2(t) = t - \dfrac{(\lambda - 1)(\lambda + 2)}{3!} t^3 + \dfrac{(\lambda - 1)(\lambda + 2)(\lambda - 3)(\lambda + 4)}{5!} t^5 + \cdots$

Section 7.2 Approximations

1. The power series solution is convergent for every t.

3. The radius of convergence is $r = 1$.

5. The radius of convergence is $r = 2$.

7. The radius of convergence is $r = \frac{1}{2}$.

9. The radius of convergence is $r = \frac{1}{2}$.

11. The radius of convergence is $r = 1$.

13. $x(t) = a_0 + a_1 t + (\frac{1}{2}a_1 - \frac{1}{2}a_0)t^2 + (-\frac{1}{3}a_1 - \frac{1}{6}a_0)t^3 + (-\frac{7}{24}a_1 + \frac{1}{6}a_0)t^4 + \cdots$
The series is convergent for all t.

15. $x(t) = a_0 + a_1 t - \frac{1}{24}a_0 t^3 - \frac{1}{48}a_1 t^4 + \cdots$
The series is convergent for t in $(-2, 2)$.

17. $x(t) = a_0 + a_1 t - \frac{1}{18}a_0 t^2 - \frac{1}{27}a_1 t^3 - \frac{1}{1944}a_1 t^4 + \cdots$
The series is convergent for t in $(-3, 3)$.

19. $\Theta(x) = a_0 + a_1 x + \frac{3}{2}a_0 x^2 + (\frac{1}{2}a_1 - \frac{1}{12}a_0)x^3 + (\frac{3}{8}a_0 - \frac{1}{24}a_1)x^4 + (\frac{3}{40}a_1 - \frac{1}{20}a_0)x^5 + (\frac{7}{180}a_0 - \frac{1}{80}a_1)x^6 + (\frac{59}{10080}a_1 - \frac{9}{1120}a_0)x^7 + (\frac{17}{6720}a_0 - \frac{3}{2240}a_1)x^8$.

Section 7.3 Regular Singular Points

1. $\tau y'' + 3y' + 3y = 0$ using the change of variables $\tau = 1 - t$

3. $\tau y'' + 3(\tau - 5)y' + (\tau - 5)y = 0$ using the change of variables $\tau = x + 5$

5. Using the change of variable $\tau = \theta + 1$, obtain $\tau r'' - 2(\tau - 1)^2(\tau - 2)r' + r(\tau - 2) = 0$.

7. 0 is not a regular singular point.

9. 0 is a regular singular point.

11. 0 is a regular singular point.

13. $x_1(t) = \sum_{n=0}^{\infty}(-1)^n \frac{1}{(3)(5)(7)\cdots(2n+1)}t^{n+1}$

$x_2(t) = t^{1/2} + \sum_{n=1}^{\infty}(-1)^n \frac{1}{2^n n!}t^{n+1/2}$
The general solution converges for all $t > 0$.

15. $v_1(t) = \sum_{n=0}^{\infty}(-1)^n \frac{1}{2^n n!(13)(17)\cdots(8n+1)}t^{2n}$

$v_2(t) = t^{-1/4} + \sum_{n=1}^{\infty}(-1)^n \frac{1}{2^n n!(7)(15)(23)\cdots(8n-1)}t^{2n-1/4}$
The general solution converges for all $t > 0$.

17. $w_1(z) = z^3$, $w_2(z) = z^{-2/3}$

The general solution converges for all $z > 0$.

19. $s_1(r) = 1 + 4r + \frac{8}{3}r^2$

$s_2(r) = r^{1/2}\left(1 + \frac{3}{2}r + \frac{3}{8}r^2 - \frac{1}{16}r^3 + \frac{3}{128}r^4 - \frac{3}{256}r^5 + \frac{7}{1024}r^6 + \cdots\right)$
The general solution converges for all $r > 0$.

21. $q_1(s) = s^{5/2} + \displaystyle\sum_{n=1}^{\infty} \frac{(3)(5)\cdots(2n+1)}{n!(7)(9)\cdots(2n+5)} s^{n+5/2}$

$q_2(s) = 1 + \dfrac{2}{3}s$

The general solution converges for all $s > 0$.

23. $r_1(\theta) = \theta^{5/2} + \displaystyle\sum_{n=1}^{\infty} \frac{(-1)^n \theta^{n+5/2}}{n! \cdot 9 \cdot 11 \cdot 13 \cdots (2n+7)}$

$r_2(\theta) = \theta^{-1}\left(1 + \dfrac{\theta}{5} + \dfrac{\theta^2}{2!(3)(5)} + \dfrac{\theta^3}{3!(3)(5)} - \dfrac{\theta^4}{4!(3)(5)}\right)$

$\qquad + \displaystyle\sum_{n=5}^{\infty}(-1)^{n+1} \frac{1}{15n!(3)(5)\cdots(2n-7)}\theta^n,$

The general solution converges for all $\theta > 0$.

25. The method of Frobenius cannot be applied because $t = 0$ is not a regular singular point; the solution is $x(t) = e^{1/t}$.

27. $q_1(t) = 1 - 10t + \dfrac{10^2}{2!6}t^2 - \dfrac{10^3}{3!(6)(11)}t^3 + \dfrac{10^4}{4!(6)(11)(16)}t^4$

$\qquad + \cdots + (-1)^n \dfrac{10^n}{n!(6)(11)\cdots(5n-4)}t^n$

$q_2(t) = t^{4/5}(1 - \dfrac{10}{9}t + \dfrac{10^2}{2!(9)(14)}t^2 - \dfrac{10^3}{3!(9)(14)(19)}t^3$

$\qquad + \dfrac{10^4}{3!(9)(14)(19)(24)}t^4 + \cdots + (-1)^n \dfrac{10^n}{n!(9)(14)\cdots(5n+4)})$

BIBLIOGRAPHY

Bell, E.T. *The Development of Mathematics.* Dover, New York, 1992.

Blanchard, P., Devaney, R.L., and Hall, G.R. *Differential Equations.* Brooks/Cole, New York, 1998.

Boyce, W.E., and DiPrima, R.C. *Elementary Differential Equations and Boundary Value Problems.* John Wiley & Sons, New York, 1992.

Derrick, W.R., and Grossman, S.I. *Introduction to Differential Equations.* West, New York, 1987.

Devaney, R.L. *Chaos, Fractals, and Dynamics.* Addison-Wesley, Reading, MA, 1990.

Diacu, F., and Holmes, P. *Celestial Encounters—The Origins of Chaos and Stability.* Princeton University Press, Princeton, NJ, 1996.

Edwards, C.H., Jr., and Penney, D.E. *Differential Equations, Computing and Modeling.* Prentice Hall, Upper Saddle River, NJ, 1996.

Goldberg, J., and Potter, M.C. *Differential Equations.* Prentice Hall, Upper Saddle River, NJ, 1998.

Guckenheimer, J., and Holmes, P. *Nonlinear Oscillations, Dynamical Systems, and Bifurcations of Vector Fields.* Springer Verlag, New York, 1983.

Hartman, P. *Ordinary Differential Equations.* Birkhäuser, Boston, 1982.

Hirsh, W.H., and Smale, S. *Differential Equations, Dynamical Systems, and Linear Algebra.* Academic Press, New York, 1974.

Katok, A., and Hasselblatt, B. *Introduction to the Modern Theory of Dynamical Systems.* Cambridge University Press, New York, 1995.

Kostelich, E.J., and Armbruster, D. *From Linearity to Chaos—Introductory Differential Equations.* Addison-Wesley, Reading, MA, 1997.

Perko, L. *Differential Equations and Dynamical Systems.* Springer Verlag, New York, 1996.

Rainville, E.D. *Intermediate Differential Equations.* Macmillan, New York, 1964.

Rainville, E.D., Bedient, P.E., and Bedient, R.E. *Elementary Differential Equations.* Prentice Hall, Upper Saddle River, NJ, 1997.

Redheffer, R. *Introduction to Differential Equations.* Jones and Bartlett, London, 1992.

Sanchez, A.D., and Allen, R.C., Jr. *Differential Equations.* Addison-Wesley, Reading, MA, 1988.

Tierny, J.A. *Differential Equations.* Allyn & Bacon, Boston, 1979.

Verhulst, F. *Nonlinear Differential Equations and Dynamical Systems.* Springer Verlag, Heidelberg, 1990.

Wiggins, S. *Introduction to Applied Nonlinear Dynamical Systems and Chaos.* Springer Verlag, New York, 1990.

Zill, G.D. *A First Course in Differential Equations.* Cole, San Francisco, 1997.

SOURCES OF PHOTOGRAPHS

INDEX